Advanced Sciences and Technologies for Security Applications

Indexed by SCOPUS

The series Advanced Sciences and Technologies for Security Applications comprises interdisciplinary research covering the theory, foundations and domain-specific topics pertaining to security. Publications within the series are peer-reviewed monographs and edited works in the areas of:

- biological and chemical threat recognition and detection (e.g., biosensors, aerosols, forensics)
- crisis and disaster management
- terrorism
- cyber security and secure information systems (e.g., encryption, optical and photonic systems)
- traditional and non-traditional security
- energy, food and resource security
- economic security and securitization (including associated infrastructures)
- transnational crime
- human security and health security
- social, political and psychological aspects of security
- recognition and identification (e.g., optical imaging, biometrics, authentication and verification)
- smart surveillance systems
- applications of theoretical frameworks and methodologies (e.g., grounded theory, complexity, network sciences, modelling and simulation)

Together, the high-quality contributions to this series provide a cross-disciplinary overview of forefront research endeavours aiming to make the world a safer place.

The editors encourage prospective authors to correspond with them in advance of submitting a manuscript. Submission of manuscripts should be made to the Editor-in-Chief or one of the Editors.

Benjamin Jacob · Edwin Michael ·
Anthony J. Masys
Editors

Tech Enabled Global Health Security

Springer

Editors
Benjamin Jacob
College of Public Health
University of South Florida
Tampa, FL, USA

Edwin Michael
College of Public Health
University of South Florida
Tampa, FL, USA

Anthony J. Masys
College of Public Health
University of South Florida
Tampa, FL, USA

ISSN 1613-5113 ISSN 2363-9466 (electronic)
Advanced Sciences and Technologies for Security Applications
ISBN 978-3-031-86996-9 ISBN 978-3-031-86997-6 (eBook)
https://doi.org/10.1007/978-3-031-86997-6

This Springer imprint is published by the registered company Springer Nature Switzerland AG
The registered company address is: Gewerbestrasse 11, 6330 Cham, Switzerland

Acknowledgements

The editors would like to acknowledge the financial support for the Malaria Seek and Destroy projects from the following donors:

Joy McCann Foundation
Tampa, FL, United States

Icel Rodriquez
Director of missions
Florida Conference
United Methodist Church
450 Martin Luther King Jr Ave
Lakeland FL 33815
United States

Russ Montgomery
President and Founder
Living in Faith
PO Box 271883
Tampa, FL 33688

Jim Manuel
Vice Chair of the Board of Trustees
Florida United Methodist Church Conference
300 E. Church St. Suite 1102
Orlando FL 32801

Contents

Possibility Thinking in the Wild: Imagination, Creativity and Innovation from Possibility Thinking to Possibility Doing

Anthony J. Masys

Abstract Rittel and Webber (Policy Sci 4:155–169, 1973) describe a wicked problem as an issue that is difficult or impossible to solve because of its complex and interconnected nature. Examples of wicked problems in Global Health relate to issues such as climate change, biodiversity loss, persisting poverty, the growing obesity epidemic, and food insecurity (Walls in Glob Health 14:34, 2018). Along with these wicked problems are equally complex events, with roots in this 'wicked space', like the COVID-19, H1N1 and H5N1 outbreaks, the emergence and spread of monkeypox, and the international spread of vector borne disease that are challenging the global health security landscape revealing an inherent fragility in our societal global health systems.

How do we design and implement solutions to the spectrum of global and public health problems that are rooted in a 'wicked space'?

This chapter presents possibility thinking/possibility doing defined here as a process and experience of creativity, imagination, and ideation to explore and generate innovative ideas and solutions to complex problems. Within this context, such thinking encourages questioning the status quo approaches and imagining alternatives to conventional thinking. In short, possibility thinking/possibility doing involves more than a methodology but a mindset that challenges assumptions and transcends traditional framing of problems.

Keywords Possibility thinking · Creativity · Imagination · Wicked problems · Ideation

A. J. Masys (✉)
College of Public Health, Global and National Security Institute (GNSI), University of South Florida, Tampa, USA
e-mail: tmasys@usf.edu

B. Jacob et al. (eds.), *Tech Enabled Global Health Security*, Advanced Sciences and Technologies for Security Applications, https://doi.org/10.1007/978-3-031-86997-6_1

1 Introduction

The conditions under which the global health security landscape exists today is characterized by an inherent Volatility, Uncertainty, Complexity, Ambiguity (VUCA) [44]. As we have experienced with the COVID-19 pandemic, our highly connected and interdependent world has given rise to vulnerabilities whereby the requirement for global efforts to manage and combat highly pathogenic infectious diseases and emerging public health crisis is at a whole new level. Such global health security issues have multidimensional roots with linkages in a complex 'wicked space' involving influences of the social and environmental determinants of health. Rittel and Weber [54] describe a wicked problem as an issue that is difficult or impossible to solve because of its complex and interconnected nature. Examples of wicked problems in Global Health relate to issues such as climate change, biodiversity loss, persisting poverty, the growing obesity epidemic, and food insecurity [69]. Along with these wicked problems are equally complex events, with roots in the 'wicked space', like the COVID-19, H1N1 and H5N1 outbreaks, the emergence and spread of monkeypox, and the international spread of vector borne disease that are challenging the global health security landscape revealing an inherent fragility in our societal global health systems.

With this in mind, global health security can be considered a wicked problem due to its complexity, interconnectedness, evolving and emerging threats and a lack of clear solutions. Global Health Security is defined as:

> '... the activities required, both proactive and reactive, to minimize the danger and impact of acute public health events that endanger people's health across geographical regions and international boundaries' (WHO[1]).

> '... the existence of strong and resilient public health systems that can prevent, detect, and respond to emerging infectious disease threats, wherever they occur in the world' (USAID[2]).

Within this 'wicked space' of global health security, possibility thinking brings a fresh approach to managing wicked problems and designing wicked solutions. Possibility thinking/Possibility doing is defined here as a process and experience of creativity, imagination, and ideation to explore and generate innovative ideas and solutions to complex 'wicked' problems. As noted by complexity theorist [53] (cited in Boulton et al. [4: 69]) '...the future can be felt in the present through the action of the imagination'. Within this context, such thinking encourages questioning the status quo approaches and imagining alternatives to conventional thinking. In short, possibility thinking/possibility doing involves more than a methodology but a mindset that challenges assumptions and transcends traditional framing of problems. We see this in the world of art where '...a similar attitude encourages artists to challenge existing assumptions and invite their audiences to do the same [26: 171] highlighting how creativity emerges out of difference [27, 28]. Similarly in the sciences, paradigm

[1] https://www.who.int/health-topics/health-security#tab=tab_1.

[2] https://www.usaid.gov/global-health/health-areas/global-health-security.

shifts (relativity, quantum theory…) are examples of breaking free from traditional views and mindsets.

Emerging and informed from over 15 years of action research and drawing upon the literature on creativity, imagination, ideation, and innovation across the arts and sciences, a framework for possibility thinking/possibility doing is presented. Within the possibility framework, experiential learning and reflective practices [56] figures prominently. As with possibility thinking, such insights derived from Action research requires one to reflect upon and challenge '…our own beliefs, values, assumptions, ways of thinking, strategies and behaviour' [10: 18] pertaining to inquiry. From the reflective practices of the action research and the continued refinement, testing and serendipitous 'playful' exploration, a 'possibility thinking/possibility doing framework' emerged both as a set of principles and a mindset.

2 Wicked Problems and Global Health

As described in Chan et al. [9: 2], the 'wickedness' of wicked problems is ubiquitous across social systems. This 'wicked space' as identified in this chapter is characterized by an inherent volatility, uncertainty, complexity and ambiguity (VUCA). As cited in Masys [44] and described in Kraaijenbrink [36], VUCA is defined as:

> **Volatility**—Volatility refers to the speed of change in an industry, market or the world in general. It is associated with fluctuations in demand, turbulence and short time to markets and it is well-documented in the literature on industry dynamism. The more volatile the world is, the more and faster things change.
> **Uncertainty**—Uncertainty refers to the extent to which we can confidently predict the future. Part of uncertainty is perceived and associated with people's inability to understand what is going on. Uncertainty, though, is also a more objective characteristic of an environment. Truly uncertain environments are those that don't allow any prediction, also not on a statistical basis. The more uncertain the world is, the harder it is to predict.
> **Complexity**—Complexity refers to the number of factors that we need to take into account, their variety and the relationships between them. The more factors, the greater their variety and the more they are interconnected, the more complex an environment is. Under high complexity, it is impossible to fully analyze the environment and come to rational conclusions. The more complex the world is, the harder it is to analyze.
> **Ambiguity**—Ambiguity refers to a lack of clarity about how to interpret something. A situation is ambiguous, for example, when information is incomplete, contradicting or too inaccurate to draw clear conclusions. More generally it refers to fuzziness and vagueness in ideas and terminology. The more ambiguous the world is, the harder it is to interpret.

The Global Health Security wicked space encompassing issues from complex humanitarian crisis and global climate change to systemic racism, poverty, food and water insecurity and widening economic inequality, require a new lens in which to see, understand and frame the problem and new thinking to 'design' wicked solutions. A cited in Chan and Xiang [9: 2], Rittel and Webber [54] '...assert that wicked problems are innately resistant to any neopositivist formulations of scientific analysis and rigid linear protocols for professional practice'. Within the context of the COVID-19 pandemic, the failure to prepare for the global pandemic is reflective of both the inconceivability of the problem as well as a lack of political will to heed the warning of those who identified this looming threat [34]. This points to the requirement that to do differently we must think differently. As described in Kawa et al. [34] 'wicked problems thus require disciplinary expertise and transdisciplinary skills but also purpose-driven commitment to problems that defy easy resolution and require stakeholders to work across ideological and epistemological differences'. This is reflected within the intelligence and security domain by facilitating alternative perspectives defined as '...the result of looking at a situation, problem, or fact through a different mindset, cultural frame, or value and belief structure' [50]. Facilitating this problem framing are the activities rooted in creativity, imagination, ideation, experimentation, learning, and innovation (defining the possibility thinking/possibility doing space). The inherent complexity of wicked problems opens a space for participative possibility thinking. As noted by Syed [63: 17] '...solutions to complex problems typically rely on multiple layers of insight and therefore require multiple points of view'. From this the possibility space is defined.

3 Overview of Possibility Thinking

Possibility thinking is grounded in a 'what if' paradigm that challenges our notions of 'what is' to imagine 'what could be' and realize 'what should be' (resonating with the appreciative inquiry approach [12]). Possibility thinking brings to the forefront ideas that can be played with, explored, developed and realized whereby what did not exist becomes real. Creativity, imagination, ideation, learning, experimentation are at the heart of possibility thinking. As noted in Cremin et al. [15]:

> 'It has been proposed that possibility thinking is at the core of creative learning [13, 14], that it can be understood from the tripartite perspective of people or agents, processes and domains and that it involves both problem finding and problem solving. Craft (ibid) also argues that possibility thinking is implicit in learners' engagement with problems, suggesting that it is exemplified through the posing, in multiple ways, of the question 'what If?' and that it involves the shift from 'what is this and what does it do?' to 'what can I do with this?'

Possibility thinking is a 'space' whereby 'we can see the world anew'. It is the experience of applying alternative and new ways of thinking to challenge our traditional notions of what is. It incorporates a 'reflective practice' [56] allowing one to consider the influences of VUCA on past, present and future circumstances. Exploring the VUCA 'wicked space' that characterizes key features of global health security requires a new perspective, a new type of thinking. Possibility thinking, within a domain of Possibility Studies, provides novel way of navigating the complexity of the Global Health Security landscape. Glaveanu [23] defines the space of Possibility Studies '…as a new and emergent multi- and trans-disciplinary field dedicated to the study of this shift of focus from being to becoming, from what is to what could be, from deterministic accounts of the world to agentic, generative and open-ended understandings'. In possibility thinking we look at '…Stretching our imagination beyond the boundaries of what is possible might turn out to be crucial to get unstuck from dominant narratives' [16: 53]. Working in this possibility space '…stimulates creativity, pushing us to explore, create, innovate, see across boundaries, disciplines…' [16: 53]. It is this very integration of creativity, imagination in the 'possibility space' that creates the possible as a field of inquiry [25]' (cited in Lin [39]).

As described by [23]:

> 'No matter whether we are talking about realistic or unrealistic possibilities, at the heart of this notion stands a commitment to the idea that the world is not yet finished, that it is in a continuous process of becoming, and that this becoming – for as much as we might anticipate it – is never entirely predictable. Beyond 'what currently is', there is room for 'what is not yet, but will be', 'what might be' and 'what can never be'. And the realm of the possible doesn't stop with the future, it also helps us imagine 'what might have been', in the past, and what exists 'as if' in the present. To engage with the possible means to infuse 'what is' with new perspectives and, in doing so, to radically transform it [20, 25]'.

> Impossible is not a fact, it is an opinion.
> Muhammad Ali

'Through possibility thinking we move beyond the here and now and reperceive, imagine, co-create and experience what is not yet. Glaveanu [24] argues that 'we try to anticipate the future, consider alternatives for our current situation, revisit the past, think in abstract terms, imagine how things could have been different and things that could never exist, and then use all of these to shape reality. Importantly, such explorations of the possible—from a quick alternative solution to a problem to a long episode of daydreaming about the future—are experienced as real'.

The chapters in this book are examples of possibility thinking/possibility doing in the wild, illustrating tech enabled solutions to many 'wicked' and complex problems. These solutions emerged from a process and experience of creativity, imagination, and ideation to explore and generate innovative ideas and solutions. Within this context, such thinking encourages questioning the status quo approaches and imagining alternatives to conventional thinking. In short, possibility thinking involves more than a methodology but a mindset that challenges assumptions and traditional framing of problems. It emerges and thrives in a culture of perseverance, grit, purpose, play, a culture that embraces failures as feedback (foundational cultural requirements to foster possibility thinking).

This chapter explores 'possibilities in the wild': from possibility thinking to possibility doing.

4 Discussion

> Every block of stone has a statue inside it, and it is the task of the sculptor to discover it.
>
> Michelangelo

We live in VUCA times where the world is experiencing unprecedented societal change. In the global health security landscape, these changes are being influenced by interdependent PESTEL factors (Political; Economic; Social; Technological; Environmental; Legal). The patterns of the past are no longer sufficient to project and predict near term events under VUCA conditions. Complex problems require a new mindset, new ways of seeing the world. Seeing and treating the world in Gaussian (normal distribution) terms is insufficient. For the world is not Gaussian. Stepping into this world requires 'possibility thinking'. Facilitating this problem framing are the activities rooted in creativity, imagination, ideation, experimentation, learning, and innovation (defining the possibility thinking space).

5 Possibility Thinking and Imagination

> Imagination is more important than knowledge. For knowledge is limited, whereas imagination embraces the entire world.
>
> Albert Einstein

Arguing that we live in a world where a 'crisis of imagination' is palpable, Finn et al. [19] point to the requirement and urgency for the application of imagination. A case in point is reflected in the 9/11 Report National Commission on Terrorist Attacks [49: 336] explicitly pointing to '...a lack of imagination and a mindset that dismissed possibilities'. Finn et al. [19: 2] argue that 'To navigate the global challenges of the twenty-first century, humans will need to develop new skills for planning, problem-solving, anticipation, and adaptation. Understanding how imagination can be fostered and deployed will determine how humans can survive and flourish in the future'.

The domain of study of imagination and applied imagination is truly interdisciplinary. As described in Standford Encyclopedia of Philosophy[3]: '*To imagine* is to

[3] https://plato.stanford.edu/entries/imagination/.

represent without aiming at things as they actually, presently, and subjectively are. One can use imagination to represent possibilities other than the actual, to represent times other than the present, and to represent perspectives other than one's own. Unlike perceiving and believing, imagining something does not require one to consider that something to be the case'.

Inherent in the human mind, Comrie et al. [11] argue that 'The ability to imagine is essential to human experience. At a broad level, imagination has a major role in human creativity, agency and everyday thoughts and actions. More specifically, humans have and express many types of imagined experiences. These include recollections, predictions, simulations, counterfactuals, fantasies, suppositions and mind-wandering'.

The creation of art and the hypothesis' within the Sciences is a function of imagination. As described by Burton [7] '…imagination enables us to form associations and connections, and thereby to apply our knowledge to real life situations. It opens up alternatives and possibilities and guides our decision-making by playing them out in our mind'. This certainly resonates with the thought experiments of Einstein.

6 Possibility Thinking and Creativity

> Creativity requires the courage to let go of certainty.
> Erich Fromm

Creativity is a multifaceted construct that has been extensively studied across various disciplines, including psychology, sciences, education, business, and the arts. The literature on creativity encompasses a wide range of theories, models, and empirical studies aimed at understanding the nature, processes, and outcomes of creative thinking and behavior. We often equate creativity to such noted accomplishments as Pablo Picasso's *Guernica* or Einstein's thought experiments pertaining to relativity.

As described in Simonton [58: 14] creativity, in the standard definition is based upon two criteria: (a) novelty or originality and (b) usefulness, value, or appropriateness [55]. However, a third criteria has emerged focusing on "surprise" or at least "nonobviousness,". With this in mind Simonton [58: 14] adopts the following three-criterion definition:

> '*An idea is creative to the extent that it is jointly original, useful, and surprising* [59]. The insertion of the qualifier "jointly" means that the definition is multiplicative rather than additive. An idea cannot possibly be creative if it is commonplace, useless, or obvious'.

Contextualizing and operationalizing the definition of creativity, Henriksen et al. [31: 104] defines it as: '—the ability to devise novel and effective ideas and solutions [55]—combined with an ability to 'see' the world, a willingness to suspend judgment and an openness for the new, is essential in addressing the uncertainty of complex global and societal challenges (e.g., climate change, wealth inequality, racial injustice, and countless others)'.

With this in mind, Possibility Thinking requires thoughtful experimentation stemming from visioning—to move beyond imagination, where you envision possibilities, to creativity, where you begin finding solutions, to innovation where the rubber meets the road. Working in this space of possibilities, requires grit and courage to take a stand that is often counter to traditional thinking and views. In carrying out research on creative individuals, Sternberg and Lubart [60, 61] argue that going against the grain and the "guts to defy the crowd" are common place and are often required. For example, the works of Picasso and Cezanne were met with distaste and disregard by the Paris art community. Similarly in the field of physics, Einsteins views on relativity were counter traditional norms and paradigms. This connects with Steinberg [62] three types of defiance in the Triangle Theory of creativity:

- Defying the crowd: defying the beliefs, values and practices of one's field
- Defying oneself: defying (moving beyond) one's earlier values, practices and beliefs
- Defying the zeitgeist: defying the often unconsciously accepted presuppositions and paradigms in one's field.

Weisberg [70: 387] presents that 'Professionals interviewed in various fields, such as artists, scientists, and businesspeople, emphasized that the creative individual went against conventional thinking, took risks, and followed through on the consequences of taking those risks'. Creative work thereby entails a "process in which ideas are generated, elaborated, evaluated, stored, discarded, or implemented into a product by individuals or groups" with the intent of developing something novel and useful' [30] often under conditions of 'friction'.

7 Possibility Thinking and Ideation

> The best way to have a good idea is to have a lot of ideas.
> Linus Pauling

A key 'space' in the design thinking process [5, 6], ideation is associated with generating unique or uncommon ideas. As described by Brown and Wyatt [6] '…The design thinking process is best thought of as a system of overlapping spaces rather than a sequence of orderly steps. There are three spaces to keep in mind: *inspiration, ideation,* and *implementation* (Fig. 1). Think of *inspiration* as the problem or opportunity that motivates the search for solutions; *ideation* as the process of generating, developing, and testing ideas; and *implementation* as the path that leads from the project stage into people's lives'.

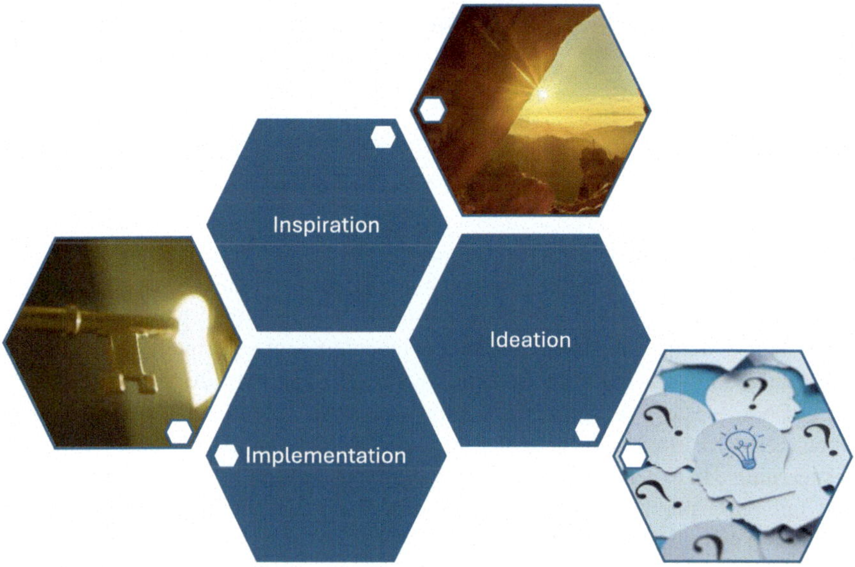

Fig. 1 Three spaces of innovation: inspiration; ideation; implementation

Within the design thinking process, 'Ideation' focuses on idea generation through divergent and convergent thinking to explore the possibility space. It is a core design thinking principle and a step in the design thinking process. As cited in Titus [65: 238] 'Divergent thinking is considered a fundamental ideation skill that facilitates the creation of an assortment of creative ideas [1, 29, 66, 67]'.

Collaborative ideation involves a process of sharing ideas that further stimulates the generation of new concepts and ideas that may not have emerged if not for a space of 'collaborative flow' akin to a jazz ensemble experience. As noted in Paulus et al. [52: 2] '...the elaboration processes involved in this building on others' ideas are key factors in enhancing the novelty of ideas in groups with diverse perspectives ... Such collaborative ideation processes require that group members carefully balance the search process of their own knowledge base with attention to the ideas shared by other group members'. The power of collaborative ideation certainly connects with and is supported by Gama [21: 100] reporting that 'Liedtka [40] describes collaborative ideation as a participatory sense-making method such as mind mapping that facilitates team-based processes to solve problems and create a shared mind amongst partners. These arguments are supported by the notion that collaborative ideation frequently uses participatory concept development techniques (i.e. collaborative prototyping, mind mapping, and causal analysis) to help address problems. Bogers and Horst [2], conducted an inductive study to investigate how collaborative activities across functional, hierarchical, and organizational boundaries can improve the overall collaborative process. The findings suggest that collaborative activities are used as a critical success factor, thereby serving as a platform for the cross-fertilization of knowledge'.

8 Possibility Thinking: Experimentation and Learning

> Research is formalized curiosity. It is poking and prying with a purpose.
> Zora Neale Hurston

Experimentation is crucial to learning within possibility thinking because it allows us to explore new ideas, test hypotheses, and discover innovative solutions. As described in Thomke [64: 125] 'All experiments, by definition, generate information, which, at a minimum, becomes an input to additional experiments or is applied to the end result—the intent of the experiment itself—or both. An experimentation process, however, can do more than generate information useful to the process itself. When well structured and integrated into an organization, experimentation generates learning'. In experimentation, the learning mindset is key in reshaping how we view failure. Experimentation supports assumption and hypothesis testing, and the exploration and discovery of new ideas.

Experimentation as a Problem Based Learning (PBL) approach, facilitates learning while engaging actively with meaningful problems. As noted in Yew and Goh [71] '…the underpinning philosophy of PBL is that learning can be considered a "constructive, self-directed, collaborative and contextual" activity. The principle of constructivism positions students as active knowledge seekers and co-creators who organise new relevant experiences into personal mental representations or schemata with the help of prior knowledge'. This resonates with theories associated with reflective practices [17, 56].

Reflective practices as described and developed by Dewey [17], Schon [56] provide a useful framework for active thinking [41] in which to operationalize the underlying processes associated with possibility thinking. Dewey [17] argues that reflection is a deliberate and active process. His basic notion is that reflection is an active, deliberative cognitive process involving a sequence of interconnected ideas that include the underlying beliefs and knowledge of an individual [46]. Schön claimed that by using reflection, practitioners can make explicit, hidden (tacit) knowledge.

9 Possibility Thinking and Innovation

> I have not failed. I've just found 10,000 ways that won't work.
> Thomas Edison

The OECD/Eurostat [51] defines Innovation as:

'… a new or improved product or process (or combination thereof) that differs significantly from the unit's previous products or processes and that has been made available to potential users (product) or brought into use by the unit (process)'.

It encompasses novelty, utility and value creation and can be considered both an activity and an outcome of an activity. As noted in Bolton [3], 'Innovation is something different that creates value. Across the application domains, innovation can be thought of in terms of outputs, processes and mindsets [33]'. An innovation mindset is a way of thinking and a way of doing that opens up a space of possibilities to explore, learn and create. Kahn [33] suggests, however, that innovation as a mindset can either mean individuals' orientation or an innovative culture. Design thinking as an innovation process and mindset are rooted in a '…set of values, cognitive and behavioral attributes constituting the design thinking mindset are suggested to be empathy, collaboration, inquisitiveness and openness to learning, acceptance of uncertainty and risk, optimism, playfulness, and tangibility…' (cited in Krohn [37: 19].

The quote from Thomas Edison certainly illustrates the salient qualities of an innovation mindset which can be described as having the following features (Fig. 2):

- Being Purpose driven
- Embracing feedback from failure
- Having courage to explore new ways of doing

Fig. 2 Innovation mindset

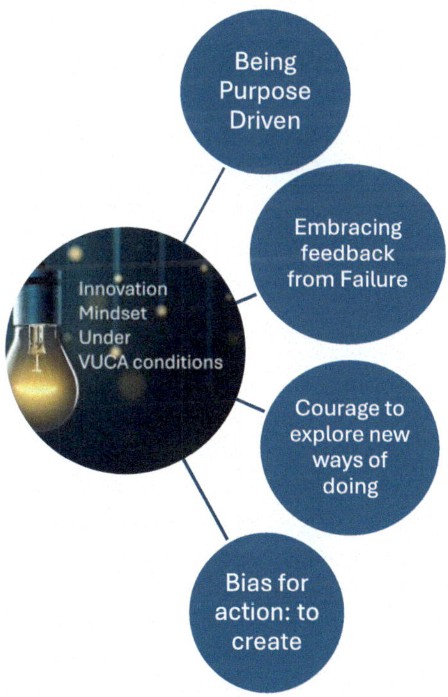

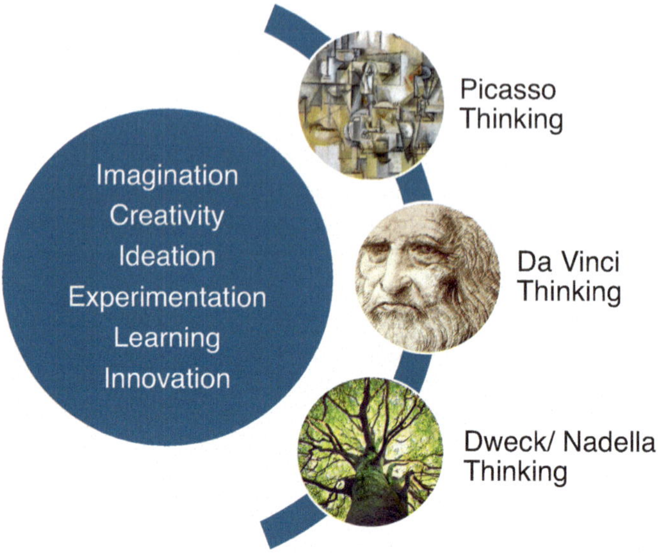

Fig. 3 Possibility thinking/possibility doing framework

- Having a bias for action: to create/co-create

10 Possibility Thinking/Possibility Doing Framework

Emerging from the analysis of the literature and from over 15 years of action research, the defining activities of the possibility thinking/possibility doing space are creativity, imagination, ideation, experimentation, learning, and innovation. These activities provide the foundation of the proposed Possibility Thinking/Possibility Doing Framework.

The framework is expressed and actualized across three mindsets (Thinking): Picasso Thinking; Da Vinci Thinking; Dweck/Nadella Thinking. Illustrated in Fig. 3.

11 Picasso Thinking

Pablo Picasso's creativity is best known for his 'courageous' innovation and ability to reinvent his artistic style. His revolutionary contributions to the Cubist movement in art is rooted in **Perspective**. As described in Masys [43], the single perspective dominated the art world until the development of Cubism during the period 1907–1911. Picasso and Braque pioneered and introduced a new visual language representing a 'revolution' in the art world. This new way of seeing challenged traditional ideas of

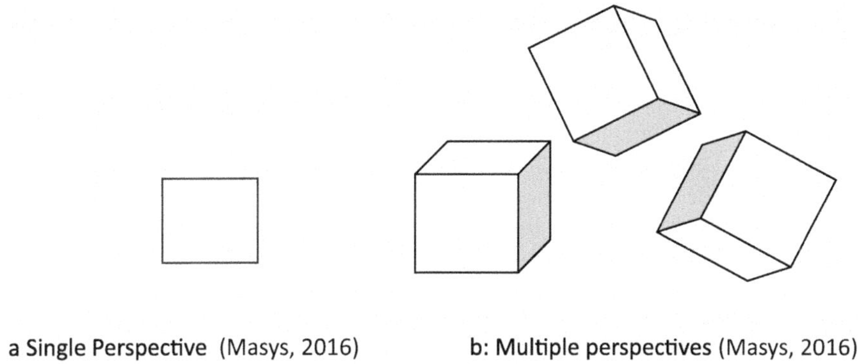

Fig. 4 **a** Single perspective [43]. **b** Multiple perspectives [43]

perspective. The notion of a single viewpoint picture was overturned challenging the rigid geometry of perspective. This 'relativistic' approach to art emerged in which images from various observations and points of view were captured. For example, as described in Masys [43], changing ones perspective when viewing an object (like walking around a cube) allows the observer to view the object from above, below and from the sides. In so doing this relativistic perspective proposed by the cubist revolution constructs and presents the object holistically (spatially and temporally) (Fig. 4a, b).

Shlain [57: 199] describes how 'Cubism breaks with Renaissance perspective. It views objects relatively: that is, from several points of view, no one of which has exclusive authority'. An excellent example of the original cubism is Picasso's Violin and Grapes [68].[4] Picasso presents the object explored from several different viewpoints thereby creating an image that was multidimensional.

We take from this the important argument that there are multiple perspectives, each adding to a more holistic impression of the possibility space. As noted by Keating et al. [35: 775], 'Shifts in vantage points, environmental conditions, or knowledge will influence perspectives of a system. It is naïve to consider there is only one system perspective that is "correct"'.

Possibility thinking and doing encompasses the work of Picasso and the Cubism revolution that challenged traditional linear thinking and gives voice to various perspectives thereby creating a more 'holistic' view of the problem space. Picasso and cubism embraces exploring the possibility landscape through perspective.

[4] https://www.metmuseum.org/art/collection/search/761944.

12 Da Vinci Thinking

Leonardo Da Vinci can be considered the actualization of creativity and possibility, hence the inclusion of Da Vinci thinking. This does certainly not negate the works of the Renaissance Masters such as Michelangelo and Raphael. Their creative and innovation mindsets are also captured under the moniker of Da Vinci Thinking. Leonardo da Vinci exemplified the Renaissance ideal of the polymath, excelling in art, science, engineering, and anatomy. His work and contributions to art, engineering and the sciences are characterized by his interdisciplinary thinking, innovative design, and his attention to observation and perspective. His creative potential was decoded in the book "**How to Think Like Leonardo da Vinci**" by Gelb [22] presents the **Seven Da Vincian Principles**:

1. **Curiosità (Curiosity)**: Embrace an insatiable quest for knowledge and continuous learning. Cultivate a curious mind by asking questions and seeking out new experiences.
2. **Dimostrazione (Demonstration)**: Commit to testing knowledge through experience, persistence, and a willingness to learn from mistakes. This principle emphasizes the importance of empirical evidence and learning from practical experience.
3. **Sensazione (Sensation)**: Sharpen your senses, particularly sight, as a means to enliven experience. Leonardo believed that refining the senses was crucial for enhancing creativity and perception.
4. **Sfumato (Going up in Smoke)**: Embrace ambiguity, paradox, and uncertainty. This principle encourages open-mindedness and the ability to navigate complex and uncertain situations without rushing to conclusions.
5. **Arte/Scienza (Art and Science)**: Develop a balance between science and art, logic and imagination. Leonardo's genius lay in his ability to integrate these seemingly disparate disciplines.
6. **Corporalità (Body)**: Cultivate grace, ambidexterity, fitness, and poise. Leonardo believed in the importance of physical health and its connection to mental well-being and creativity.
7. **Connessione (Connection)**: Recognize and appreciate the interconnectedness of all things. This holistic approach encourages seeing patterns and relationships between different fields and ideas [22].

The contributions of Da Vinci in the sciences, arts, engineering and his influence even today over 500 years later point toward a Da Vinci mindset described by the **Seven Da Vincian Principles** that is exemplified and realized across the activities of the possibility thinking space: creativity, imagination, ideation, experimentation, learning, and innovation. These principles of Da Vinci thinking become a foundation to support possibility thinking/possibility doing.

13 Dweck/Nadella Thinking

Carol Dweck, the Lewis and Virginia Eaton Professor of Psychology at Stanford University and renowned author of Mindset: The New Psychology of Success [18], introduced the concept of 'growth mindset' to describe an individual's belief in learning and intelligence. The growth mindset argues for the position that a person's abilities can be developed through perseverance and effort. It is argued that '…organizations and teams that hold a growth mindset…have employees who are more motivated and committed, more mutually supportive, and more creative and innovative' [48: xi]. The relationship between growth mindset and innovation has been well articulated in Krskova and Breyer [38] and is well demonstrated by Microsoft.

Satya Nadella was appointed as CEO of Microsoft in 2014 with a vision of transforming Microsoft into a 'growth mindset culture' [48: xv]. Satya Nadella, cited in McKinsey [45] argues that in reimaging innovation at Microsoft '… you will have to overemphasize [creating] the right culture so that you can continue to cultivate new capabilities and new concepts'. This is rooted in shifting Microsoft from a "know-it-all" culture to embrace a "learn-it-all" curiosity paradigm. Nadella credits the work of Dweck [18] for shaping the future of Microsoft. As reflected in Ibarra and Rattan [32] 'Dweck divides the world between learners and non-learners," explained Nadella, "demonstrating that a fixed mindset will limit you and a growth mindset can move you forward." Microsoft's culture change, he concluded, would centre on "the belief that everyone can grow and develop, potential is nurtured, not predetermined, and anyone can change their mindset'. They would shift from being 'know-it-alls' to 'learn-it-alls". This new growth mindset culture has been operationalized and reflected in the Microsoft company mission: "To empower every person and every organisation on the planet to achieve more."

Emerging from the operationalization of a 'growth mindset culture', Dweck/Nadella thinking can be characterized by several key attributes that fuel creativity, imagination, innovation, and ideation:

1. **Embracing Challenges**: The growth mindset encourages individuals, teams and organizations to see challenges as opportunities to learn and grow leading to innovative solutions and creative breakthroughs.
2. **Persistence**: At the foundation of the growth mindset is persistence. In the face of obstacles, persistence fosters a resilient approach to problem framing and solution navigation, often leading to more imaginative and innovative outcomes.
3. **Empathy and Purpose**: As noted by Satya Nadella 'The reason I talk about empathy is that I believe this is the leading indicator of success. Innovation comes only when you are able to meet unmet, unarticulated needs—and this comes from a deep sense of empathy we all have' [45].
4. **Inclusivity, diversity and Collaboration**: Collaboration is at the foundation of innovation. Sharing diverse perspectives and ideas within a team can spark creativity and lead to innovative solutions. As presented in Cardenas [8], by championing a workplace where every voice mattered and every individual felt valued, employee morale improved and innovation flourished, by leveraging the

broad range of experiences and ideas. This growth mindset culture fueled individuals and teams to expand their creative and imaginative capacities, leading to fresh ideas and innovative approaches.
5. **Risk-Taking**: A growth mindset is a key driver for innovation as it embraces failure as a learning opportunity thereby encouraging individuals to take risks and experiment with new ideas.

These attributes combine to create a unique mindset that not only fosters innovation but also inspires others to think creatively and pursue ambitious goals.

These three mindsets: Picasso, Da Vinci, Dweck/Nadella reveal, leverage and operationalize the principles of the Possibility Space: creativity, imagination, ideation, experimentation, learning, and innovation. Picasso's rule-breaking and constant evolution, da Vinci's interdisciplinary genius and empirical observation, and Dweck/Nadella growth mindset and intense focus and a lean forward/learning posture each offer unique insights and together reify into a framework of Possibility Thinking/Possibility Doing.

The chapters in this book are instantiations of Possibility Thinking/Possibility Doing in the 'wild'. They reflect the application of Picasso Thinking (Perspective), Da Vinci Thinking (interdisciplinarity, experimentation, reflection) and Dweck/Nadella Thinking (growth mindset). Each employing the principles of the possibility space: creativity, imagination, ideation, experimentation, learning, and innovation. Of note, during the foundational research and development of the chapters, collaborative ideation, co-creation and even applied Red Teaming [42, 43] were observed.

14 Conclusion

The social reality of wicked problems is palpable and has become the norm in complex societal systems. As described by Moritz and Kawa [47], 'Many problems facing humanity are so daunting they seem impossible to solve'. Wicked problems, such as those within the global health security domain, require disciplinary expertise and transdisciplinary skills but also purpose-driven commitment. The possibility thinking/possibility doing framework opens a space and opportunities to solve wicked problems.

Possibility thinking/possibility doing becomes a key enabler in addressing such wicked problems. The defining activities of the possibility thinking/possibility doing space are creativity, imagination, ideation, experimentation, learning, and innovation. These activities provide the foundation of the proposed Possibility Thinking/Possibility Doing Framework actualized across three mindsets (Thinking): Picasso Thinking; Da Vinci Thinking; Dweck/Nadella Thinking.

References

1. Baer JS (1993) Creativity & divergent thinking: a task-specific approach. Erlbaum, Hillsdale, NJ
2. Bogers M, Horst W (2014) Collaborative prototyping: cross-fertilization of knowledge in prototype driven problem solving. J Prod Innov Manag 31(4):744–764
3. Bolton, R. (2019) Back to basics: what is innovation? https://www.forbes.com/councils/forbesbusinesscouncil/2019/12/30/back-to-basics-what-is-innovation/
4. Boulton JG, Allen PM, Bowman C (2015) Embracing complexity: strategic perspectives for an age of turbulence. Oxford University Press, Oxford
5. Brown T (2008) Design thinking. Harv Bus Rev
6. Brown T, Wyatt J (2009) Design thinking for social innovation. Stanford Soc Innov Rev 8(1):31–35. https://doi.org/10.48558/58Z7-3J85
7. Burton N (2018). https://neelburton.com/2018/11/02/the-psychology-and-philosophy-of-imagination/
8. Cardenas B (2024) Transformative leadership: unveiling Satya Nadella's three revolutionary strategies at Microsoft. https://worthyleadership.com/transformative-leadership-unveiling-satya-nadellas-three-revolutionary-strategies-at-microsoft/#:~:text=Another%20cornerstone%20of%20Nadella's%20leadership,innovation%2C%20engagement%2C%20and%20inclusivity.
9. Chan JKH, Xiang W-N (2022) Fifty years after the wicked-problems conception: its practical and theoretical impacts on planning and design. Socio-Ecol Pract Res 2022(4):1–6
10. Coghlan D, Brannick T (2010) Doing action research in your own organization, 3rd edn. SAGE Publishing, London
11. Comrie Alison E, Frank Loren M, Kenneth K (2022) Imagination as a fundamental function of the hippocampus. Phil Trans R Soc B37720210336
12. Cooperrider DL, Whitney D, Stavros JM (2008) Appreciative inquiry handbook: for leaders of change, 2nd edn. Crown Custom Publishing, Inc., Ohio, USA
13. Craft A (2000) Creativity across the primary curriculum: framing and developing practice. Routledge Falmer, London
14. Craft A (2001) Little c creativity. In: Craft A, Jeffrey B, Liebling M (eds) Creativity in education. Continuum, London, pp 45–61
15. Cremin T, Burnard P, Craft A (2006) Pedagogy and possibility thinking in the early years. Int J Think Skills Creat 1(2):108–119
16. Damhof L, Gulmans J (2023) Imagining the impossible: an act of radical hope. Possibility Stud Soc 1(1–2):51–55
17. Dewey J (1933) How we think: a restatement of the relation of reflective thinking to the educative process. D.C. Heath and Company, New York
18. Dweck CS (2006) Mindset: the new psychology of success. Random House Publishing Group, New York
19. Finn E, Torrejon Capurro C, Bennett MG, Wylie R (2023) Applied imagination. Front Psychol 14:1275942
20. Gaggioli A (2020) Transformative cognition. In: Glăveanu V (ed) The Palgrave encyclopedia of the possible. Palgrave Macmillan. https://doi.org/10.1007/978-3-319-98390-5_63-1
21. Gamma F (2019) Managing collaborative ideation: the role of formal and informal appropriability mechanisms. Int Entrep Manage J 15:97–118
22. Gelb MJ (2004) How to think like Leonardo da Vinci: seven steps to genius every day. Delta Trade Paperback
23. Glăveanu VP (2023) Possibility studies: a manifesto. Possib Stud Soc 1(1–2):3–8
24. Glăveanu VP (2020) The possible: a sociocultural theory. Oxford University Press
25. Glaveanu VP (2018) The possible as a field of inquiry. Eur J Psychol 14(3):519–530
26. Glaveanu VP (2017) Creativity and wonder. J Creat Behav 53(2):171–177
27. Glăveanu VP, Gillespie A, Valsiner J (eds) (2015) Rethinking creativity: perspectives from cultural psychology. Routledge, London

28. Glăveanu VP, Beghetto RA (2017) The difference that makes a 'creative' differ-ence in education. In: Beghetto RA, Sriraman B (eds) Creative contradictions in education. Springer, Cham, pp 37–54
29. Guilford JP (1950) Creativity. Am Psychol 5:444–454
30. Harrison S, Rouse E, Fisher C, Amabile T (2022) The turn toward creative work. Acad Manage Collect 1(1):1–15
31. Henriksen D, Richardson C, Gruber N, Mishra P (2022) The uncertainty of creativity: opening possibilities and reducing restrictions through mindfulness. In: Beghetto RA, Jaeger GJ (eds) Uncertainty: a catalyst for creativity, learning and development, creativity theory and action in education, vol 6. https://doi.org/10.1007/978-3-030-98729-9_7
32. Ibarra H, Rattan A (2019) Microsoft: instilling a growth mindset. https://herminiaibarra.com/microsoft-instilling-a-growth-mindset/
33. Kahn KB (2018) Understanding innovation. Bus Horiz 61(3):453–460. https://doi.org/10.1016/j.bushor.2018.01.011
34. Kawa NC, Arceño MA, Goeckner R, Hunter CE, Rhue SJ, Scaggs SA, Biwer ME, Downey S, Field JS, Gremillion K, McCorriston J, Willow A, Newton E, Moritz M (2021) Training wicked scientists for a world of wicked problems. Humanit Soc Sci Commun 8(1). https://doi.org/10.1057/S41599-021-00871-1
35. Keating CB, Kauffmann P, Dryer D (2001) A framework for systemic analysis of complex issues. J Manage Dev 20(9):772–784
36. Kraaijenbrink J (2018) What does VUCA really mean? Forbes. https://www.forbes.com/sites/jeroenkraaijenbrink/2018/12/19/what-does-vuca-really-mean/?sh=6fdda5c917d6
37. Krohn M (2022) The crucial role of mindsets in innovation efforts opening the black box in the context of frugal innovation. Springer
38. Krskova H, Breyer YA (2023) The influence of growth mindset, discipline, flow and creativity on innovation: introducing the M.D.F.C. model of innovation. Heliyon 18;9(3):e13884
39. Lin Y (2021) Possibility thinking. In: The palgrave encyclopedia of the possible. Palgrave Macmillan, Cham. https://doi.org/10.1007/978-3-319-98390-5_156-1
40. Liedtka J (2015) Perspective: linking design thinking with innovation outcomes through cognitive bias reduction. J Prod Innov Manag 32(6):925–938
41. Machost H, Stains M (2023) Reflective practices in education: a primer for practitioners. CBE—Life Sci Educ 22:es2, 1–11
42. Masys AJ (2012) Black Swans to Grey Swans-Revealing the Uncertainty. Int J Disaster Prev Manage 21(3):320–335
43. Masys AJ (2016) Counter-terrorism and design thinking: supporting strategic insights and influencing operations. In: Masys AJ (ed) Disaster forensics: understanding root cause and complex causality. Springer Publishing
44. Masys AJ (2021) The security landscape—systemic risks shaping non-traditional security. In: Masys AJ (ed) Sensemaking for security, advanced sciences and technologies. Springer Publishing
45. McKinsey and Company (2018) Microsoft's next act CEO Satya Nadella talks about innovation, disruption, and organizational change. McKinsey Quarterly March 2018. https://www.mckinsey.com/industries/technology-media-and-telecommunications/our-insights/microsofts-next-act/
46. Mohamed M, Rashid RA, Alqaryouti MH (2022) Conceptualizing the complexity of reflective practice in education. Front Psychol; Sec Educ Psychol 13. https://doi.org/10.3389/fpsyg.2022.1008234
47. Moritz M, Kawa NC (2022) The world needs wicked scientists. Am Sci 110:212–217
48. Murphy MC (2024) Cultures of growth: how the new science of mindset can transform individuals, teams, and organizations. First Simon & Schuster Hardcover Edition. Simon & Schuster, New York
49. National Commission on Terrorist Attacks (2004) The 9/11 commission report. WW Norton https://govinfo.library.unt.edu/911/report/911Report.pdf

50. NATO (2017) The NATO alternative analysis handbook, 2nd Edn. https://www.act.nato.int/wp-content/uploads/2023/05/alta-handbook.pdf
51. OECD/Eurostat (2018) Oslo manual 2018: guidelines for collecting, reporting and using data on innovation. In: The measurement of scientific, technological and innovation activities, , 4th edn. OECD Publishing, Paris. https://doi.org/10.1787/9789264304604-en
52. Paulus PB, Baruah J, Kenworthy JB (2018) Enhancing collaborative ideation in organizations. Front Psychol 9:2024. https://doi.org/10.3389/fpsyg.2018.02024
53. Prigogine I (2001) Creativity change and time's arrow. In: Nair R (ed) Mind matter and mystery. Sristi Publishers, New Delhi
54. Rittel HW, Webber MW (1973) Dilemmas in a general theory of planning. Policy Sci 4:155–169. https://doi.org/10.1007/BF01405730
55. Runco M, Jaeger GJ (2012) The standard definition of creativity. Creat Res J 21:92–96
56. Schön D (1983) The reflective practitioner: how professionals think in action. Basic Books, NewYork
57. Shlain L (1991) Art and physics: parallel vision in space, time and light. Harper Perennial, New York
58. Simonton DK (2022) From everyday creativity to eminent cases of creative achievement in professional domains. In: Lubart T et al (eds) Homo creativus, creativity in the twenty first century. Springer Publishing. https://doi.org/10.1007/978-3-030-99674-1_2
59. Simonton DK (2013) Creative thought as blind variation and selective retention: why sightedness is inversely related to creativity. J Theor Philos Psychol 33:253–266
60. Sternberg RJ, Lubart TI (1996) Investing in creativity. Am Psychol 51:677–688
61. Sternberg RJ, Lubart TI (1995) Defying the crowd. Free Press, New York
62. Sternberg RJ (2018) A triangular theory of creativity. Psychol Aesthet Creat Arts 12(1):50–67
63. Syed M (2021) Rebel ideas: the power of diverse thinking. Flatiron Books, New York
64. Thomke S (2014) Accelerating learning by experimentation. In: Gassmann O, Schweitzer F (eds) Management of the fuzzy front end of innovation. Springer International Publishing Switzerland, pp 125–140
65. Titus PA (2018) Exploring creative marketing thought: divergent ideation processes and outcomes. Psychol Mark 35:237–248
66. Torrance EP (1962) Guiding creative talent. Prentice Hall, Englewood Cliffs, NJ
67. Torrance EP (1974) Torrance tests of creative thinking: norms and technical manual. Scholastic Testing Press, Bensenville, IL
68. Violin and Grapes (1912) https://www.metmuseum.org/art/collection/search/761944
69. Walls HL (2018) Wicked problems and a 'wicked' solution. Glob Health 14:34
70. Weisberg RW (2020) Two confluence theories of creativity. In: Rethinking creativity: inside-the-box thinking as the basis for innovation. Cambridge University Press
71. Yew EHJ, Goh K (2016) Problem-based learning: an overview of its process and impact on learning. Health Prof Educ 2(2):75–79

Training an Artificial Intelligent iOS App Using a GPU Cluster with Four Nodes of GeForce GTX TITAN X, with 12 GB of Memory for Optimizing a Regional Convolutional Neural Network VGG-16 Model with a Robotic Arm for Real-Time Implementation of Seek and Destroy Malaria Mosquito Larval Source Management Tactics in Nkolondon Cameroon

Benjamin G. Jacob, Ricardo Izurieta, Hugues Nana Djeunga, Philippe Nwane, Jesse Casonova, Andre Domche, Arnauld Efon-Ekangouo, Donald Fezeu-Fankam, Namit Choudhari, Alexander Gambrell, Antonio-Nkondjio Christophe, Martin Rono, Joseph Mwangangi, Edwin Micheal, and Joseph Kamgno

Abstract Malaria continues to be endemic across Cameroon, causing thousands of deaths every year. With the ever-increasing mosquito resistance to insecticides, it becomes more important to consider the addition of other control strategies to help prevent the disease. Larval source management (LSM), in particular, continues to show significant decreases in transmission while still being low-cost, but it has yet

---

B. G. Jacob (✉) · R. Izurieta · A. Gambrell · E. Micheal
College of Public Health, University of South Florida, Tampa, USA
e-mail: bjacob1@usf.edu

H. N. Djeunga · A. Domche · A. Efon-Ekangouo · D. Fezeu-Fankam · J. Kamgno
The Center for Research On Filariasis and Other Tropical Diseases, Yaounde, Cameroon

P. Nwane
The University of Yaoundé I, Yaounde, Cameroon

J. Casonova
USF Health International University of South Florida, Tampa, USA

N. Choudhari
School of Geosciences, University of South Florida, Tampa, USA

A.-N. Christophe
Organisation de Coordination Pour La Lutte Contre Les Endémies en Afrique Centrale—OCEAC, Yaoundé, Cameroon

B. Jacob et al. (eds.), *Tech Enabled Global Health Security*, Advanced Sciences and Technologies for Security Applications, https://doi.org/10.1007/978-3-031-86997-6_2

to see wide-scale implementation in Cameroon. The proposed Seek and Destroy (S&D) program is an LSM strategy that combats malaria through the prediction and identification of malaria mosquito habitats followed by destruction and continuous suppression of said habitat utilizing real-time artificial intelligence, [AI] machine learning, [ML], unmanned aerial vehicle [UAV], or intelligent, smartphone application [app] technology, Python, ArcGIS Pro, and training of local abatement vector control officers. Once these habitats are identified, vector control officers can continuously suppress them, leading to significant decreases in vector density. These habitats can range from commercial roadside ditches (~1 m × 0.5 m) to agroecosystem pastureland swamps (~10 m x 10 m), which can be identified geolocated and treated with our proposed real-time, S&D, Artificial Intelligent [AI]-ML, pipeline workflow. We have built a customized, intelligent iOS app that allows for a drone to be flown over areas of suspected habitats with a built-in multispectral [Red, Green and Blue (RGB)] camera that can generate capture point signatures that can be interpolated using gridded, satellite visible and near-infra red [NIR] reflectance emissivity data [e.g., WorldView-2 satellite data which provides commercially available imagery of 0.46 m(m) resolution, image wavebands]. By so doing, real-time unmanned aerial vehicle [UAV] real time identification of potential georeferenceable, [GPS indexable], hyper/hypoendemic, aggregation/non-aggregation-oriented [i.e., hot/ cold spots] of sentinel site larval habitat, breeding sites can be optimally conducted, and field data recorded to make the AI–ML algorithm more accurate continuously. The S&D real time program was executed during a baseline collection in Cameroon to display the capability of malarial reduction in a hyperendemic malarious location. Baseline collection was carried out in two vegetable farming areas: the intervention site and a control site. The program's efficiency was measured based on weekly entomological surveillance through Human landing catches (HLC) and Pyrethrum spray catches (PSC) and biting rates. After four weeks post-intervention, there was a significant decline in indoor resting adult anopheline mosquitos and a steady decline in biting rates at the intervention site. The control site showed the same number of indoor resting adult anopheline mosquitoes and a similar biting rate. These results indicate the immense potential for malarial suppression by employing the S&D, real time, AI-ML assisted program alongside ongoing malaria control efforts. To prevent the reemergence of the destroyed habitats, S&D vector control officers were trained in habitat surveillance. This research revealed that investment in community health systems enables the longevity of AI-ML real time malaria control by training local officials in habitat identification, treatment, and surveillance. This ensures continual vector density suppression in combination with facilitating program sustainability in an entomological, study site, and intervention area. We have set out to achieve four aims with the completion of the LSM program. Our first aim is to evaluate the efficiency of S&D real-time tools to detect mosquito breeding habitats and deliver optimal microbial loads to suppress the development of malaria mosquitoes in different spatio temporal landscapes. Our second aim is to

M. Rono · J. Mwangangi
Kenya Medical Research Institute, Centre for Geographic Medicine Research-Coast, Kilifi, Kenya

conduct field surveys in control versus intervention settings to evaluate the impact of combining real time S&D with ongoing malaria control and elimination strategies. The third aim is to construct data-informed transmission models coupled with the effectiveness and costs of AI-ML S&D-driven malaria control strategies to determine the optimal mix of interventions using S&D to eliminate malaria in different endemic settings. The final aim is to develop a protocol to institute S&D alongside ongoing control measures with the Ministry of Health to guide elimination, monitoring, and decision-making.

Keywords Malaria · Seek and destroy · Artificial intelligence · Machine learning · Cameroon

1 Introduction

Cameroon remains among 11 countries that account for 92% of the malaria infection in sub-Saharan Africa (SSA). The disease is endemic in Cameroon, and the level of endemicity varies from one ecological zone to another. The Ministry of Public Health in Cameroon reported that malaria causes 30 to 35% of overall mortality and 67% of annual childhood mortality [1]. Current WHO control strategies implemented by the Ministry of Health and local stakeholders are based on mass preventive chemotherapy, targeted chemotherapy, and reactive strategies focused on drug treatments of infected individuals. In addition, indoor residual spraying (IRS) of insecticide to homes, long-lasting insecticide-treated nets, and rapid diagnostic tests. While these strategies have had a positive effect in reducing the burden of malaria, they are insufficient to achieve malaria control and elimination. The persistent issues with sustainability-related rising resistance of malaria parasites to Artemisine -based Combined Treatments [ACTs], increases vector resistance to insecticides, including to pyrethroids used in long-lasting bed nets, and increased invasiveness of new vectors, such as *Anopheles stephensi*, into new geographies mean that these strategies will not lead to a long-term solution for malaria control and would leave malaria to continue as the most widespread vector-borne disease (VBD) in Cameroon [5]. The disease would thus continue to disproportionately affect children under 5, pregnant women, and immunocompromised individuals living in remote rural, urban poor, and conflict settings throughout the country. Plasmodium falciparum is the predominant malaria parasite among the four reported nationwide. *Anopheles gambiae* s.l. and *An. funestus* s.l are the primary vectors responsible for transmission [2–4].

These challenges have renewed interest in innovative measures that can target malaria mosquitoes in different (non-chemical) ways in the outdoor environment to overcome the above barriers to malaria elimination. Routine entomological surveillance, for example, is needed in Cameroon to monitor and evaluate the coverage, effectiveness, and quality of the deployed vector control interventions. Only two-thirds of malaria-endemic countries in sub-Saharan Africa [SSA] monitor insecticide resistance annually, and even fewer countries routinely track and monitor mosquito larval habitats as it is operationally challenging to conduct frequently and across large eco–geographical areas [6–11]. Low-cost, scalable, and sustainable tools for real time entomological surveillance are needed in Cameroon to urgently guide the targeting and timing of larval source management (LSM). This targeted treatment of mosquito larvae breeding sites continues to show substantial decreases in transmission, up to 70–90%, especially when combined with long-lasting impregnated nets [12]. To date, LSM has not been widely implemented in Cameroon, yet targeting the immature, aquatic stages of the mosquito, when mosquitoes are clustered with limited ability to evade or escape treatment, may accelerate progress towards malaria elimination [11]. **This baseline collection is significant because it seeks to reduce the burden of malaria by improving entomological surveillance and larval control in Cameroon.**

Artificial intelligence, [AI] and machine learning [ML], object-based detection algorithms, Python, ArcGIS Pro, real-time, drone sensing technology, and geospatial analyses have been used to create maps and predict the geo-spatiotemporal abundance, occurrence, and distribution of larval habitats of vectors of malaria and other VBDs [12–16]. Large semi-permanent and permanent larval habitats (e.g., lakes, swamps, and drainage ditches) can easily be identified using remote sensing satellite data's coarse spatial and temporal resolution [17]. However, small or shallow transient larval habitats (e.g., puddles, tire tracks, roof gutters, hoof prints, vehicle ruts, and artificial water containers) are more challenging to identify [17, 18]. Unmanned aerial vehicles (UAV) commonly provide high spatial resolution images with temporal flexibility and at a relatively low cost [19]. Drone-based mapping has been used to identify aquatic larval habitats of different mosquito species, including *Anopheles, Aedes,* and *Culex* [8, 9, 20–22]. Improvements in the battery life and payload capacity of drones have resulted in using drones to deploy larvicides [23–25] Real-time drone-based larval habitat identification and treatment overcomes the challenges of the traditional approaches that are labor-intensive, time-consuming, costly, and limited to a small geographic scope [8]. ***This baseline collection is significant as it l*****everages novel technology for larval habitat identification and treatment in Cameroon.**

To keep continuous identification and elimination of the larval habitats, local people were trained to run the program themselves after initial implementation. We trained 15 people for each health district on implementing the Seek and Destroy [S7D] program for the maintainable control of malaria in Cameroon. During the baseline collection, 11 people were trained in Nkolondom (Fig. 1a and b) on using this novel real time technology for larval habitat identification and destruction.

Fig. 1 **a** Farmland-ecosystem homesteads and **b** Peri-urban market junction in Nkolondon, Cameroon

1.1 *AI-ML Technological Tools for* Anopheline *Larval Habitat Identification to Support Real-Time Data Collection, Automated Data Processing and Analysis, and Immediate Decision-Making in Cameroon*

In this baseline drone-based real time mapping and entomological collection, we employed different approaches to process acquired Worldview 2, satellite visible and near-infra-red [NIR] data of Nkolondon and classified potential georeferenceable [GPS indexable], malaria, oviposition, mosquito *Anopheles* larval habitats (e.g., technology-assisted digitization, Python code generation, ArcGIS unsupervised and supervised classification, etc.); in multiple intervention capture point, sentinel sites. We leveraged real time drone technology, AI-ML and deep learning algorithmic methods, high-performance computing smartphone dashboard platforms, at the intervention locality.

First, we employed a drone with a multi-angular, multispectral, RGB camera sensor to acquire visible, NIR, and short-wave infrared (SWIR) images. There are currently six species of *Anopheles* mosquitoes implicated in Cameroon [5]. The SWIR sensors offer incremental improvements in differentiating larval habitats from other land use land cover (LULC) types [8, 9]. To further enhance image classification, we fused the multispectral UAV imagery with WorldView-2 satellite 0.46 imagery employing deep learning.

Second, we employed a Mask Region-based Convolutional Neural Network (Mask R-CNN), which is an object-detection ML algorithm based on the architecture of Faster R-CNN and combined a Region Proposal Network (RPN) algorithms into a single, end-to-end-unified-network for real-time, capture point, sentinel site, *Anopheles* habitat, identification. A more sophisticated deep learning approach was selected for the benefits of computational speed and efficiency in real time habitat object detection and segmentation, improving flexibility in the algorithms workflows, and eco-geographic scalability.

Third, we infused our complementary AI-ML algorithmic methods and techniques in an interactive, configurable iOS dashboard application[app], which we built using

Unity game engine software and Vuforia 10. 22 software development kit (SDK). Vuforia Studio is a web-native and easy-to-use tool for authoring domain and task-specific experiences. We installed Vuforia Studio Software I Vuforia Studio I PTC into the iOS database. Vuforia offered a range of tools and features which created georeferenced, capture point, sentinel site georeferenceable, *Anopheles,* larval habitat, LULC image object, real-time environment, and text recognition for implementing S &D (See Fig. 2). The app provided an integrated view of digital and physical product data, real-time alerts with 2D, 3D, augmented realities of capture point *Anopheles* larval habitats as in Jacob et al. (2021) [38]. Running the software application on handheld mobile devices (e.g., iPad) allowed the display of interpolated RGB signature, sentinel site LULC classified map data of identified larval habitats, and multi-directional aerial footage in the Nkolondom intervention site. The local vector control officers of Nkolondom were trained on these mobile devices and the use of the drone technology to ensure continuous seasonal larval habitat identification after the baseline collection.

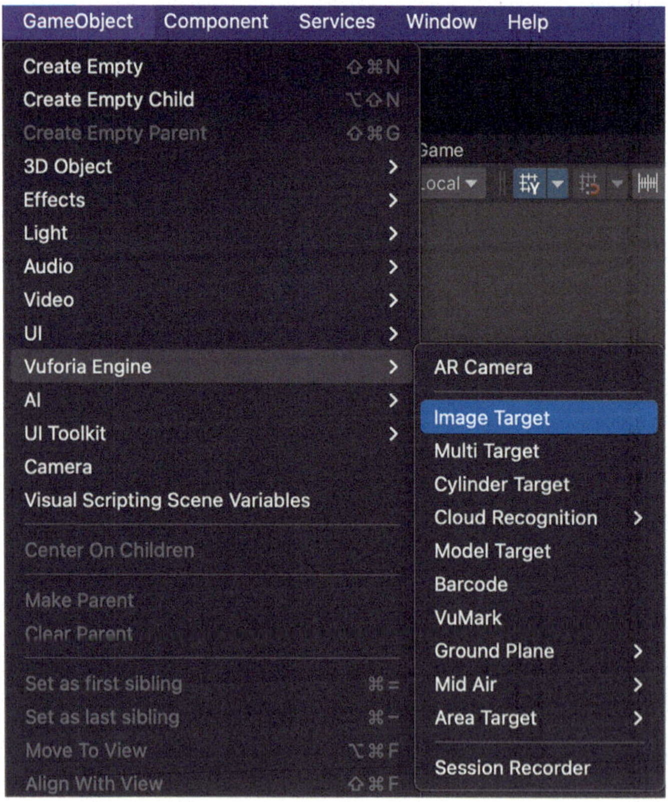

Fig. 2 Vuforia engine menu and select image target

Finally, despite the potential of real-time, drone-based, larval habitat identification and treatment in Nkolondom, a critical question in this baseline collection was the approach's effectiveness on real time retrievable entomological indicators (e.g., number of larval habitats, larval abundance, adult mosquito abundance). To help answer this question, we employed a quasi-experimental design to estimate the intervention effect.

2 Methodology

See (Fig. 3).

2.1 Study Site

The study was conducted in two vegetable farming areas, namely Nkolondom (3∘57,018″ N, 11∘29,036″ E) and Nkolbisson (3∘52,046″ N, 11∘25,055″ E), located around Yaoundé the capital city of Cameroon. Nkolbisson is located 8 km from the city west of Yaoundé, while Nkolondom is 16 km north of Yaoundé. Farming activities sustained by the permanent use of chemical insecticides are carried out year-round, and crops are manually watered extensively. Nkolondom is under the

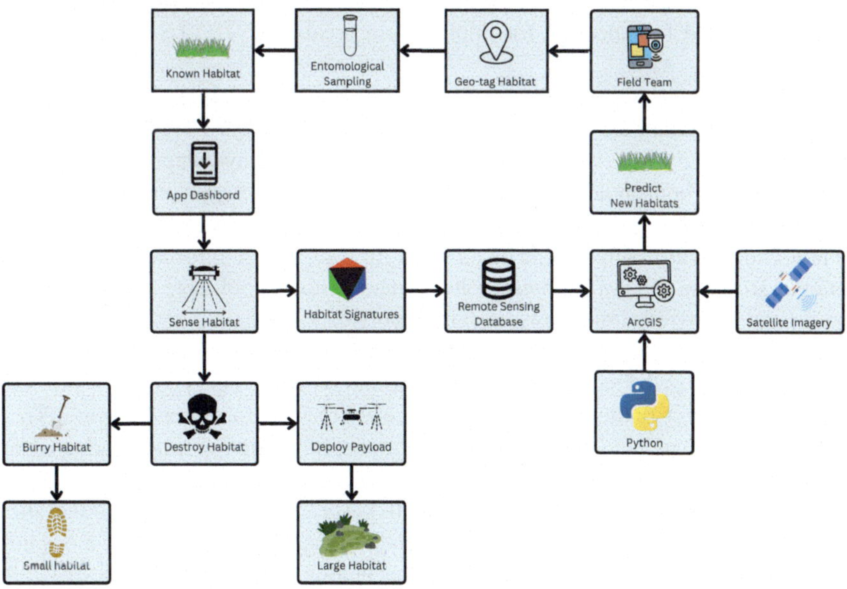

Fig. 3 Proposed Seek and Destroy malaria habitat prediction and treatment workflow

influence of the equatorial climate, with average yearly rainfall above 1,500 mm spread out over 4 seasons: 2 dry seasons (December-February and July–August). Malaria transmission is perennial, with a high annual entomological inoculation rate. This location is a significant production site for spices, condiments, and vegetables that supply households in the city of Yaoundé. It is a marshy site whose exploitation for seasonal crops favors the creation of mosquito breeding sites, especially malaria vectors. Two anopheline species, *An. gambiae* and *An. coluzzii* are found in sympatry although at variable densities over the year. These species are resistant to all four prominent insecticide families (OC, OP, CB, and Pyr), which are recommended for vector control. Resistance mechanisms inducing insecticide target site mutations such as Vgsc-L1014F, Vgsc-L1014S, Vgsc-N1575Y and Ace-1-G119S and overexpression of metabolic genes including CYP6m2, CYP9k1, CYP6p4, CYP6z1, GSTe2 and CYP4g16 genes found in these local malaria vectors [27–31]. Nkolondom was considered a test intervention site, and Nkolbission was considered a control site. The distance between the two villages is 13 km.

Prior to the onset of this study, households in Nkolondom were enumerated and mapped, which was used to generate a real time sampling frame for the entomology surveys. All households enumerated during the survey were assigned a unique number. A random sample of 25 households was selected to generate a list of households to be approached for recruitment into the entomology survey. From the list, households were selected for participation in the human landing catches, pyrethrum spray catches [PSCs], exit trap collections, and environmental measures. A separate list of random households was chosen to generate a list of households to be approached for recruitment into the study being conducted under a separate protocol. The households of all local villagers 16–22 recruited into the cohort study were approached for selection for implementing real-time S7D-LSM strategies in Nkolondom (Fig. 4).

Randomly selected houses were sprayed using an aerosol of non-residual pyrethroids with a piperonyl butoxide synergist. These sprays were combined with exit traps placed over the windows of the houses to capture any escaping mosquitoes. In each site, 10 households were randomly selected for the spray collections from the entomology recruitment list generated from the enumeration database. The same 10 households were sampled one day a week for four weeks. Informed consent was obtained from the head of the household or an adult household representative before conducting the pyrethrum spray and exit trap collections.

No other activities were carried out at the control site apart from collecting adult mosquitoes during the same periods as in the intervention site. The study site was surveyed for water bodies. Site coordinates were recorded using a Garmin e-Trex 10 Worldwide Handheld GPS Navigator. Purposeful sampling was conducted to maximize the collection of the aquatic stages of mosquitoes using a 350-ml dipper (Clarke Mosquito Control Products, Roselle, IL). At each georeferenced sentinel site, *Anopheline,* marine habitat, ten dips were made in places likely to harbor mosquito larvae, such as around tufts of submerged vegetation or substrate, edges of water bodies, and floating debris. In extensive water bodies, dipping was carried out over a 100-m walk. The larvae were classified either as *Anophelines* or *Culicines.*

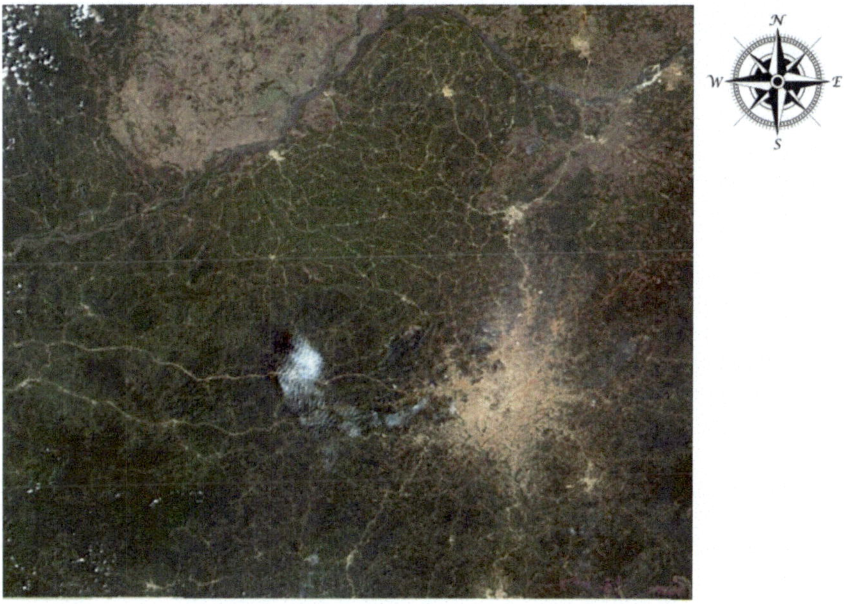

Fig. 4 Nkolondom intervention study site map

2.2 *Sensed Real Time Data*

Drone surveys were carried out using a DJI Phantom 4 Pro (DJI, Shenzhen, China) quadcopter fitted with a DJI 4 K camera (8.8 mm/24 mm; f/2.8; 1" CMOS; 20 MP) for conventional RGB signature, capture point, LULC imagery collection, A 3DR Solo (3D Robotics, California, US) quadcopter fitted with a Parrot Sequoia sensor (Parrot, France) composed of single-band cameras (Green, Blue, Red, and NIR of 1.2 MP) was used for multispectral imagery collection. The flight plan was programmed with the Pix4D Capture app in an iPad Mini 4 (Apple, California, US). The connection between the controller and DJI Phantom 4 Pro and 3DR Solo was set up using the DJI GO 4 app and 3DR Solo app, respectively. For approximately 30 min, the drone flew over the entomological intervention site using the high-end, radio-controlled, and camera-equipped for agro-village, and peri-urban pastureland explorations. The drone was integrated with handheld devices to enhance its aerial footage capabilities. The multangular drone camera within the kit recorded, stored, and managed the capture point, seasonal, georeferenced, sentinel site, and RGB signature datasets in the dashboard spectral library. A copy of the georeferenced, identified sentinel site, *Anopheline*, larval habitat LULC data, (Fig. 5) and spectral imagery was stored in the on-flight computer and concurrently transmitted down to the ground stations via Wi-Fi communication in real-time employing the cloud-based DroneDeploy™ platform DroneDeploy software. 2.8. Orthomosaic.

Fig. 5 Real-time UAV identified *Anopheles* habitat capture point objects

The photogrammetric processing of the *Anopheline,* larval habitat, capture point, reflectance, surface measurements was based on drone sentinel site, imagery which was conducted in AgiSoft PhotoScan Pro (https://www.agisoft.com). The resulting UAV imagery was imported into PhotoScan and processed to construct an orthomosaic (i.e., georeferenced, overlapped, sentinel site, land cover georeferenced image with correction for topographic distortions) for the entomological intervention site. The position of the drone at the time of image capture for each aquatic, *Anopheles*, larval habitat, sentinel site photo was recorded automatically by the onboard GPS; hence, the orthomosaic was georeferenced without the need for Ground Control Points (GCPs). The standard procedure used was: photo-alignment (accuracy: highest; generic preselection active, reference preselection active; KeyPoint limit: 80,000; adaptive camera model fitting active); (2) dense cloud building (quality: high; depth filtering: aggressive); (3) elevation model (geographic projection using; resolution of 0.1 m and 0.02 m per pixel for the multispectral images respectively; interpolation: extrapolated; all georeferenced, *Anopheline*, larval habitat, capture point classes to generate digital surface model); (4) orthomosaic building (input surface, blending mode mosaic, resolution of 0.1 m and 0.02 m per pixel for the multispectral, georeferenced, aquatic, larval habitat, sentinel site, land cover, UAV sensed, images respectively.

Python configurable apps provided a suite of app templates that allowed creating a web app from the signature sentinel site LULC maps and the UAV scenes without having to write code. We could interact with the UAV real-time maps with the field-sampled entomological data by leveraging an app template and choosing a few options. For testing, we flew the UAV over the sentinel sites. During these wayward

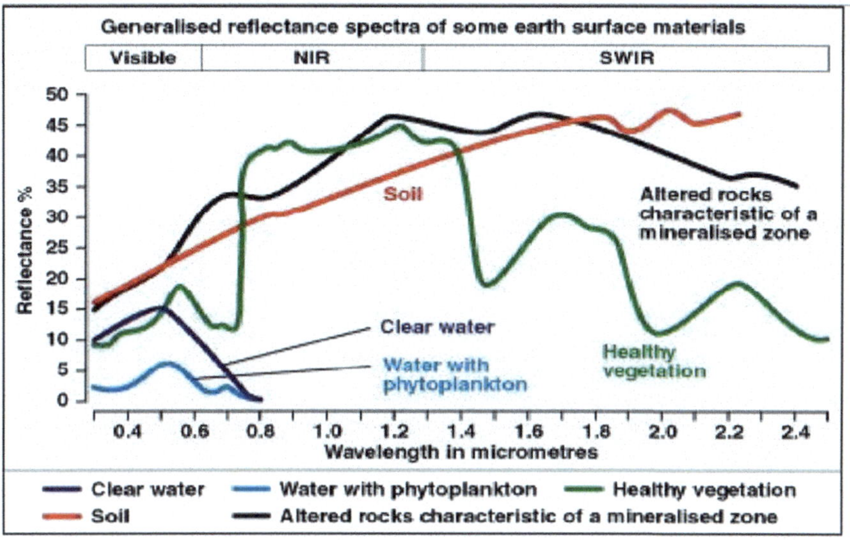

Fig. 6 Generalized reflectance spectra of an *An. gambiae* s.l. habitat materials

flights, 12 videos totaling 28 min were collected. The total number of frames extracted was 1,126, with 87% containing at least one potential *Anopheles* larval habitat.

The features were identified around the predicted district-level aquatic habitats in the app (Fig. 6). This tool required a model definition file containing trained model information. The real-time, UAV-sensing, signature, sentinel site model was trained using third-party training software [i.e., Py Torch] to capture the Anopheles habitat signature and implement the drone larvicide portion of S&D.

Anopheline larval sampling sites were characterized by biotic factors visible in the sentinel site real-time drone imagery and generalized linear mixed (GLM) models were constructed to determine their association with larval presence. The statistical methods also constructed regression, analysis of variance, analysis of covariance, multivariate analysis of variance, and partial correlations employing the sampled, larval habitat, seasonal, entomological, and empirical data in Python after the intervention to compare results. These quantitative methods were used to test the Seek and Destroy intervention in a Nkolbisson, the non-intervention pastureland site.

Once the drone signatures were overlaid onto commercial high-resolution satellite data (46 cm spatial resolution, Worldvew-2, visible and NIR wavebands) (Fig. 7) it was real-time transferred into the ArcGIS-AI (i.e., geo-AI) Interface Kit™, where a Maximum-likelihood Classification using Raster ran the signature over the entire intervention site to identify unknown larval habitat aquatic foci at the entomological, intervention site (Figs. 6, 8).

This real-time scaling up of the signature, sentinel site, Anopheline, larval habitat, capture point, RGB-LULC signature mapping was based on the Faster R-CNN algorithm being applied to real-time georeferenced, sample, estimator sentinel site

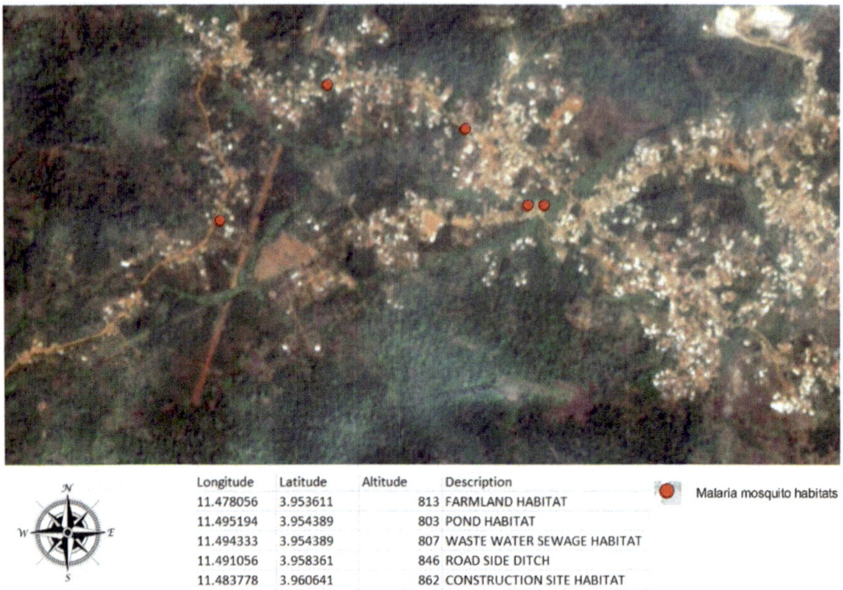

Fig. 7 Drone signature GPS *Anopheles* habitat capture points overlaid onto World View 2 satellite data of Nkolondom

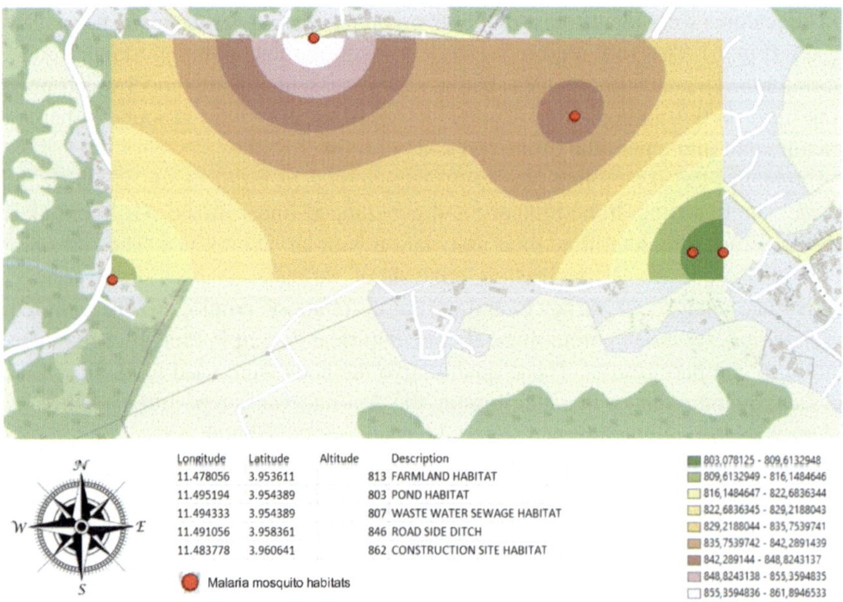

Fig. 8 Signature interpolation map of construction site *Anopheles* habitats in Nkolondom

datasets which included differentially corrected GPS indexed component video data. The analog signatures and priority information extracted from the collection points were used to optimize seasonal, field control, sentinel sites, imaging, and entomological field sampling operations. For example, when a user (e.g., trained local district-level vector control officer) submitted a query of Anopheles larval habitats and video clip, the system retrieved a ranked list of visually similar district, aquatic, and larval habitat types with GPS coordinates in real-time (Table 1).

The water depth of an aquatic, sentinel site, *Anopheles,* larval habitat was measured from separate places depending on the size of the habitat using a meter stick, and the average depth was taken. The distance to the nearest homestead was measured using a tape measure of less than 100 m and estimated to be more than 100 m. The distance was then categorized into four classes: (1) $\leq$ 100 m, (2) 101 to 200 m, (3) 201 to 300 m, and (4) 301 to 400 m. Surface debris, the presence of algae, and emergent plant coverage were determined based on visual observation. Vegetation cover was visually observed and expressed as open (no vegetation), tree (for the presence of large trees within a range of 10–15 m where shade and foliage could reach), and shrub (woody plants more petite than a tree within 10–15 m). Habitat perimeter was measured using a tape measure and classified as < 10 m, 10–100 m, and > 100 m.

Larval breeding habitats and several immature *Anopheles* mosquitoes sampled were described using tables. Correlation analysis investigated the relationship between pH, temperature, and water depth to the *Anopheles* larval density. The density weights were determined by dividing the number of Anopheles larvae (early or late) by the number of dips taken from each larval habitat. Larval density was log-transformed log10 (x + 1) to improve the normality of distribution. Multiple regression analysis was employed to identify the environmental variables associated with *Anopheles* larvae. Mann–Whitney U test was employed to compare samples with the variables: the presence of algae (presence or absence), habitat permanency (temporary or permanent), surface debris (present or absent), the intensity of light (sunlit or shaded), and water movement (still or flowing). The Kruskal–Wallis H test was used to compare samples with more than two groups: water turbidity, water

Table 1 Shows the results of larval sampling and the types of larval habitats productive in the study area. Six sentinel habitat types were identified in the entomological intervention study site: pastureland habitats, rain pools, construction site habitats, wastewater sewage habitats, swamps, and commercial roadside ditches

Sentinel habitat types	Intervention site	Control site
Pastureland	2	3
Rain pools	11	13
Construction site	6	10
Wastewater sewage	6	9
Swamps	3	1
Roadside ditches	12	7

perimeter, distance to the nearest house, canopy cover, emergent plant coverage, habitat type, and substrate type. These non-parametric tests were used to compare larval densities from sites with different habitat characteristics. Data were analyzed using Python version 3.10. Values were considered significantly different if $p < 0.05$ for all the tests. We employed the package sci-kit-learn, a widely used Python library for machine learning built on the top of NumPy. The package provided the means for preprocessing the entomological data by reducing dimensionality, implementing regression and classifying, clustering of the sampled georeferenced Anopheles larval aquatic habitats.

We connected to Cloud Recognition databases on the Vuforia Developer Portal. We introduced the Web API employing Python into the mobile real-time UAV dashboard iOS app for adding and meshing the drone and remote sensed, capture point, *Anopheles,* and larval habitat data. Python library for the Vuforia Web Services (VWS) API and the Vuforia Web Query API were installed into the dashboard. We employed Cloud Targets Web API and the Cloud Query Web API to manage and query various Cloud targets. We employed the API to upload habitat imaged targets to a Cloud recognition database and query Cloud Database, which created model target [centroid coordinates of RGB signatures of potential, hyper/hyperendemic, aggregation/non-aggregation, oriented *Anopheles* habitats] and generated VuMarks. [i.e., printed target that can be placed on any planar surface or imprinted on an object in a discreet size]. Our real-time app performed tasks such as authentication location-aware functions and provided targeted content and communication for all the collaborating researchers [i.e., end users] at the University of Yaounde 1, University of South Florida, and the Center for Research on Filariasis and Other Tropical Diseases) [i.e., CRFilMT]. For Vuforia targets, we employed an Optimize Tracking mode to improve real-time UAV tracking for habitat sites [e.g., peri-urban elevated farmlands with built environment].

In this proposal, we introduced a Regional Proposal Network [RPN] that shared full-image convolutional features with the real-time detection retrieval and transfer network, thus enabling nearly cost-free region proposals using Tensor Flow object detection. The TensorFlow Object Detection API is an open-source framework built on TensorFlow. It makes constructing, training, and deploying object detection Anopheles larval habitat capture point, sentinel site, RBG, signature, and interpolated models easy. The RPN was trained end-to-end to generate high-quality region proposals, which Faster Regional Convolutional Neural Network [R-CNN] employed for real-time detection retrieval and transfer of the dashboard *Anopheles* habitat information, data and capture points.

Our object detection system [i.e., Faster R-CNN] comprised two modules. The first module was a deep, fully convolutional network that proposed regions for UAV Anopheles larval habitat signature acquisition, and the second module was the Fast R-CNN detector that used the interpolated signatures to determine habitat geolocations in the proposed regions. The entire system was a single, unified network for object detection. Using the recently popular terminology of neural networks with 'attention' mechanisms, the RPN module tells the Fast R-CNN module where to look (Ren 2015). For potential, hyper/hyperendemic, aggregation/non-aggregation-oriented foci, we introduce the designs and properties of the network for region proposal in the real-time dashboard.

We develop ML algorithms for training both modules with the capture point, sentinel site, georeferenced, Anopheles, and larval habitat attribute features. We modeled this process with a fully convolutional network. Because our ultimate goal was to share computation with a Fast R-CNN object detection network, we assumed both nets would share a standard set of convolutional layers. In our experiments, we investigated Zeiler and Fergus 2014. We built a VGG-16, which had 13 shareable convolutional layers. To generate the region proposals, we slide a small network over the convolutional feature output by the last shared convolutional layer in the real-time, smartphone, intelligent AI-ML dashboard. This small network employed an input n × n spatial window of the capture point, sentinel site, georeferenced, convolutional habitat, data feature attributes. Each sliding window was geospatially associated with a lower-dimensional feature, 512-d for VGG, with Rectified Linear Unit [ReLU] in Pytorch as in Jacob et al. (2024).

We introduce the use of rectified linear units (ReLU) as the classification function in a deep neural network (DNN). Conventionally, ReLU is used as an activation function in DNNs, with SoftMax function as their classification function. We accomplished this by taking the activation of the penultimate layer hn −1 in a neural network, then multiply it by weight parameters θ to get the raw scores oi. Afterwards, we thresholded the raw scores oi by 0, i.e. f (o) = max (0, oi), where f (o) was the ReLU function. We provided class predictions yˆ through arg max function, i.e. arg max f (x). Keras with Google TensorFlow backend was used to implement the deep learning algorithms in this study, with the aid of other scientific computing libraries: matplotlib, numpy and scikit-learn.

ReLU is an activation function which [38] was demonstrated to further improve training of deep neural networks for real time imaging georeferenced, potential, hyper/hypoendemic, aggregation/non-aggregation-oriented, *Anopheles,* larval aquatic habitats. It worked by thresholding values at 0, i.e. f (x) = max (0, x). The Rectified Linear Unit (ReLU) activation function rendered 0 as an output when x < 0 ['cold spot'], and then produced a linear with slope of 1 when x > 0. [hot spot] We employed ReLU not only as an activation function in each hidden layer of the Ios real time, interactive intelligent AI-ML neural network, but also as the classification function at the last layer of a network. Here, the predicted class for ReLU classifier was yˆ, yˆ = arg max i ∈ 1,…,N max. Many real time neural networks use the SoftMax cross-entropy function to learn the weight parameters θ. In this experiment, we implemented the fore prior mentioned loss function, but with the distinction of using the ReLU for the prediction units. The θ parameters were then learned by backpropagating the gradients from the ReLU classifier. To accomplish this, we differentiated the ReLU-based cross-entropy function for the activation of the penultimate layer, $\ell(\theta) = -\tilde{O}\, y \cdot$ loд.

The ReLU function, a key component of our neural network, was subsequently applied elementwise on an input tensor using torch.nn. ReLU (). This function played a crucial role in our system, as it replaced all the negative elements in the input tensor with 0 (zero) in the real time intelligent UAV-iOS app dashboard, and all the non-negative elements were left unchanged. The real time, UAV- iOS dashboard, supported the real-valued input tensors, benefitted from this function. ReLU was used as an activation function in the neural network, enhancing the system's performance and efficiency.

Syntax
Relu = torch.Nn. ReLU ()
Output = relu(Input)

We imported the required library [i.e., Python library was torch].

Import Torch
Import Torch. Nn as Nn

We subsequently defined the input tensor and printed it.

Input = torch.Randn(2,3)
Print ("Input Tensor:", Input)

We defined a ReLU function relu using torch.in.ReLU().

Relu = torch.Nn.ReLU()

We applied the above-defined ReLU function rely on the input tensor in the dashboard and optionally assigned the output to a new UAV sampled, Anopheles, larval habitat, capture point, sentinel site, and signature variable.

Output = relu(Input)
We printed the tensor containing ReLU function values.
Print ("ReLU Tensor:", Output)

We then imported the required library.

Import Torch
Import Torch. Nn as Nn
Relu = torch.Nn. ReLU ()
Input = torch. Tensor ([[-1., 8., 1., 13., 9.],
[0., 1., 0., 5., −5.],
[3., −5., 8., −1., 5.],
[0., 3., −1., 13., 12.]])
Print ("Input Tensor:", Input)
Print ("Size of Input Tensor:", Input. Size ())

We computed the ReLU) function elementwise.

Output = relu(Input)
Print ("ReLU Tensor:", Output)
Print ("Size of ReLU Tensor:", Output. Size ())

An Output/Input Tensor was generated:

Tensor([[−1., 8., 1., 13., 9.],
[0., 1., 0., 5., −5.],
[3., −5., 8., −1., 5.],
[0., 3., −1., 13., 12.]])

The size of the Input Tensor was determined:

Torch. Size ([4, 5]

Finally, the ReLU Tenso was created:

Tensor ([[0., 8., 1., 13., 9.],
[0., 1., 0., 5., 0.],
[3., 0., 8., 0., 5.],
[0., 3., 0., 13., 12.]])

The final sizc of ReLU Tensor was: **Torch.Size([4, 5])**

We noted that the harmful elements in the input tensor were replaced with zero in the output tensor. This feature was fed into two sibling fully connected layers—a box-regression layer (reg) and a box-classification layer (cls). We used n = 3, noting that the effective receptive field on our input entomological information data capture points. Their explanatory variables had a maximum threshold of 171 and 228 pixels for VGG. Max Pooling moved a window (224 × 64) across the input sentinel site, drone feature *Anopheles* larval habitat, capture point map to 1 × 1 × 1000). In each window, Max Pooling selected the maximum mapped habitat signature values. A VGG-16 model CNN architecture consisting of 16 layers, including 13 convolutional layers and 3 fully connected layers was built to achieve strong performance on various computer tasks, including real-time classification and object recognition of archived retrieved and or communicated Anopheles larval habitat RGB signature capture point sentinel site data (Fig. 9).

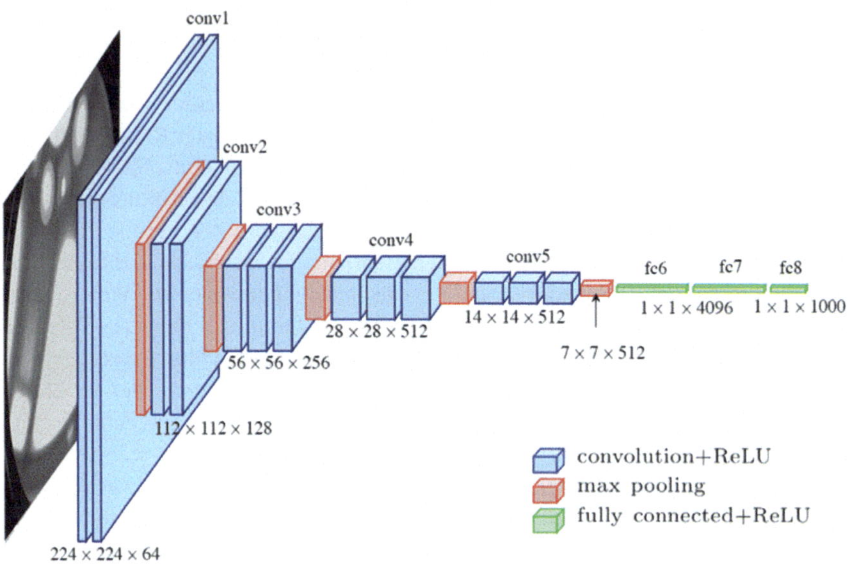

Fig. 9 A VGG-16 model CNN architecture was constructed in the smartphone dashboard

Note that because the mini-network operated in a sliding-window fashion, the fully connected layers were shared across all spatial locations at the Cameroonian intervention study sites in the neural architecture, which was implemented with an n × n convolutional layer followed by two siblings 1 × 1 convolutional layers (for reg and cls, respectively). At each sliding-window location, we predicted multiple region proposals, where the number of maximum possible proposals for each location diagnostic, stratified, explanatory variable was denoted as k. The k proposals were parameterized relative to k reference boxes. Subsequently, we unified the RPNs with Faster R-CNN object detection networks, which employed a scheme that alternated between fine-tuning the RPN task for fine-tuning the capture point, sentinel site *Anopheles,* aquatic, larval habitat, real-time UAV information while keeping the real-time, drone sensed, data fixed in the real-time iOS dashboard. We transformed a pre-trained ResNet-50 network into a Faster R-CNN object detection network by adding an ROI pooling layer, a bounding box regression layer, and a RPN. The Faster R-CNN network was trained using trainFasterRCNNObjectDetector.

We trained our models on an 8-GPU implementation, ensuring their robustness and reliability. The adequate mini-batch size became 7 for RPN (1 per GPU) and 14 for Fast R-CNN (2 per GPU). The RPN and Fast R-CNN steps were trained for 280 k iterations with a learning rate of 0.004 and then for 60 k iterations with 0.0002. We modified the learning rates (starting with 0.003 instead of 0.001) as the mini-batch size was changed. We employed three aspect ratios and four scales (adding 64) for the anchors, mainly motivated by handling small habitat objects in the UAV image, capture point, and sentinel site dataset. In addition, in our Fast R-CNN step, the negative samples were defined as those with a maximum IoU with ground truth in the interval of (0, 0.5), ensuring the accuracy and reliability of our models.

We validated the intelligent AI-ML smartphone app using a GPU cluster with four nodes of GeForce GTX TITAN X, each with 12 GB of memory for generating a real-time, interactive app-generated Faster R-CNN model. Faster R-CNN network [i.e., 2-step deep learning object detector] using a pre-trained convolutional neural network model which extracted and transferred convolutional data capture points and diagnostic variables from the last layer of a trained network in real-time (Table 2). Using the pre-trained convolutional neural network model, we validated the real-time sensed capture point RGB-indexed sentinel site and signature datasets. We extracted the convolutional feature map from the last layer of the trained network. This step enabled the neural network to understand the critical features within the actual time, UAV sampled, entomological, larval, *and Anopheles habitat data that separated multiple classes of objects within the capture point, and* sentinel site data. To train towards feature recognition in real-time, the right neural network for real-time UAV sensing, capture point, sentinel site, and *Anopheles* mosquito habitats was used and adapted. This study employed Inception V2 as the base pre-trained convolutional network for extracting the capture point, sentinel site, RGB signature, *Anopheles*-related habitat sampled data. Inception V2 is a complex deep learning architecture (Szegedy et. al. 2016) that we employed using clever factorization methods in the intelligent AI-ML-infused dashboard to make the convolutions efficient in terms of computational complexity.

Table 2 Validated interpolated *Anopheles* habitats at the entomological intervention sites

Total durationof all videos	32 min
Total number of frames	1,889
Total number of frames with breeding habitats	1,100
Total number of frames without breeding habitats	789

The robotic appendage is a powerful mechanism used in industries to carry out pick and place tasks with great speed and accuracy [github.com]. A development team at COPH designed, produced, and attached two high-performance, completely original-axis robotic arms to the DJ-4 drone. The appendages provided the drone with the versatility to perform across a wide range of at the cameroonian intervention study sites for larval habitat surveillance, for example, to extract larval habitat water samples, clear canopy debris surrounding foci, transport field equipment, etc.]. The appendage used its intelligence to complete the task [i.e. real time entomological sampling of the georeferenced, capture point, sentinel site, habitats]. The robotic arm clipped onto the DJI Phantom drones to provide aerial manipulation capability. The arm included a backpack controller clipped onto the DJI remotes' back. The interactive smartphone dashboard adjusted two slider controls. We employed one arm to deploy the carbon fiber-reinforced appendage. We used the other to close the carbon fiber talon-like claws to grasp and carry habitat objects. The appendage was installed or removed without tools. Specifications of the drone with the robotic arm post-testing was Motor-to-Motor Length: 1450 mm (57 in.)Maximum Arm Payload: 10 kg (22 lbs.), Airframe Weight: 20 kg (44 lbs.), Maximum Speed: 60 km/h (37 mph), Total Height: 1300 mm (51 in.), Maximum Operating Wind Speed: 10 m/s (22 mph), Propeller Diameter: 27 inches Maximum Flight Time: 30 min Battery: 22.2v/ 16000mAh × 2,Maximum Operating Altitude(Pressure Altitude): 5,000 m Water Resistance: All-Weather Capable (Fig. 10).

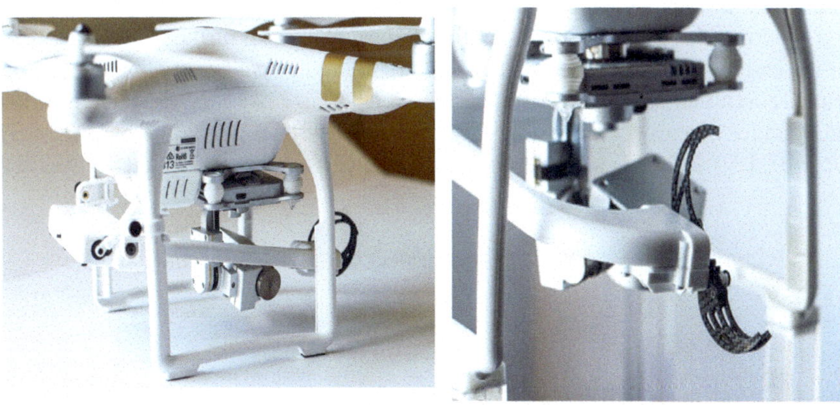

Fig. 10 **a** Robotic Arm Kit mounted on a DJ-4 **b** programming, mechanical claw

The problem with an arm which lacks intelligence is that with changing circumstances they cannot acclimatize well, the arm needs human intervention to reprogram as per requirement and perform only as per instructed [github.com]. However, the intelligent robotic arm we employed was capable enough to make its own decision for the entomological tasks. Some prevailing projects in this domain were majorly based on the drone and satellite sensors for the classification and detection of the georeferenced, capture point, sentinel site, *Anopheles,* larval habitat objects [e.g., larval sampling, earth-bottom irrigation canals, within-canopy pigments (e.g., chlorophyll, zeathinins etc.). The arm employed an RGB signature color-based sensor for detection of the *Anopheles* habitats. We employed the Faster R-CNN, ML model to implement the arm for automatically sorting the red, green and blue bounding boxes in the drone images by capturing their signatures with the multispectral camera which was processed on a Raspberry Pi 3 B + to detect and classify the capture points accurately. The mechanical arm movement ensured a precise inverse kinematic model, which helped pick and segregate potential, hyper/hyperendemic, aggregation/non-aggregation-oriented habitats. Both accuracy and real-time computing performance were good in the experimental findings of the Faster RCNN model. The precision we achieved using the Faster RCNN model was 98.8% for detecting and sorting the red, green and blue boxes. Hence, the Faster RCNN for smart robotic arm successfully resulted in good outcomes [real time, capture point, sampling geolocations of potential, hyper/hypendemic, aggregation/non-aggregation-oriented *Anopheles* sentinel site habitats]. This architecture helped in extracting the real time, entomological, UAV and field sampled, larval habitat data in prioritization order based on georeferenced Euclidean distances to human habitation to localize the habitats in the real time drone images.

According to [38] an anchor is a predefined bounding box varying in scale and aspect ratio, scanned the image for potential hyper/hypoendemic, aggregation/non-aggregation-oriented, *Anopheline* habitats We employed the Faster Region-based CNNs generated to predefine anchor aspect ratios and scale capture point variable feature data attribute in the real time dashboard. Anchors, the rectangular boxes were used to scan habitat objects were emplaced during training of the neural network. For the case of our, real time sampled, capture point, *Anopheles* larval habitat RGB signature, sentinel site data we set the base anchor size as (256, 256) pixels, with scaling ratios and aspect ratios as (0.25, 0.5, 0.75, 1.0, 1.50, 2.0, 2.5) and (0.25, 0.5, 0.75, 1.0, 1.50, 2.0, 2.5) respectively. The width and height of each anchor was set as

$$\text{width anchor} = \text{scale}[i] * \text{sqrt (aspect ratio}[i]) * \text{base anchor}[0],$$

$$\text{height anchor} = \text{scale}[i] * \text{base anchor } [1]/\text{sqrt (aspect ratio}[i])$$

where i was the index of the matrices of scales and aspect ratios. In total, 300 anchors was generated per capture point for training. We fed our anchors and feature maps to RPN. Here, the task was to train the network to identify those boxes that were indicative of, real time detected, retrieved and transmitted information data capture points and diagnostic habitat, explanatory variables. To do so, we field verified each box prior to training as a potential georeferenceable, hyper/hyperendemic, aggregation/non-aggregation-oriented information data capture point and diagnostic variable (s) and the classified LULC background. The RPN consisted of small convolutional layers, which was trained to identify the anchors having the relevant objects of interest (in our case, a potential, *Anopheles* larval habitat capture point) with an "objectness" score and returned the ones that most likely contain the information within the UAV-iOS dashboard based on the score along with the center coordinates, width and height of all anchors. We resized the feature data of the anchors so as to optimize differentiating habitat types learned from the above step into fixed sized feature maps, which served as input to two branches in the iOS dashboard as a) classifier to identify potential, hyper/hyperendemic, aggregation/non-aggregation-oriented data capture points within the anchor; and (b) regression to tighten the anchor for improved accuracy. All steps were performed in parallel. We fine-tuned the classifier by varying learning rates. We employed the stochastic gradient descent (SGD) solver for 50,000 iterations with a base learning rate of 2e-5, then another 25,000 iterations by reducing the base learning 2e -6 and the rest 25,000 with 2e-7 for faster convergence. The learning rate for each iteration is shown in (Figs. 1, 11).

The real-time high-risk entomological larval habitat data was archived in the real-time dashboard memory. We created classified datasets and converted the real-time UAV image capture point, and sentinel site entomological data to TF. Tensor objects which were extracted from tensor slices. A code embedded the larval habitat data feature attributes and labelled arrays in a TensorFlow graph as TF. constant() operations in the dashboard.

According to Jacob et al. (2024), during training, a critical criterion for the correctness of UAV real-time sensor data is the loss function. Briefly, our loss function in the dashboard measured the learning ability of a neural network architecture. It ranged from 0 to 1 where 0 meant perfect learning and 1 meant no teaching. The goal of training the real-time detected, retrieved, and habitat-related data was to minimize

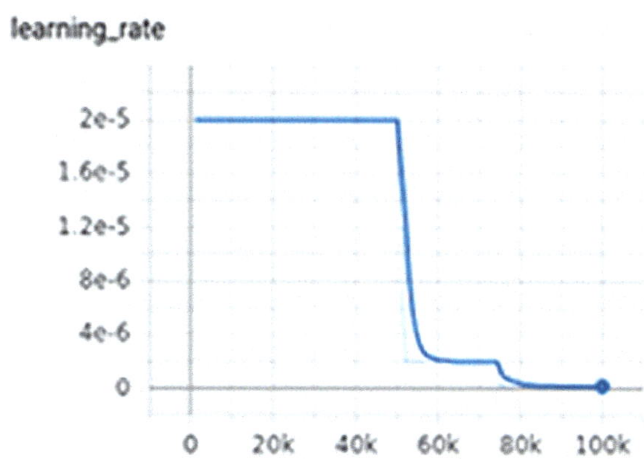

Fig. 11 The learning rate for each iteration in the *Anopheles* habitat, an interactive AI-ML infused smartphone app

the loss of training data. To reduce the loss, we applied a novel focal loss function that penalized instances of false negatives. If the classifier was biased towards the background class in the real-time Anopheles model, this replaced in biased learning Focal loss is an improvement over the more standard cross-entropy loss since it operates by lowering the loss of well-classified cases and emphasizing the misclassified ones [37]. We considered the following equation: where, y denoted the ground truth of the class (+1 for the Anopheles RGB, signature-related capture point, and −1 for LULC classified background), which was the model estimate probability for the class label. The Focal loss (FL) for each anchor k was defined as FL(k) =−(1−Q) log(Q). Here, y denoted the ground truth of the class (+1 for a potential hyper/hypoendemic, aggregation/non-aggregation-oriented, aquatic, breeding site foci and −1 for LULC background), and p ϵ [0,1], which was the app-generated real-time model estimated probability for the class label y = + 1. We quantified the focal loss in the real-time UAV sampled, capture point, sentinel site, and entomological data. We added modulating factor $(1 - Q)^{\lambda}$ to the above loss equation for tunability, where $\lambda > = 0$. The focal loss (FL) was subsequently defined as $FL(Q) = -(1-Q)^{\lambda} log(Q)$ in the dashboard neural architecture (Fig. 12). Here, l was a tunable parameter.

After applying focal loss, we determined that overall minimal loss was achieved in the iOS, AI-ML infused, real-time iOS dashboard based on an uncertainty-oriented eigen-Bayesian semi-parametric, non-Gaussian Markovian model (See [38]). Essentially, when the loss in training data is similar to the loss in validation data (here, the validation data was also used to train the model), we assumed that the UAV and field sampled data could not decrease any further, and the process of training was complete. Training and validating the capture point, sentinel site, *Anopheles*, larval habitat, prognosticative, signature models was iterative.

It is emphasized here that we also trained the capture point, sentinel site, larval, *Anopheles*, habitat-related, real-time, dashboard app-generated Faster R-CNN model using the annotated data to detect and locate sources of solutions (meshing georeferenced, distant measurements from potential hyper/hypoendemic, aggregation/non-aggregation-oriented, *Anopheles* larval habitat to a human habitat). A testing team

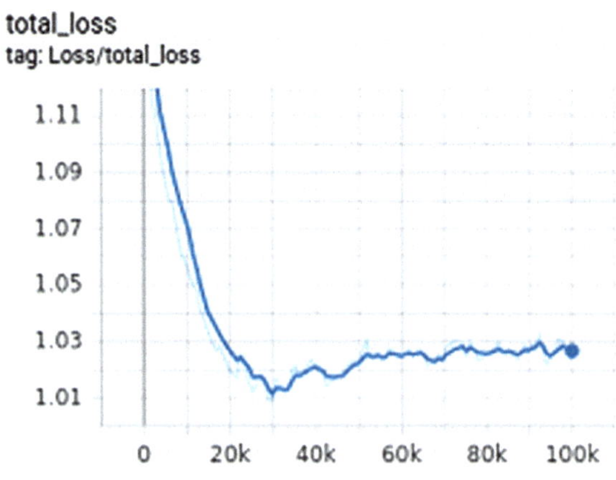

Fig. 12 Total loss graph with focal loss

trained and validated the smartphone app, real-time, UAV, and field-sampled entomological data using a GPU cluster as in Jacob et al. (2024). The cluster had four GeForce GTX TITAN X nodes, each with 12 GB of memory. Each capture point sentinel site update was directed into previous models constructed for classification and localization, which strengthened our expected results using the GPU. We conducted early stopping during iteration. We monitored the model's performance on validation data during the training process while stopping the training whenever the validation error did not improve anymore. We also employed noise injection to overcome any over fitting issues. This process consisted of adding noise to the inputs or the outputs of hidden layers during the training to make the model more robust and prevent it from a weak generalization. Furthermore, we added a penalty to the loss function based on the size of the RGB habitat signature-related information, real-time, UAV, and field. Entomological sampled, data capture point predictions weights. Data augmentation was also conducted to artificially increase the size and diversity of the training dataset by applying random transformations [e.g., rotation, scaling, flipping, or cropping] to the input georeferenced, potential, hyper/hypoendemic, aggregation/non-aggregation-oriented, *Anopheles,* capture point, sentinel site data from video analog data.

We noted that the default binary cross entropy loss function increased after 15,000 iterations. The probabilities of detection by the real-time mobile app neural network (e.g., the confidence in the forecast model in making decisions such as whether a potential, hyper/hypoendemic, aggregation/non-aggregation-oriented, *Anopheles,* capture point, sentinel site) was close to 0.99. Meanwhile, the number of anchors containing the real-time UAV sampled capture points was smaller than that of anchors that contained the LULC background class. In the real-time dashboard, when a potential, RGB, signature capture point, or diagnostic variable was misclassified as LULC background, and Q was small, the modulating factor (1−Q) l was close to 1, which did not affect loss. When Q tended to 1, the modulating factor was close to 0, and loss for well-classified examples was down weighted. The focal loss (FL) was definable as $FL(Q) = -(1-Q)^{\lambda} log(Q)$ in the real-time neural app dashboard architecture. The real-time dashboard's final set of capture points, sentinel site, and Anopheles habitat parameters was subsequently minimized due to the data loss criteria. We computed the True Positives, False Positives, and False Negatives for classification in the validation set which allowed for integrating a final mAP in the iOS mobile app.

The model definition file was an Esri model definition JSON file (.emd), which contained the path to the Python raster function, which called to process each sentinel site, aquatic, *Anopheles* larval habitat, and the trained binary deep learning model file in the configurable, intelligent, smartphone app. We trained the deep learning models using sensitive (5-min maximum wayward UAV flights and processing of sentinel site, capture point signature) over a seasonal sampled, georeferenced, *Anopheles,* aquatic, larval habitat, capture point and derived information products [e.g., forecasted, district-level, prognosticated, maps of potential, unknown hyper/hypoendemic, aggregation/non-aggregation-oriented, seasonal breeding site aquatic foci] using the ArcGIS Pro and ArcGIS API for Python and scaled up the processing of ArcGIS Image Server in the app. Detection geoprocessing tools [e.g., convolutional neural networks (R-CNN, Region Based Convolutional Neural Networks], Fast R-CNN and Deep Learning tools in ArcGIS Pro along with other infused geo-AI remote sensing, object-based algorithms in the app were able to to precisely, forecast, signature map seasonal, *Anopheles,* aquatic, larval habitat geolocations by scaling up a database of UAV imaged, sentinel site, capture point, georeferenced signatures using Worldview-2, visible and NIR satellite data.

The network first processed the whole satellite image with several convolutional (conv) and max pooling layers to produce a conv feature map. Then, for each obect proposal a region of interest [i.e., georeferenced, capture point, sentinel site *Anopheles* habitat] a (RoI) pooling layer extracted a fixed-length feature vector from the feature map. Each feature vector was fed into a sequence of fully connected (fc) layers that finally branched into two sibling output layers: one that produced probability estimates over K object classes plus a catch-all "LULC background" class and another layer that output four real-valued numbers for each of the K object classes. Each set of 4 values encodes refined the bounding-box positions for one of the K classes.

The RoI pooling layer used max pooling to convert the RGB signature habitat features into a small feature map with a fixed spatial extent of H × W (e.g., 7 × 7), where H and W were layer hyper-parameters that were independent of any particular RoI. In this baseline collection. RoI was a rectangular window on a conv feature map. Each RoI was definable by a four-tuple (r, c, h, w) that specified its top-left corner (r, c) and its height and width (h, w).. An input UAV capture point, sentinel site image and multiple *Anopheline* habitat RoIs were input into a fully convolutional network. Each RoI was pooled into a fixed-size feature map and then mapped to a feature vector by fully connected layers (FCs). The network had two output vectors per RoI: SoftMax probabilities and per-class bounding-box regression offsets. The Faster R-CNN architecture was trained end-to-end with a multi-task loss. RoI max pooling worked by dividing the h × w RoI window into an H × W grid of sub-windows of approximate size h/H × w/W and then max pooling the values in each sub-window into the corresponding output grid cell. Pooling was applied independently to each habitat feature map channel, as in standard max pooling. We validated a subset of the forecasted habitats.

We experimented with three pre-trained ImageNet networks, each with five max pooling layers and between five and thirteen conv layers. A pre-trained network initialized the R-CNN network which underwent three transformations. First, the last max pooling layer was replaced by a RoI pooling layer that was configured by setting H and W to be compatible with the net's first fully connected layer (e.g., H = W = 7 for VGG16). Second, the network's last fully connected layer and SoftMax (which were trained for 1000-way ImageNet classification) and replaced with the two sibling layers (a fully connected layer and SoftMax over K + 1 categories and category-specific bounding-box regressors). Third, the network was modified to map the RGB interpolated *Anopheles* habitat capture point data inputs.

A total of 25 potential *Anopheles* larval breeding sites were predicted. Of the 25 sites expected to be suitable *Anopheles* habitats by the UAV model, 23 (92%; 95%

Fig. 13 Validated *Anopheline* mosquito habitat site as defined in the red box

confidence interval (CI), 81–100%) contained larvae. In contrast, just 2/10 (20%; 95% CI 0–45%) sites examined, which were not predicted to represent Anopheles aquatic habitat by the model, were found to contain larvae (Fig. 13). The scaled-up signature model thus exhibited a sensitivity of 80% and a specificity of 92% when applied in the intervention site in Nkolondum, a statistically significant performance ($p < 0.0001$; Fisher's Exact test).

Based on weekly entomological surveillance using Human landing catches and Pyrethrum spray catches (PSC) performed indoors, our preliminary results conducted in the Nkolondom intervention site showed a drastic decline in the number of resting adult anopheline mosquitoes from week 1 to week four after the intervention (Fig. 14). After implementing the Seek and Destroy intervention, there was a steady decline in biting rates with reduction rates varying from 50 to 100%. In the control site of Nkolbisson, this trend was not observed during the two rounds of entomological surveys. Indeed, the number of mosquitoes collected varies from week-to-week collection over the two sampling periods.

Given increasing insecticide resistance, LSM is essential in Cameroon. Our approach can be adapted to other VBDs where VBDs pose an increasing public health threat in Cameroon, and there is a critical need to expand vector surveillance and improve outbreak preparedness. The entomological baseline data acquired in Yaoundé shows a significant reduction in malaria mosquitos found within homes [15a and b]. Having a program that can decrease the mosquito population drastically combined with other malaria control efforts will lead to a remarkable reduction in malaria transmission. This combination of programs will be long-term preventive measures whose upkeep will be in the hands of local programs after initial

Fig. 14 Typical *Anopheles* habitats in Nkolondom

implementation. Overall, the addition of the Seek and Destroy program can give malaria control one of the final pieces of the puzzle and help the population and the economy of Cameroon. These programs cannot work independently, and each area of transmission must be dealt with to eliminate malaria, especially at the host site (Fig. 15).

Innovation for this baseline proposal stems from use of real time, innovative technology, sophisticated analytical methods for geospatial data to improve on the current practice for larval habitat identification and treatment in malaria endemic countries.

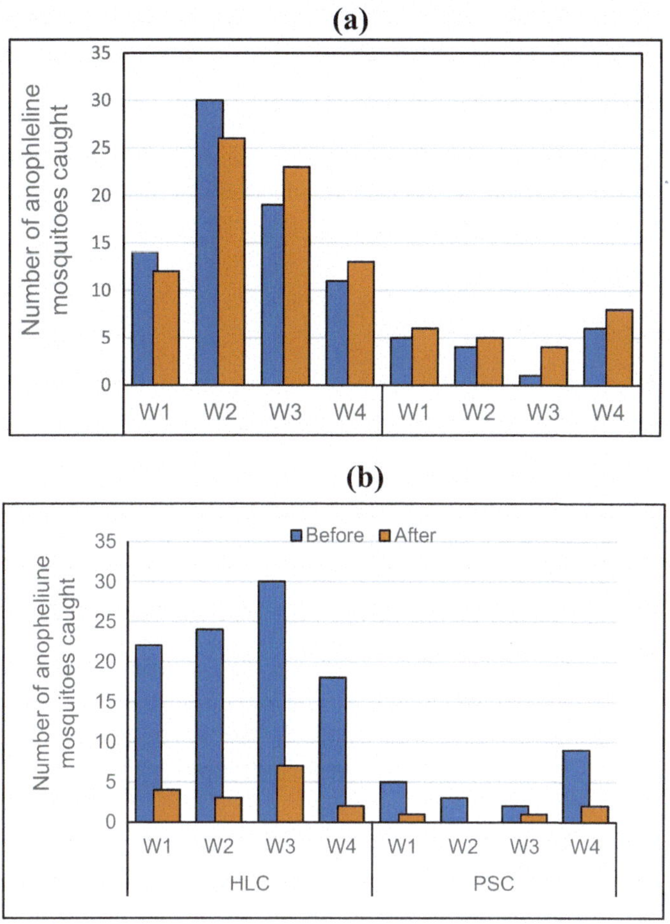

Fig. 15 **a** An *Anopheline* mosquito collection by HLC and PSC in the Nkolondom intervention site and in the Nkolbisson control site **b** Baseline (Blue) and follow up data (Orange) during the implementation of the "Seek and Destroy" project

- First, this baseline collection is technologically innovative in its use of real time drone technology, hyperspectral technology, and an iOS configurable, web-friendly, interactive, smartphone application.
- Second, the proposal is conceptually innovative. While larval habitat identification and treatment with LSM are not novel concepts, our approach is innovative as we focus on accuracy, timeliness, geographic scalability and leveraging tools.
- Third, previous drone-based mapping studies only hint at potential operational use, real-time identification, and scalability, while this baseline collection directly shows it.
- Our proposal bridges the gap between development of tools and operational use as our proposal develops and validates a tool for real-time detection and scalability.

Our work further builds on prior work by considering the effectiveness of such a tool on entomological monitoring and surveillance.

In summary, this baseline collection data challenges and seeks to shift the current paradigm of entomological surveillance treatment of larval habitat using novel machine learning techniques to advance our knowledge of larval ecology of malaria vectors in a low-resource Cameroonian setting characterized by climate change, urbanization, and habitat fragmentation.

3 Discussion

A key goal of the WHO Global Technical Strategy (GTS) for Malaria 2016–2030 is to eliminate malaria in most currently endemic countries by 2030. Approaches based on mass preventive chemotherapy, targeted chemotherapy and reactive strategies focused on drug treatments of infected individuals and indoor application of insecticide to homes of these individuals and their neighbors constitute the mainstay of the proposed strategy to achieve this goal.

While the GTS strategy is comprehensive, critical challenges are the financial resources required to implement the strategy successfully, especially in Africa [32, 33] where rising resistance of malaria parasites to ACTs, increased vector resistance to insecticides, including to pyrethroids used in long-lasting bed nets, and increased invasiveness of new vectors, such as *Anopheles stephensi*, into new geographies occur [34, 35]. In addition, short-term seasonal and long-term trends of climate and land-use change can alter not only biological traits but also the contact between mosquitoes and humans, and hence, transmission trajectories [36, 37].

These challenges have renewed interest in innovative measures that can target malaria mosquitoes in different (non-chemical) ways in the outdoor environment to overcome the above barriers to malaria elimination. One such tool, increasingly being discussed is the active control of aquatic stages of the mosquito through LSM, with innovative approaches based on application of microbials, such as Bacillus thuringiensis israeliensis (Bti) and B. sphaericus (Bs). These products kill only larvae since their toxins must be ingested and lead to starvation [12]. One such tool is the detection of larval habitats followed by delivery of these microbials through real time, UAV technology as part of a Seek and Destroy real time LSM program. Recent work carried out by us and others have shown how these UAVs can rapidly detect and treat many permanent, temporary, or transient mosquito breeding sites over large areas at low cost [38], highlighting their potential for enhancing contemporary malaria control and elimination efforts.

Before the discovery of the involvement of anopheline mosquitoes in malaria transmission to humans by Ronald Ross, vector control based on environmental management was already practiced as the primary tool of malaria vector control [39]. This strategy has been put at half-mast with progress made in pesticide research where insecticides have taken the lead in vector control. The "Seek and Destroy" approach of malaria vector control tested in this pilot study is reminiscent of the successes that were achieved in malaria vector control during the construction of the railway in Italy in 1899–1900 [40, 41] and the building of the Panama Canal [42]. Thus, evidence of the effectiveness of the physical control strategy against disease vectors has been known since the dawn of time and documented in the scientific literature [43–46]. Physical vector control targeting aquatic stages aims to reduce the vector population density by permanently removing standing water, for example by draining or filling land; making temporary changes to mosquito habitats to disrupt breeding sites.

The "Seek and Destroy" vector control approach implemented in this pilot study is in line with this trend towards the destruction of mosquito breeding sites. Preliminary results obtained during this study show a significant reduction in vector population density in the intervention site compared to the control site. In sites such as Nkolondom and Nkolbisson where malaria vectors show resistance to public health recommended insecticides, this approach would be best to not only significantly reduce vector population density and consequently malaria transmission, but also indirectly contribute to resistance management. In addition, with insecticide pressure from the mass use of impregnated mosquito nets and indoor residual spraying in the community, the vectors opt for a change in behavior for their adaptation in the community environment [47, 48]. This change in adult mosquito behavior has implications for the role of currently recommended control tools (bed nets and indoor residual spraying) and malaria transmission. Combining the "Seek and Destroy" approach with currently existing malaria control strategies in endemic areas would be a starting point towards eliminating the disease through prevention and treatment.

4 Seek and Destroy Pilot Program Proposal

The objective of this proposal is four-pronged. First, to assess how efficiently the real time, AI-based drone-driven "Seek and Destroy" interactive, intelligent platform can detect mosquito breeding habitats and deliver microbials to suppress the development of malarial mosquitoes in endemic settings, and measure how long this suppression would last. The second goal is to evaluate the impact this tool and ongoing malarial control activities will have in reducing malaria cases and clinical disease. Third, to use the impact data and costs of various interventions for modeling the prospects of malaria transmission elimination using the combination of S&D with different malaria control activities to determine the best cost-effective mix of elimination strategies in different settings. Finally, we will evaluate protocols to best embed the tool within Ministries of Health so that S&D using AI and drones can be used by health officials for monitoring impacts and for adaptive decision making.

We propose the following specific aims:

Aim 1: To evaluate the efficiency of S&D tool to detect mosquito breeding habitats and deliver optimal microbial loads to suppress the development of malaria mosquitoes in different spatio-temporal landscapes.

The current S&D technology has shown that it can significantly reduce adult malaria mosquito populations significantly additional to the effects of using ITNs alone which can be ineffective against exophagic and exophilic mosquitoes. Expanded field experiments are now required to: (1) evaluate the effectiveness of the real time tool to detect permanent, temporary, or transient, mosquito breeding sites over the range of the major malaria ecosystems in Cameroon, and (2) determine the most effective payload sizes for reducing and suppressing larval and emerging adult mosquitoes in each of these ecosystems. We propose to undertake these evaluations in 4 villages each from a total of five malarial zones—the far north Sahelian region, the northern Soudanian/Guinean zone, the northwestern humid savannah, the south-western forest zone, and from the coastal region. 2 villages will form the intervention sites alongside 2 controls in each zone. Drone spraying with 2 different Bacillus payloads will be tested and duration of effects assessed weekly over 2 months. In each site, mosquito larval/pupal densities and adult emergence counts will be conducted daily using standard WHO dippers and emergence traps, respectively. Field workers will be recruited in each village and trained to perform these collections, including larval and adult species identification using taxonomic keys—provided by Dr Phillpe Nwane lecturer and entomologist from Yaoundé 1 university.

Aim 2: To conduct field surveys in control versus intervention settings to evaluate the impact of combining "S&D" with different on-going malaria control and elimination strategies.

Apart from the effectiveness studies above on habitat larval source reduction, we will also use these sites to conduct surveys of impacts of habitat management using AI-empowered drones on adult biting mosquitoes and case prevalence. Here, the control villages will continue with current malaria intervention strategies (ITNs, chemotherapy), while the intervention villages will additionally include habitat management using S&D. The marginal reduction in adult biting mosquitoes (to be caught using human landing catches (HLCs) and via indoor aspiration) and case prevalence will then allow calculation of the added contribution of the drone-based habitat larval source management tool. These surveys will be carried out during different malarial transmission seasons or at monthly intervals in year 1 to determine the optimal timing, duration, and frequency of the drone intervention strategy that will best suppress malarial mosquitoes to bring about control or eradication.

Aim 3: To construct data-informed transmission models coupled with the effectiveness and costs of S&D-driven malaria control strategies to determine the optimal mix of interventions using S&D to eliminate malaria in different endemic settings.

We will develop a mathematical model to estimate the basic reproductive number, Ro, in the different study villages (and ultimately for all villages in Cameroon) based on EIR rates that can be quantified in the study villages directly from adult mosquito catches and their measured sporozoite rates, and from the published literature on EIRs for West Africa), which will be used to create interpolated smooth maps of EIR distribution across Cameroon for extracting the corresponding values for every village of the country. Similarly, PR, the case prevalence of malaria, will be assembled for each study village directly and for all villages in Cameroon from records held on malaria incidence from the MOH. The malaria Indicator survey data on IRS and ITN usage and treatments delivered will also be compiled either directly from the study villages or for all villages in Cameroon from both the MOH and from the USAID DHS program that will be interpolated using geostatistical methods to obtain values for each village. This data will be inputted into the Ro expression to obtain the respective Ro values for each village. Year 1 will focus on developing this Ro base data for Cameroon. The data on the impacts of larval resource management in reducing the number of mosquitoes in relation to humans and the effects of ITNs and IRS usage and treatments provided to deflect, prevent biting, and kill mosquitoes via reduced biting and increased mortality will then be used together with data on the fraction of habitats removed and ITN/IRS/treatment coverages per village to determine the relative effects of current strategies (ITN/IRS/treatments) versus integrated malaria control combining these interventions with LSM to reduce and suppress Ro below 1.0. In so doing, transmission was either controlled sustainably or will be broken. Monitoring malarial mosquito numbers and PR as well as ITN/IRS/treatment coverages in the integrated versus current control villages from year 2 will then allow: (1) evaluating the levels to which these variables will need to be reduced sustainably to keep Ro below 1, implying control or elimination of malaria, and (2) determining the mix of interventions also to reduce and maintain Ro < 1 in all endemic villages of Cameroon. Once Ro is reduced to below 1, the absence of malaria infection in communities can be used to validate if the proposed interventions break or control malaria transmission. We expect to be able to create the Ro map in year one and validate our predictions by year 4. Once validated, our methods could form the basis of malaria surveillance for achieving and maintaining sustained malaria control in the country. A locally installable dashboard built using Shiny R will be developed for use by malaria program managers at the MOH to facilitate input of data on changes in biting mosquito numbers per human and in PR plus habitat coverage (which can be monitored using the drones) and ITN/IRS/treatment coverages from Cameroonian villages undergoing S&D so that values of Ro are automatically generated for monitoring if values are being kept <1 in these villages (Fig. 16). The app will be constructed in year two, tested with data, and used by district-level health officials for decision-making in years 3 and 4. A hierarchical system will also be developed along with maps to aid surveillance at various levels of the health administration, from district to national, by year 4.

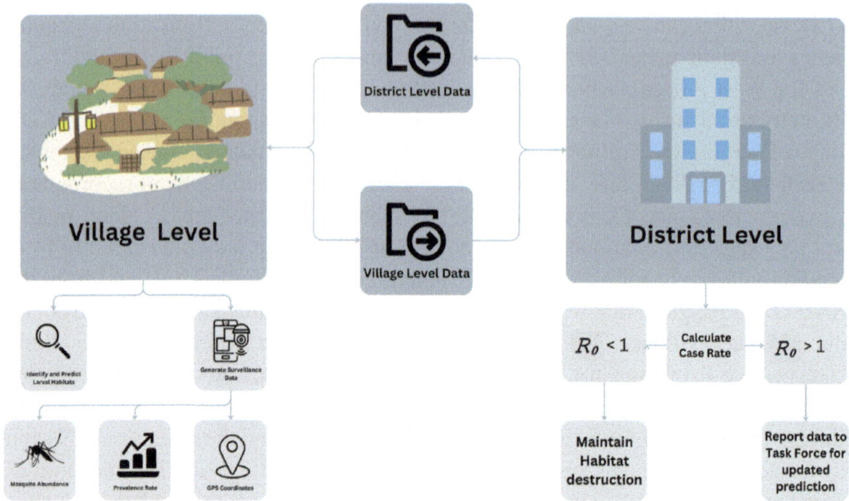

Fig. 16 Ro testing sequence

Aim 4: To Develop a Protocol to Institute S&D Alongside Ongoing Control Measures with the Ministry of Health to Guide Elimination Monitoring and the Making of Stopping Decisions.

The S&D program will work alongside existing malaria control programs and integrate existing campaigns (Fig. 17). To accomplish this, the program will work alongside the Ministry of Health through the National Malaria Control Program, PMI, Peace Corps, and civil society organizations and plans to create a Seek and Destroy Taskforce to review available data and make informed decisions on how to adapt S&D initiatives to meet the changing built environment continuously. This task force will meet at least once monthly to review data and constantly make any necessary adjustments to S&D.

We intend to train field task forces at the village level to implement habitat management activities and conduct larval/emerging adult abundance surveys, including identifying mosquito species using taxonomic keys. This will be achieved using five drones as we aim to establish the program in four villages spanning five transmission zones. Thus, we propose to investigate the total impact of 20 villages. Our partners will include President Malaria Incentive, The Ministry of Health, Yaoundé University, and local village/district leaders.

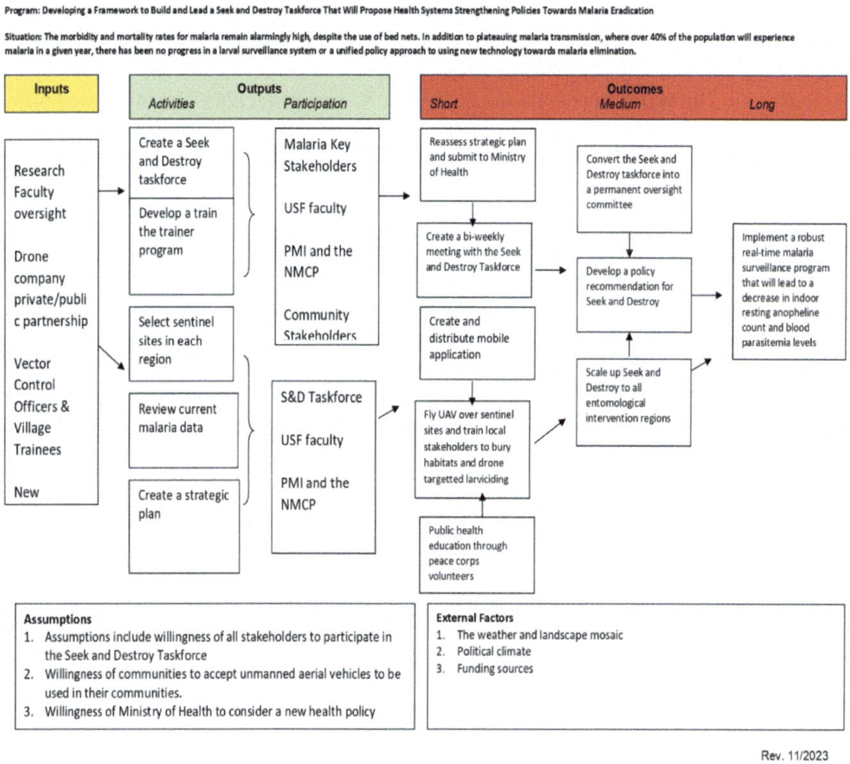

Fig. 17 Program flowchart

References

1. WHO, Malaria Factsheet (2017) Available at: OBJ
2. Bates I, Fenton C, Gruber J, et al. (2004) Vulnerability to malaria, tuberculosis, and HIV/AIDS infection and disease. Part 1: Determinants operating at individual and household level. Lancet Infect Dis. 4(5):267–277. https://doi.org/10.1016/S1473-3099(04)01002-3
3. Alirol E, Getaz L, Stoll B, Chappuis F, Loutan L (2011) Urbanisation and infectious diseases in a globalised world. Lancet Infect Dis 11(2):131–141. https://doi.org/10.1016/S1473-3099(10)70223-1
4. Jamieson DJ, Theiler RN, Rasmussen SA (2006) Emerging infections and pregnancy. Emerg Infect Dis 12(11):1638–1643. https://doi.org/10.3201/eid1211.060152
5. World Health Organization (WHO) (2022) World Malaria Report 2022. World Health Organization
6. World Health Organization (WHO). Malaria surveillance, monitoring & evaluation: a reference manual. (2018)
7. Clennon JA, Kamanga A, Musapa M, Shiff C, Glass GE (2010) Identifying malaria vector breeding habitats with remote sensing data and terrain-based landscape indices in Zambia. Int J Health Geogr 9(1):58. https://doi.org/10.1186/1476-072X-9-58
8. Stanton MC, Kalonde P, Zembere K, Hoek Spaans R, Jones CM (2021) The application of drones for mosquito larval habitat identification in rural environments: a practical approach for malaria control? Malar J 20(1):244. https://doi.org/10.1186/s12936-021-03759-2

9. Sarira TV, Clarke K, Weinstein P, Koh LP, Lewis M (2020) Rapid identification of shallow inundation for mosquito disease mitigation using drone-derived multispectral imagery. Geospat Health. 15(1). https://doi.org/10.4081/gh.2020.851
10. Burkot TR, Farlow R, Min M, Espino E, Mnzava A, Russell TL (2019) A global analysis of National Malaria Control Programme vector surveillance by elimination and control status in 2018. Malar J 18(1):399. https://doi.org/10.1186/s12936-019-3041-2
11. Maienfisch P (2016) Malaria vector control. Chimia (Aarau). 70(10):681. https://doi.org/10.1016/j.ijid.2014.03.720
12. Fillinger U, Lindsay SW (2011) Larval source management for malaria control in Africa: Myths and reality. Malar J 10(1):353. https://doi.org/10.1186/1475-2875-10-353
13. Kanyangarara M, Mamini E, Mharakurwa S et al (2016) High-Resolution Plasmodium falciparum Malaria Risk Mapping in Mutasa District, Zimbabwe: Implications for Regaining Control. Am Soc Trop Med Hyg. 95(1):141–147
14. Louis VR, Phalkey R, Horstick O et al (2014) Modeling tools for dengue risk mapping - a systematic review. Int J Health Geogr 13(1):50. https://doi.org/10.1186/1476-072X-13-50
15. Estallo EL, Benitez EM, Lanfri MA, Scavuzzo CM, Almiron WR (2016) MODIS environmental data to assess chikungunya, dengue, and zika diseases through Aedes (Stegomia) aegypti Oviposition Activity Estimation. IEEE J Sel Top Appl Earth Obs Remote Sens. 9(12):5461–5466. https://doi.org/10.1109/JSTARS.2016.2604577
16. Palaniyandi M (2012) The role of remote sensing and GIS for spatial prediction of vector-borne diseases transmission: A systematic review. J Vector Borne Dis 49(4):197–204
17. Zou L, Miller SN, Schmidtmann ET (2006) Mosquito Larval Habitat Mapping Using Remote Sensing and GIS: Implications of Coalbed Methane Development and West Nile Virus. J Med Entomol 43(5):1034–1041. https://doi.org/10.1093/jmedent/43.5.1034
18. Kiang R (2009) Modeling and surveillance: benchmark report
19. Berni JAJ, Zarco-Tejada P, Suárez L, González-Dugo V, Fereres E (2009) Remote sensing of vegetation from UAV platforms using lightweight multispectral and thermal imaging sensors. Int Arch Photogramm Remote Sens Spat Inform Sci. 38(6) p 6
20. Byrne I, Chan K, Manrique E, et al. (2021) Technical workflow development for integrating drone surveys and entomological sampling to characterise aquatic larval habitats of anopheles funestus in Agricultural Landscapes in Côte d'Ivoire. Nolan M, ed. J Environ Public Health. 2021:1–14. https://doi.org/10.1155/2021/3220244
21. Hardy A, Makame M, Cross D, Majambere S, Msellem M (2017) Using low-cost drones to map malaria vector habitats. Parasit Vectors 10(1):29. https://doi.org/10.1186/s13071-017-1973-3
22. Valdez-Delgado KM, Moo-Llanes DA, Danis-Lozano R et al (2021) Field Effectiveness of drones to identify potential Aedes aegypti breeding sites in household environments from Tapachula, a Dengue-Endemic City in southern Mexico. Insects 12(8):663. https://doi.org/10.3390/insects12080663
23. Amenyo J-T, Phelps D, Oladipo O, et al. (2014) MedizDroids Project: Ultra-low cost, low-altitude, affordable and sustainable UAV multicopter drones for mosquito vector control in malaria disease management. In: IEEE global humanitarian technology conference (GHTC 2014). 590–596
24. Mechan F, Bartonicek Z, Malone D, Lees RS (2023) Unmanned aerial vehicles for surveillance and control of vectors of malaria and other vector-borne diseases. Malar J 22(1):1–11
25. Mukabana WR, Welter G, Ohr P et al (2022) Drones for area-wide larval source management of malaria mosquitoes. Drones. 6(7):180
26. Ohrt C, Roberts KW, Sturrock HJW, Wegbreit J, Lee BY, Gosling RD (2015) Information systems to support surveillance for malaria elimination. Am J Trop Med Hyg 93(1):145
27. Nwane P, Etang J, Chouaïbou M, Toto JC, Mimpfoundi R and Simard F (2013) Multiple insecticide resistance mechanisms in Anopheles gambiae s.l, populations from Cameroon, Central Africa. Parasites & Vectors, 6: 41
28. Kimbatsa FG, Mahoungou E, Ofoueme YB (2018) L'importance de l'horticulture dans la lutte contre l'insécurité alimentaire, la pauvreté et la protection de l'environnement à Brazzaville (République du Congo). Études caribéennes, (39–40)

29. Piameu M, Nwane P, Toussile W, Mavridis K, Wipf NC, Kouadio PF et al (2021) Pyrethroid and Etofenprox Resistance in Anopheles gambiae and Anopheles coluzzii from Vegetable Farms in Yaoundé, Cameroon: Dynamics, Intensity and Molecular Basis. Molecules. janv 26(18):5543
30. Tchouakui M, Assatse T, Mugenzi LMJ, Menze BD, Nguiffo-Nguete D, Tchapga W, et al. (2022) Comparative study of the effect of solvents on the efficacy of neonicotinoid insecticides against malaria vector populations across Africa. Infect Dis Poverty. 25 avr 11(1):35
31. Sonhafouo-Chiana N, Nkahe LD, Kopya E, Awono-Ambene PH, Wanji S, Wondji CS, et al. (2022) Rapid evolution of insecticide resistance and patterns of pesticides usage in agriculture in the city of Yaoundé, Cameroon. Parasites & Vectors. 15(1):186
32. Griffin JT, Bhatt S, Sinka ME, Gething PW, Lynch M, Patouillard E, Ghani AC (2016) Potential for reduction of burden and local elimination of malaria by reducing Plasmodium falciparum malaria transmission: a mathematical modelling study. Lancet Infect Dis 16(4):465–472
33. Mwenesi H, Mbogo C, Casamitjana N, Castro MC, Itoe MA, Okonofua F, Tanner M (2022) Rethinking human resources and capacity building needs for malaria control and elimination in Africa. PLOS Global Public Health 2(5):e0000210
34. Oladipo HJ, Tajudeen YA, Oladunjoye IO, Yusuff SI, Yusuf RO, Oluwaseyi EM, El-Sherbini MS (2022) Increasing challenges of malaria control in sub-Saharan Africa: Priorities for public health research and policymakers. Annals of Medicine and Surgery 81:104366
35. Srinivasan S, Ghosh C, Das S, Thakare A, Singh S, Ganesh A, Subramani S (2022) Identification of a TNF-TNFR-like system in malaria vectors (Anopheles stephensi) likely to influence Plasmodium resistance. Sci Rep 12(1):19079
36. Sun XD, Zhao YL, Lin ZR, Zhao Y, Zhou YW, Li SG, Zheng Z (2023) Implementing a novel capture and ligation probe-PCR method in mass screen and treatment to support malaria elimination efforts in the China-Myanmar border region. Malar J 22(1):1–10
37. Bousema T, Drakeley C (2011) Epidemiology and infectivity of Plasmodium falciparum and Plasmodium vivax gametocytes in relation to malaria control and elimination. Clin Microbiol Rev 24(2):377–410
38. Jacob BG, Loum D, Kaddumukasa M, Kamgno J, Djeunga HN, Domche A, Mubangizi A (2021) Geospatial Artificial Intelligence Infused into a Smartphone Drone Application for Implementing'Seek and Destroy'in Uganda. Am J Entomol 5(4):92–109
39. Ross R (1897) On some peculiar pigmented cells found in two mosquitoes fed on malarial blood. BMJ 18:1786–8
40. Celli A (1901) The new preventative treatment of malaria in Latium. In Collected Papers on Malaria. Angelo Celli, 1899–1912. London: London School of Hygiene and Tropical Medicine
41. Lindsay SW, Emerson PM, Charlwood JD (2002) Reducing malaria by mosquito-proofing houses. Trends Parasitol 18:510–4. https://doi.org/10.1016/s1471-4922(02)02382-6. PMID: 12473368
42. Le Prince JA, Orenstein AJ (1916) Mosquito control in Panama; the eradication of malaria and yellow fever in Cuba and Panama. Putnam, New York/London
43. Wilson AL, Courtenay O, Kelly-Hope LA, Scott TW, Takken W, Torr SJ et al (2020) The importance of vector control for the control and elimination of vector-borne diseases. PLoS Negl Trop Dis 14(1):e0007831. https://doi.org/10.1371/journal.pntd.0007831
44. Steverding D (2008) The history of African trypanosomiasis. Parasit Vectors 1(1):1–8
45. Chernin E (1983) Sir Patrick Manson's studies on the transmission and biology of filariasis. Rev Infect Dis 5:148–66. https://doi.org/10.1093/clinids/5.1.148. PMID: 6131527
46. Reed W, Carroll J, Agramonte A, Lazear JW (1900) The etiology of yellow fever—a preliminary note. Public Health Pap Rep 26:37

Acellular Pertussis Vaccine Behind the Reemergence of the Whooping Cough

Jeegan U. Parikh, Miguel Reina Ortiz, Benjamin Jacob, and Ricardo Izurieta

Abstract *Bordetella pertussis*, a highly contagious organism, has reemerged in the United States ever since the switch from the whole-cell pertussis vaccine (DTwP) to acellular pertussis vaccine (DTaP) was implemented partially in 1992 and completely in 1997. The DTwP was associated with a higher number of adverse effects which prompted the change to acellular pertussis vaccine despite it having lower reactogenicity and durability to immunity. The switch in the vaccination has been associated with the increase in the number of cases with periodic outbreaks in 2010, 2012 and 2014. In addition to increase in cases, since 2000 there is a change in the epidemiological profile of cases, from childhood cases to cases in adolescents and vaccinated children. The introduction of the booster dose "Tdap" produced a mild initial decline but has not prevented cyclical outbreaks which reached a peak of 48,277 in 2012 with 30–35% of those in the age group of 11–19 years. The reasons behind this change in epidemiological profile might be related to either waning immunity, which makes the adolescents more susceptible to the infectious agent, or genotypical changes in the bacteria leading to a selection of strains without vaccine antigens. The trend analysis done suggest that the earlier waning of immunity after acellular vaccination may have led to the emergence of *B. pertussis* with non-vaccine strains by selective vaccination pressure and this combined effect may have led to an increase in the number of cases and change in epidemiological pattern of the disease. Further research is needed to study the immunological phenomenon behind the reemergence of disease and the creation of a more tolerable whole-cell vaccine. In addition, a possible revision of immunization schedule should be considered, which includes first priming with the whole-cell vaccine (first 3 doses) and then booster with acellular vaccines.

Keywords *Bordetella pertussis* · DTwP · DTaP · Immunity · Epidemiological profile

---

J. U. Parikh · M. R. Ortiz · B. Jacob (✉) · R. Izurieta
Department of Global Health, College of Public Health, University of South Florida, Tampa, USA
e-mail: bjacob1@usf.edu

B. Jacob et al. (eds.), *Tech Enabled Global Health Security*, Advanced Sciences and Technologies for Security Applications, https://doi.org/10.1007/978-3-031-86997-6_3

1 Introduction: Background and Purpose of the Project

Pertussis or whooping cough caused by *Bordetella pertussis,* is a highly contagious respiratory illness with very severe consequences in young children. Pertussis has a secondary attack rate of 11–18, requiring a vaccination coverage of 95% to achieve herd immunity [20]. Ever since the introduction of whole-cell Pertussis vaccine (DTwP) in the 1940s, the incidence of the disease decreased from 120,718 in 1950 to 4,083 in 1992 in the United States [6]. However, due to the increasing adverse effects of pain and fever, many of the parents elected to avoid getting their children vaccinated [12, 17]. These resulted in the introduction of acellular pertussis vaccines in 1992 which had shown similar efficacy as a whole-cell vaccine in vitro [3]. The current recommended immunization schedule for children mandates a prime series of 3 doses of DTaP at 2, 4, and 6 months and 2 booster dose at 15–18 and 48–60 months [8]. However, there has been some controversy around this switch from whole-cell to acellular pertussis vaccine, as research suggests that DTaP leads to lower duration of protection than that offered due to whole-cell vaccine or natural infection [18, 22, 27]. The number of cases of Pertussis has increased from 5137 in 1995 to 17,912 in 2016 with the number of cases as high as 48,277 in 2012 and 32,971 in 2014 [6]. The immunization coverage of the 3 doses or more of DTaP in the United States was 95% in 2016 which is similar to the vaccination coverage in 1995 [28]. These have led to questions being raised on the efficacy of the DTaP. The waning of immunity post DTaP vaccination may be responsible for the increase in the incidence of Pertussis. This increase in the incidence of pertussis has been associated with the change in the epidemiological profile of the disease with a high number of cases in the age group of 11–19 years [10, 25]. These increased number of cases of Pertussis lead to the introduction of a single dose of "Tdap" in 2005 as booster for those aged 11 or 12 years. Along with the waning of immunity, antigenic variation in the pertussis bacteria may have been responsible for the re-emergence of Pertussis [19, 23]. In addition, there is a possibility that the vaccination coverage could be low in certain groups or in pockets due to lack of access to health care or the recent anti-vaccination movement, which has resulted in the increase in the number of cases of Pertussis along with better surveillance measures. The purpose of these project is to look at the available data to come to the inference regarding the possible reasons behind the reemergence of Pertussis.

2 Literature Review

Pertussis or Whooping cough is caused by *Bordetella pertussis*, a fastidious, aerobic, gram-negative bacterium [4]. The bacterium produces multiple active products like pertussis toxin, pertactin, tracheal cytotoxin, agglutinogens and filamentous haemagglutinin that produce the signs and symptoms of the infection. It is a highly contagious disease, transmitted from person to person by respiratory droplets. Pertussis or Whooping cough develops in all the age groups, with the disease and associated complications being much more severe among infants or younger children than adults [4]. Before the availability of the vaccine, Pertussis was one of the most common diseases of childhood and childhood mortality in the United States.

A whole-cell vaccine with an aim to reduce the infection and severity of the disease was introduced in the 1940s, which helped in reducing the number of cases of the disease. The incidence of disease reduced by 75% from 120,718 cases in the 1950s to 4,083 in 1992 [4]. The coverage of DTP-3 vaccine in 1992 was 95%, which is very close to the vaccine coverage required to achieve herd immunity for pertussis [14, 28]. However, the administration of whole-cell vaccine was associated with adverse effects of pain, fever, erythema, higher crying from babies which was leading to increasing number of parents refusing to vaccinate their children [12, 17]. This prompted a shift from whole-cell pertussis vaccine to acellular vaccine despite acellular pertussis having lower vaccine reactivity and a small period of providing protection (waning immunity) [22].

Acellular pertussis vaccines which consisted of 5 purified antigens (pertussis toxin, filamentous hemagglutinin, pertactin and fimbriae types 2 and 3), were developed to address the concerns about the safety of the whole-cell pertussis vaccine and was later introduced in the recommended immunization schedule in 1992 in United States [3]. The Acellular vaccines replaced the whole-cell vaccine first as 4th and 5th dose in 1992 and then for the whole childhood series in 1997 [11]. It had a similar in vitro efficacy of 60–89% as a whole-cell vaccine but with less severe adverse effects. However, since the introduction of the vaccine there has been an increase in the number of cases of pertussis with a changing epidemiological pattern [10].

The vaccination coverage of 3 doses of DTaP remained around 93.7 (92.8–94.5%) in 2016 with a significant disparity in coverage among black children and those below poverty level who were uninsured or covered by Medicaid compared to Private insurance [14]. The coverage of DTaP was 94.6% in White (non-Hispanic), 93.3% in Hispanic, 96.6% in Asian, 93.6% in American Indian, 91.4% in Native Hawaiian and 90.8% in black, non-Hispanics. Similarly, the coverage was 95.9% among those who had private insurance, 92.5% among Medicaid and only 80% among those uninsured [14]. Compared to the required vaccination coverage for herd immunity, there are certain sections among the population which has a lower level of vaccination coverage and could be responsible for the increasing cases of pertussis [14, 20].

The demography of the cases suggests older adults vaccinated with 4 doses of DTaP were being affected by disease compared to the children in 6 months to 6 years of age group being predominantly affected before the introduction of vaccine [10, 26]. Initially, the increase in the number of cases of Pertussis was considered to be due to the better surveillance system and laboratory testing.

The highly contagious nature of the disease and the presence of pertussis in the adult's pre-vaccination era suggests that natural infection did not confer a lifelong immunity [26]. Even in the early days of whole-cell vaccine, cyclical outbreaks of pertussis were observed suggesting a waning of immunity and while the vaccine may have prevented the disease, transmission of the infection continued [10, 26]. However, the epidemiological trend of the disease remained the same with incidence being lower with increasing age. This trend of disease changed in the 2000s with adolescents responsible for 30–35% of the cases in 2004 and 2005 [10, 25]. Further research into the causes has suggested waning immunity as one of the factors to be responsible for this resurgence of disease and shift in the age group with the burden of disease. To address this issue of waning immunity, a single booster dose of Tdap was introduced in 2005–2006. Despite the introduction of booster, the cases of pertussis reached as high as 48,277 in 2012 and 32,971 in 2014, despite a DTaP-3 vaccination coverage of approximately 95% with significant number of cases in the age group of 7–19 years of age [6, 28]. During the outbreak of 2012, the rates of pertussis were greatly elevated among children aged 7–10 and 13–14 and most were fully vaccinated [9]. In 2016, out of 17,972 cases, 12,631 (70.7%) cases were found in the population above 7 years of age with 11–19 years of age group having the highest number of cases. Even among the age group of 6 months to 6 years, 3913 cases of Pertussis were reported and 1808 (46%) out of them had received complete primary series of DTaP vaccine (3-doses) [9]. Comparing the trends between 1992 and 2016, the incidence of disease per 100,000 population increased from 38.41 to 51.41 among <1 year age group, 3.93–13.65 among 1–6 years age group, 1.95–14.88 among 7–10 years of age group, 1.99–16.31 among 11–19 years of age and 0.27–1.68 among 20+ years [7]. This reemergence of the disease among highly vaccinated children and adolescents, suggests that the acellular vaccines may have a lower durability of protection. A case-control study in California showed that after 5 doses of DTaP, 98% of children among 4–10 years had protective immunity after one year, however, it reduced to <90% after 3 years and <71% after 5 years [18]. A study published in American Academy of Pediatrics in June 2013, showed that teenagers who had received whole-cell vaccine were more protected than those who received acellular vaccines [15].

In addition to these, there has been a change in the antigenic and genotypic features of the circulating *B. pertussis* strains with selective pressure from vaccination as a possible reason for this rapid emergence [19, 23]. The emergence of strains without the vaccine antigens pertactin and pertussis toxin and the ability to cause disease among vaccinated population has raised doubts over the effectiveness of the current vaccines [1, 2, 21].

The changing epidemiological pattern of pertussis with increasing susceptibility among adolescents and vaccinated children along with the emergence of *B. pertussis* strains without vaccine antigens suggests an urgent need to understand the immunologic basis for vaccine failure [10]. Data is increasingly emerging that children receiving whole-cell vaccines are at lower risk than those vaccinated with acellular vaccines and priming initially with the whole-cell vaccine may lower the risk of disease [16, 24]. A coordinated research effort needs to be developed to understand the problem and development of more effective and tolerable whole-cell vaccine.

3 Methods

To find an effective correlation and infer whether the resurgence of cases is due to the waning of immunity after acellular pertussis vaccination we will use the PICO (Population, intervention, compare, outcome) format.

3.1 Study Population

There are 3 significant events in the history of Pertussis vaccination, the introduction of whole-cell vaccine in 1940s, acellular pertussis vaccine (DTaP) in 1992, and Tdap in 2005. To compare the effect of whole-cell and acellular vaccine on the population, we will compare the number of cases before and after the introduction of the vaccine. The effective population under study would be US population between the age group of 1–20 years. As the basic reproduction rate of the pertussis is very high (11.0) and its mode of transmission being airborne, 93% of the cases were found to be in the population of age group <10 years. Scientific studies done regarding the persistence of immunity after the vaccination has suggested earlier waning of immunity after acellular pertussis vaccine than due to the whole-cell vaccine or natural infection [13]. This has also resulted in a switch in the epidemiological profile with most cases being reported in the age group of 0–6 years pre-DTaP and 7–19 years post-DTaP. Even the introduction of Tdap has not stopped the changing pattern of pertussis epidemiology. These factors let us to the conclusion that most of the cases of pertussis in the pre-vaccination and post vaccination would happen in the age group till 20 years of age.

3.2 Data Source and Analysis

The data for number of the cases was obtained from the "Nationally Notifiable Disease Surveillance System (NNDSS)", Surveillance reports, and the Centers for Disease Control and Prevention. The Vaccination coverage of 3rd and 4th doses of DTP or DTaP were obtained from World Health Organization database on immunization coverage in United States, Centers for Disease Control and Prevention and doing a search in PubMed for "DTP", "DTaP" and "Pertussis vaccine". In addition, to understand the effect of the switch of vaccines on the epidemiological profile of the disease, the incidence numbers were obtained from the "NNDSS" with the incidence numbers divided in the age group of <1 year, 1–6 years, 7–10 years, 11–19 years and +20 years. The data were compared based on the introduction of the vaccine in 3 phases before introduction of vaccine (1990, 1991), after introduction of DTaP but before the introduction of Tdap (1992–2005) and after the introduction of Tdap (2005–2016). The results in all the 3 periods were compared by trend analysis to come to the outcome of our research question "if the switch from whole-cell pertussis vaccine to acellular pertussis vaccine has resulted in re-emergence of the disease and change in epidemiological profile".

4 Ethical Considerations

As most of the data to be used in the study is freely available on the internet, it is implied that use of data for further analysis is permitted, with the acknowledgment of the ownership of the original data. As the data obtained may have not been collected for the current research question, the data will be evaluated for accuracy, the appropriate period of data collection and content of data. The data will be kept only for the duration necessary for the study. The soft copies of data will be kept in encrypted files in computers and cloud to prevent it from unauthorized use and accidental destruction. Along with this, the appropriate analysis of the data must be conducted to prevent the rise of ethical issues with the study.

5 Results

When the chart of the number of cases of Pertussis was analyzed it showed a decline in the number of cases from 107,473 in 1922 to 17,972 cases in 2016 (Fig. 1) [6]. However, when the trend of the overall decline in cases was done it showed an interesting pattern. Initially, in the period of 1922–1948, the number of cases of pertussis continued to remain above 100,000 per year with the peak of 265,269 cases in 1934 (Fig. 1) [6]. However, with the introduction of the vaccine in 1940s, the number of cases continued to decline from 120,718 in 1950 to as low as 1,730 cases per year in 1980 [6]. If the trend had continued the disease should have been eliminated by 2000, but we saw a resurgence of the cases in the 90s when the incidence started to rise (Fig. 1) [6].

An analysis of the number of cases versus coverage of DTP-3 vaccine showed the coverage at 96% in 1980 with a sudden decline in late 80 s with the coverage being as low as 69% in 1991 (Table 1 and Fig. 2) [6, 29]. The decrease in the coverage of DTP3 vaccination can be directly correlated to the increase in the number of cases of pertussis. As the acellular pertussis vaccine (DTaP) replaced whole-cell vaccine (DTwP) in 1992, the vaccination coverage increased with 83% in 1992 to as high as 96% in 1999 (Table 1 and Fig. 2) [6, 29].

However, the increase in the vaccination coverage didn't stop the increase in the incidence of pertussis, with the number of cases crossing 10,000 in 2003 despite an immunization coverage of 96% (Table 1 and Fig. 2) [6, 29]. As can be seen from the graph, the number of cases of pertussis continued to rise with periodic outbreaks in 2010, 2012–2014 when cases were approximately above 30,000 per year (Fig. 3) [6,

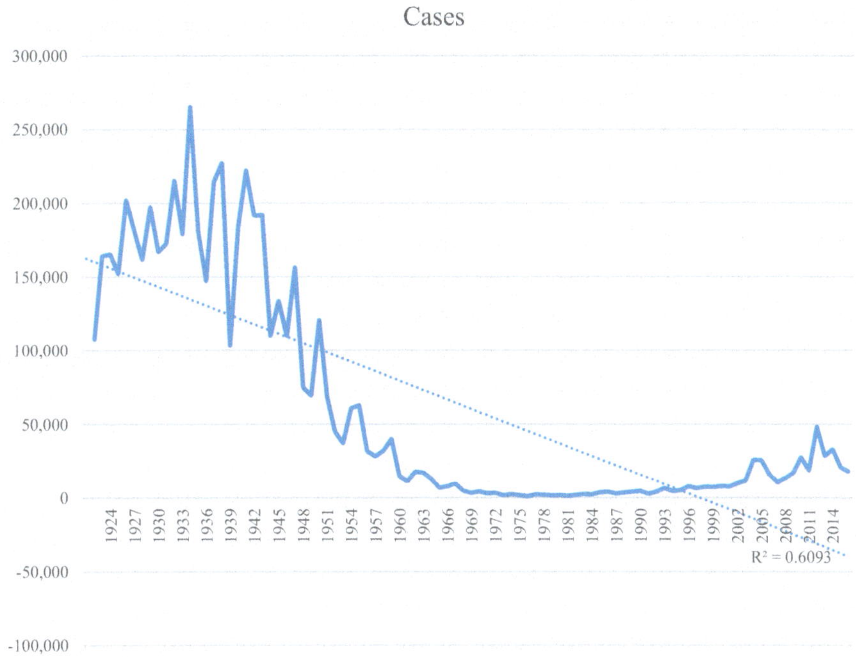

Fig. 1 Cases of pertussis per year in United States, 1922–2016

29]. The coverage of DTP-4 was 59% in 1992 which improved to 81.1 in 1996 with a peak of 85.7 in 2005. In 2016, the DTP-4 coverage was 83.4% in the population (Table 2 and Fig. 4) [5].

An epidemiological study of the incidence of pertussis cases among the population based on age group lead to a disturbing trend. While the number of cases of pertussis was highest in <1 year of age group, the steady increase in the pertussis cases in the 7–19 years of age group raises questions over the efficacy of the immunization program (Fig. 5) [7]. The incidence per 100,000 population increased from 38.41 in 1992 to 96.47 in 2005 and then decreased to 51.41 in 2016 in the <1 year of age group (Table 3 and Fig. 6) [7]. In the age group of 7–10 years and 11–19 years, the incidence per 100,000 cases were 14.84 and 16.31 in 2016 compared to 1.95 and 1.99 in 1992 respectively (Table 3, Figs. 6 and 7) [7]. In the outbreak years of 2010, 2012, 2013 and 2014 the incidence was as high as 31.78, 58.52, 30.61 and 34.04 in 7–10 years of age group (Table 3, Figs. 6 and 7) [7]. Percentage of adolescents in the age group of 11–19 years with pertussis cases were 13.3, 38.02, 21.27, and 29.57 in 2010, 2012, 2013 and 2014 respectively (Table 3, Figs. 6 and 7) [7]. The incidence in the age group of 1–6 years and 20+ years increased from 3.93 and 0.27 in 1992 to 13.65 and 1.68 in 2016 (Table 3 and Fig. 6) [7]. The increase in the number of cases in vaccinated age group lead to the introduction of a single booster of Tdap vaccine in 2005. The introduction of Tdap decreased the incidence of pertussis in

Table 1 Pertussis cases per year versus DTP-3 vaccination, 1980–2016

Year	Pertussis cases	DTP-3 (%)
1980	1,730	96
1981	1,248	96
1982	1,895	96
1983	2,463	97
1984	2,276	97
1985	3,589	96
1986	4,195	97
1987	2,823	97
1988	3,450	97
1989	4,157	
1990	4,570	
1991	2,719	69
1992	4,083	83
1993	6,586	88
1994	4,617	94
1995	5,137	95
1996	7,796	95
1997	6,564	96
1998	7,405	96
1999	7,298	96
2000	7,867	94
2001	7,580	94
2002	9,771	94
2003	11,647	96
2004	25,827	96
2005	25,616	96
2006	15,632	96
2007	10,454	96
2008	13,278	96
2009	16,858	95
2010	27,550	95
2011	18,719	96
2012	48,277	94
2013	28,639	94
2014	32,971	95
2015	20,762	95
2016	17,972	95

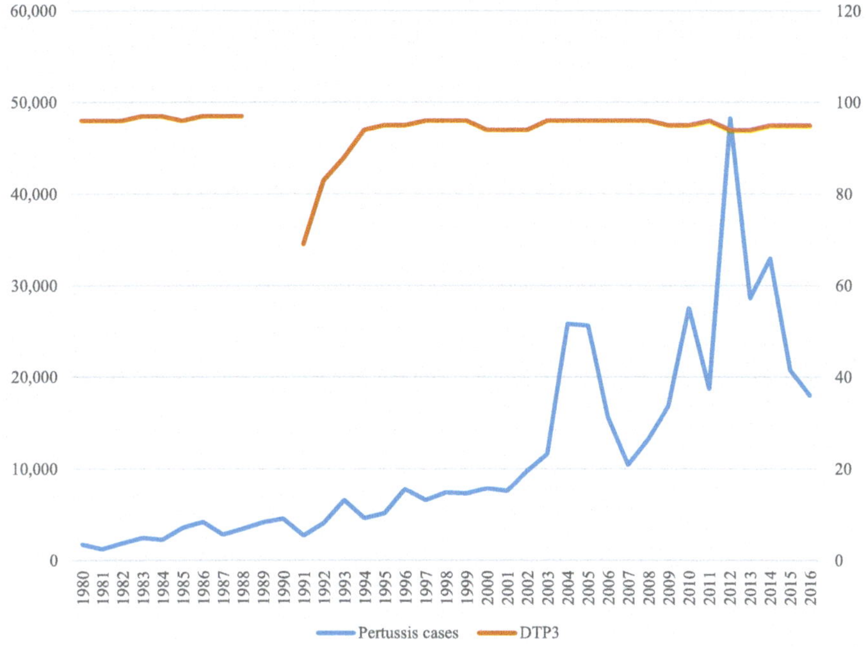

Fig. 2 Two axis graphs showing the number of pertussis cases versus DTP3 coverage (1980–1992 DTP3, 1992–2016 DTaP3), 1980–2016

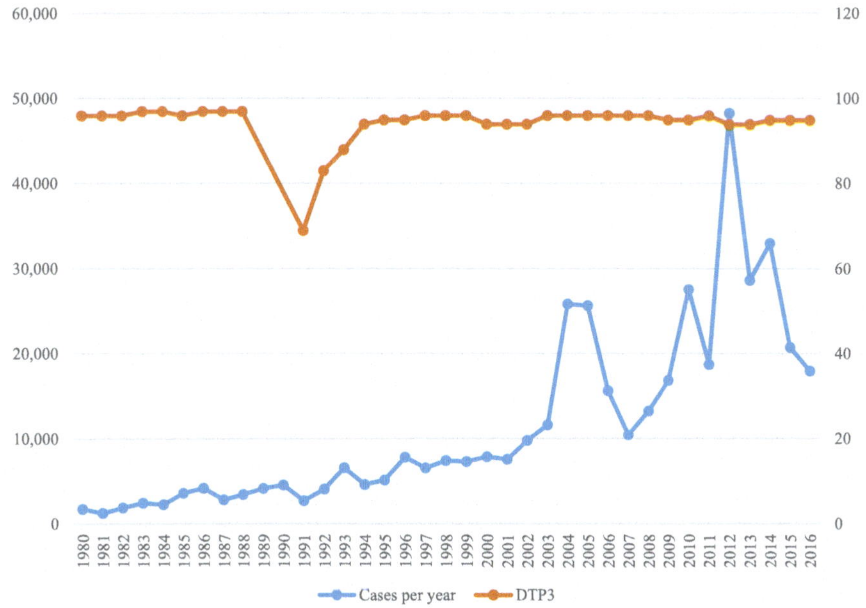

Fig. 3 Pertussis cases per year versus DTP3 coverage line scaled, 1980–2016

Table 2 DTP-4 coverage in the population

Year	DTP-4
1992	59
1993	72
1994	78
1995	78.5
1996	81.1
1997	81.5
1998	83.9
1999	83.3
2000	81.7
2001	82.1
2002	81.6
2003	84.8
2004	85.5
2005	85.7
2006	85.2
2007	84.5
2008	84.6
2009	83.9
2010	84.4
2011	84.6
2012	82.5
2013	83.1
2014	84.2
2015	84.6
2016	83.4

11–19 years of age group between 2005 and 2009, but it had an exponential increase in the outbreak years of 2012 and 2014 (Table 3) [7].

6 Discussion

The basic reproduction rate of pertussis is 11 and thus it requires an approximate level of 93–94% of immunization coverage to confer the herd immunity in the entire population [13]. The vaccination coverage of DTP-3 has been consistently above 95% in the population despite this, the trend of the incidence of the disease has reversed since the 90s and reached as high as 48,277 during the outbreak of 2012 [6, 29]. The reversal in the trend has been seen since the introduction of acellular pertussis vaccine

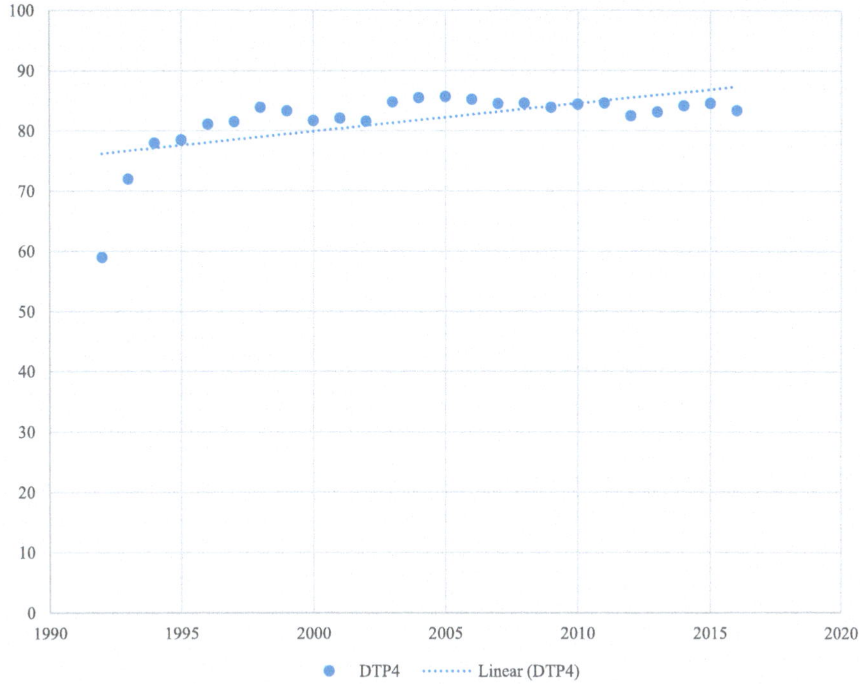

Fig. 4 DTP-4 coverage in the population per year, 1992–2016

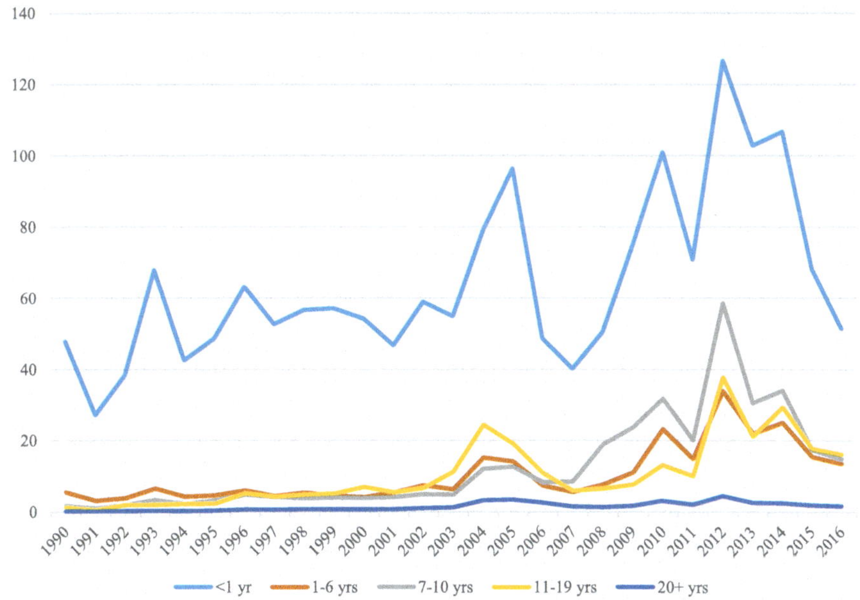

Fig. 5 Incidence of pertussis cases depending on age group per 100,000 persons, 1990–2016

Table 3 Incidence of pertussis per 100,000 persons by age group

Year	<1 year	1–6 years	7–10 years	11–19 years	20+ years
1990	47.89	5.57	1.68	1.38	0.19
1991	27.27	3.2	0.96	0.58	0.14
1992	38.41	3.93	1.95	1.99	0.27
1993	68	6.59	3.26	2.08	0.33
1994	42.61	4.35	2.26	2.3	0.27
1995	48.74	4.71	3.17	2.41	0.42
1996	63.23	6.07	4.81	5.25	0.72
1997	52.73	4.49	4.23	4.33	0.66
1998	56.71	5.45	3.8	4.88	0.78
1999	57.12	4.57	3.98	5.17	0.74
2000	54.23	4.15	3.9	7.08	0.74
2001	46.84	5.49	4.09	5.58	0.79
2002	59.01	7.53	4.94	6.8	1.08
2003	54.98	6.41	4.81	11.34	1.32
2004	79	15.28	12.07	24.5	3.29
2005	96.47	14.15	12.64	19.17	3.5
2006	48.78	7.47	8.32	11.05	2.63
2007	40.25	5.64	8.48	6.02	1.57
2008	50.54	7.67	18.85	6.58	1.35
2009	75.23	11.16	23.76	7.75	1.73
2010	100.9	23.27	31.78	13.3	3.15
2011	70.89	15.02	20.05	10.26	2.15
2012	126.65	34.09	58.52	38.02	4.51
2013	102.77	22.09	30.61	21.27	2.61
2014	106.68	25.14	34.04	29.57	2.5
2015	68.1	15.6	17.45	17.9	1.9
2016	51.41	13.65	14.84	16.31	1.68

in 1992 (Fig. 1). The introduction of acellular pertussis was warranted because of the complaint of the parents regarding the side effects of vaccine which resulted in the reduction of vaccination coverage in the second half of the 80s [12, 17, 29]. While the data regarding the immunization coverage is not available for 1989 and 1990, the vaccination coverage (DTP-3) of only 69% in 1991, a sharp fall from 97% in 1988 resulted in the recommendation of acellular pertussis vaccine with similar in vitro efficacy as whole-cell vaccine and lesser side effects (Table 1) [3, 29]. The shift to acellular vaccine helped in overcoming the fears and increased the immunization coverage in the population which reached the required 93–95% for herd immunity in 1994 (Table 1 and Fig. 2) [29]. Despite the high coverage of DTP-3 immunization,

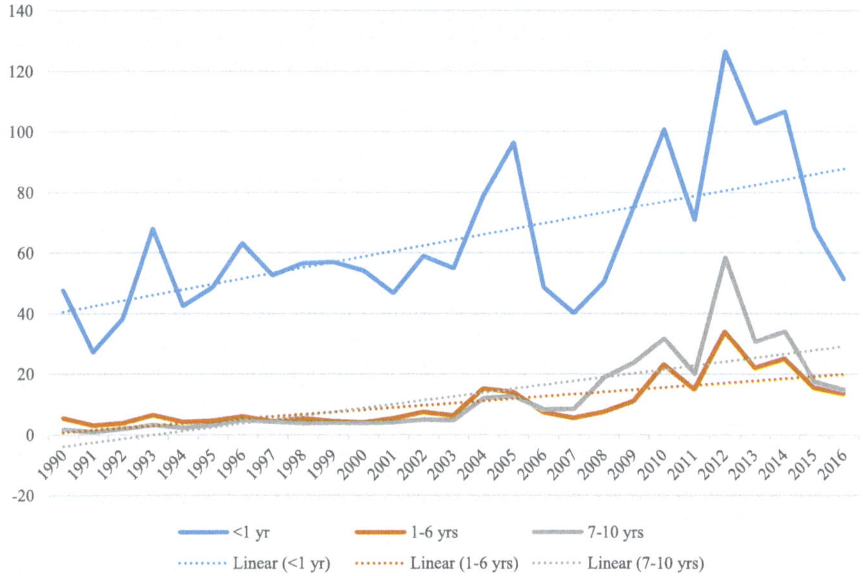

Fig. 6 Incidence of pertussis cases per 100,000 persons with trendline in <1 year, 1–6 years and 7–10 years age group, 1990–2016

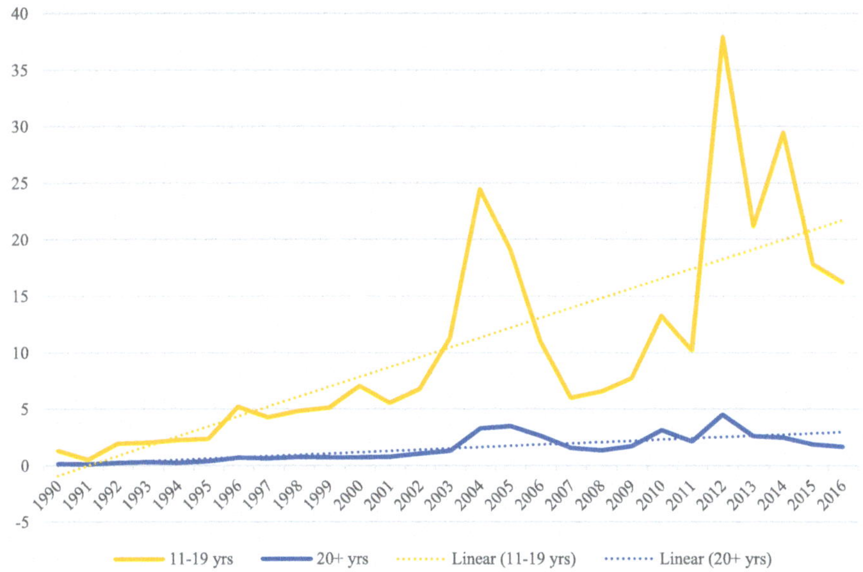

Fig. 7 Incidence of pertussis cases per 100,000 persons with trendline in 11–19 years and 20+ years age group, 1990–2016

the coverage of DTP-4 has plateaued and continue to remain in mid-80s, which suggests a certain deficiency in the system that fails to educate the parents about the importance of booster doses (Fig. 4) [5].

The expected trend in the cases would have been a reduction in the incidence of disease with increasing coverage. However, the reversal of trend started in the early 90s continued throughout 2000 with cases rising to five figures since 2003 (Table 1 and Fig. 2) [6]. Initially, the reversal of the trend in pertussis cases was explained due to better surveillance and diagnostic techniques. However, that does not explain the exponential growth in cases with periodic outbreaks. In 2015, Gambhir et al. proved that the waning of immunity after acellular pertussis vaccine (DTaP) was higher than after whole-cell pertussis vaccine (DTwP). The natural infection with pertussis results in loss of immunity at the rate of 3×10^{-5} $year^{-1}$, and results in nearly lifelong immunity [13]. The whole-cell vaccine series has a rate of loss of immunity similarly to the natural infection and thus can lead to nearly lifelong protection. The acellular vaccine results in the loss of immunity at the rate of 0.018 $year^{-1}$, resulting in the waning of immunity after some time [13]. Scientific studies have suggested that the immunity after acellular pertussis vaccine starts to wane after 2 years compared to the 7 years of whole-cell vaccine or natural infection. The mutation in the pertussis bacteria with the emergence of strains without the vaccine antigens pertactin and pertussis toxin was held responsible for the increase in the cases [1, 2, 21]. The period of DTwP was also associated with a cyclical increase in cases but the increase in incidence was much lower than the outbreaks after the switch to acellular pertussis (Fig. 1). The faster waning of immunity after the acellular pertussis vaccine may have resulted in the selection of non-vaccine antigen strains of pertussis bacterium [19, 23]. Along with the faster waning of immunity, plateauing of the DTap-4 at 84–85% results in 15% of the population with only partial immunity against bacterium. This partial immunity could be one of the factors in the emergence of the *B. pertussis* bacterium with non-vaccine antigens [19, 23].

The epidemiological profile of pertussis natural infection included many cases in the age group of <1 year of age. This continued throughout the phase of whole-cell pertussis vaccination phase. The reversal in the trend of pertussis cases was accompanied with the change in the epidemiological profile of the disease with 30–35% of the cases among the 7–19 years of age group. The introduction of Tdap as the booster dose in 2005, to counter this shift in epidemiological profile was met with selective success with decrease in the incidence in the age group of 11–19 years between 2006 and 2009 (Table 3) [7]. However, the increase in the number of cases in this age group during the outbreaks of 2010, 2012 and 2014 suggest the Tdap vaccine has a similar issue of waning immunity as DTaP vaccine.

The waning of immunity and mutation in strains were hypothesized to be responsible for the increase in the number of pertussis cases in the community and changing epidemiological profile of the disease [19, 23]. However, our trend analysis of incidence of cases versus vaccination coverage, previous scientific studies on waning of immunity and change in the epidemiological profile of disease has led us to come to a hypothesis that the switch to acellular pertussis vaccine may have resulted in earlier waning of immunity which put selective pressure on the emergence of non-vaccine antigen strains of pertussis bacteria. The emergence of non-vaccine strains of pertussis bacteria put the previously vaccinated people at risk and could be responsible for the change in the epidemiological profile of the disease with 30–35% of the cases in 7–19 years of age group. The failure of the Tdap booster dose to reverse the trend could also be explained by the emergence of mutated pertussis strain with non-vaccine antigens.

Despite this, our study has certain limitations, as it cannot prove the temporality of the association between waning of immunity and the emergence of new strains. Further immunological and microbiological studies are required to explain the correlation between the acellular vaccine and the emergence of new strains of bacteria which could explain the increase in the cases of the disease. Our study also does not take into account pockets of the unvaccinated population which could be inflating the number of pertussis cases in the entire population [14, 20].

7 Recommendations

The prevention of Whooping cough has become a challenge with the periodic outbreaks of cases including cases in the vaccinated population [9]. Further scientific studies are required to understand the immunologic basis of the failure of vaccination, steps to be taken to reverse this trend and bring the number of cases to the level of the 1980s. In addition, a widespread epidemiologic study is required to find if pockets of the unvaccinated population exist that could be lowering the immunity of the community and are responsible for the rise in the incidence of pertussis. Lack of access to health care due to racial factors or fear of vaccines could be the factor that could create these pockets of unvaccinated population and requires further resources to fill this gap. It is also important to inform the parents about the importance of the completion of vaccination series. More resources should be put in place to increase the coverage of DTaP-4 and Tdap booster doses.

A new method of priming with the first 3 doses with whole-cell vaccine and then a booster with acellular pertussis vaccine should be considered to overcome the emergence of strains of bacteria which lack acellular vaccine antigens. The children who went priming with the acellular pertussis vaccine in 1997 had a much higher rate of pertussis cases than those who went priming with whole-cell vaccine between 1992 and 1997 [16, 24]. Parents need to be educated regarding the safety profile of the vaccine and possible side-effects of whole-cell vaccine to prevent the refusal of vaccines by parents as established in 1990. Additional efforts should be made to make parents educated about basic hygiene measures, respiratory hygiene, cough etiquette and hand washing that could help in the prevention of pertussis and other disease transmitted in the same manner. The recommendations for Pertussis immunization should be made considering the balance between vaccine side effects and effectiveness. The current recommendations of complete immunization with DTaP and Tdap should be continued in a timely manner until further effective vaccines develop [16, 24].

It is particularly important for the developing countries with lower vaccination coverage to continue to use whole-cell vaccine rather than switch to acellular pertussis vaccine to prevent the pertussis epidemic in those countries. Pediatricians in these countries should educate the parents regarding the importance of vaccine and its safety profile to prevent the cases of rejection of vaccination.

It is a failure of the public health community that a disease which was thought to be under control has seen such a reversal in trend. The re-emergence of pertussis suggests the need for continuous environmental surveillance to understand the effect of immunization on the genotype of bacteria. In addition to further research, co-operation between different agencies and leadership is necessary to gain control and success in preventing the transmission of pertussis.

References

1. Barkoff A-M, Mertsola J, Guillot S, Guiso N, Berbers G, He Q (2012) Appearance of Bordetella pertussis strains not expressing the vaccine antigen pertactin in Finland. Clin Vaccine Immunol: CVI 19(10):1703–1704. https://doi.org/10.1128/CVI.00367-12
2. Bodilis H, Guiso N (2013) Virulence of pertactin-negative Bordetella pertussis isolates from infants, France. Emerg Infect Dis 19(3):471–474. https://doi.org/10.3201/1903.121475
3. Center for Disease Control and Prevention (1997, March 28) Pertussis vaccination: use of acellular pertussis vaccines among infants and young children recommendations of the advisory committee on immunization practices (ACIP). https://www.cdc.gov/mmwr/preview/mmwrhtml/00048610.htm. Accessed 01 Nov 2018
4. Center for Disease Control and Prevention (2017, August 7) Pertussis | Whooping Cough | Causes and Transmission | CDC. https://www.cdc.gov/pertussis/about/causes-transmission.html. Accessed 01 Nov 2018
5. Center for Disease Control and Prevention (2018) Pinkbook—appendix E vaccine coverage levels—United States 1962–2016, Atlanta, GA. https://www.cdc.gov/vaccines/pubs/pinkbook/downloads/appendices/e/coverage-levels.pdf
6. Center for Disease Control and Prevention (2018, February 7) Pertussis | Whooping Cough | Surveillance | Cases by Year | CDC. https://www.cdc.gov/pertussis/surv-reporting/cases-by-year.html. Accessed 01 Nov 2018
7. Center for Disease Control and Prevention (2018, July 11) Reported pertussis incidence by age group and year. CDC. https://www.cdc.gov/pertussis/surv-reporting/cases-by-age-group-and-year.html. Accessed 01 Nov 2018
8. Center for Disease Control and Prevention (2018, August 24) Vaccine information statement. Diphtheria Tetanus Pertussis. VIS, CDC. https://www.cdc.gov/vaccines/hcp/vis/vis-statements/dtap.html. Accessed 01 Nov 2018
9. Center for Disease Control and Prevention, National Center for Immunization and Respiratory Diseases (2018, February) 2016 Final pertussis surveillance report. https://www.cdc.gov/pertussis/downloads/pertuss-surv-report-2016.pdf. Accessed 01 Nov 2018
10. Clark TA (2014) Changing pertussis epidemiology: everything old is new again. J Infect Dis 209(7):978–981. https://doi.org/10.1093/infdis/jiu001
11. Clark TA, Messonnier NE, Hadler SC (2012) Pertussis control: time for something new? Trends Microbiol 20(5):211–213. https://doi.org/10.1016/j.tim.2012.03.003

12. Cody CL, Baraff LJ, Cherry JD, Marcy SM, Manclark CR (1981) Nature and rates of adverse reactions associated with DTP and DT immunizations in infants and children. Pediatrics 68(5):650–660
13. Gambhir M, Clark TA, Cauchemez S, Tartof SY, Swerdlow DL, Ferguson NM (2015) A change in vaccine efficacy and duration of protection explains recent rises in pertussis incidence in the United States. PLoS Comput Biol 11(4):e1004138. https://doi.org/10.1371/journal.pcbi.1004138
14. Hill HA, Elam-Evans LD, Yankey D, Singleton JA, Dietz V (2016) Vaccination coverage among children aged 19–35 months—United States, 2015. MMWR. Morb Mortal Wkly Rep 65(39):1065–1071. https://doi.org/10.15585/mmwr.mm6539a4
15. Klein NP, Bartlett J, Fireman B, Rowhani-Rahbar A, Baxter R (2013) Comparative effectiveness of acellular versus whole-cell pertussis vaccines in teenagers. Pediatrics 131(6):e1716–e1722. https://doi.org/10.1542/peds.2012-3836
16. Liko J, Robison SG, Cieslak PR (2013) Priming with whole-cell versus acellular pertussis vaccine. N Engl J Med 368(6):581–582. https://doi.org/10.1056/NEJMc1212006
17. Long SS, Deforest A, Smith DG, Lazaro C, Wassilak GF (1990) Longitudinal study of adverse reactions following diphtheria-tetanus-pertussis vaccine in infancy. Pediatrics 85(3):294–302
18. Misegades LK, Winter K, Harriman K, Talarico J, Messonnier NE, Clark TA, Martin SW (2012) Association of childhood pertussis with receipt of 5 doses of pertussis vaccine by time since last vaccine dose, California, 2010. JAMA 308(20):2126–2132. https://doi.org/10.1001/jama.2012.14939
19. Mooi FR, Van Der Maas NaT, De Melker HE (2014) Pertussis resurgence: waning immunity and pathogen adaptation—two sides of the same coin. Epidemiol Infect 142(4):685–694. https://doi.org/10.1017/S0950268813000071
20. Plans-Rubió P (2012) Evaluation of the establishment of herd immunity in the population by means of serological surveys and vaccination coverage. Hum Vaccin Immunother 8(2):184–188. https://doi.org/10.4161/hv.18444
21. Queenan AM, Cassiday PK, Evangelista A (2013) Pertactin-negative variants of Bordetella pertussis in the United States. N Engl J Med 368(6):583–584. https://doi.org/10.1056/NEJMc1209369
22. Salmaso S, Mastrantonio P, Tozzi AE, Stefanelli P, Anemona A, Ciofi degli Atti ML et al (2001) Sustained efficacy during the first 6 years of life of 3-component acellular pertussis vaccines administered in infancy: the Italian experience. Pediatrics 108(5):E81
23. Schmidtke AJ, Boney KO, Martin SW, Skoff TH, Tondella ML, Tatti KM (2012) Population diversity among Bordetella pertussis isolates, United States, 1935–2009. Emerg Infect Dis 18(8):1248–1255. https://doi.org/10.3201/eid1808.120082
24. Sheridan SL, Ware RS, Grimwood K, Lambert SB (2012) Number and order of whole cell pertussis vaccines in infancy and disease protection. JAMA 308(5):454–456. https://doi.org/10.1001/jama.2012.6364
25. Skoff TH, Cohn AC, Clark TA, Messonnier NE, Martin SW (2012) Early impact of the US Tdap vaccination program on pertussis trends. Arch Pediatr Adolesc Med 166(4):344–349. https://doi.org/10.1001/archpediatrics.2011.1093
26. Wendelboe AM, Van Rie A, Salmaso S, Englund JA (2005) Duration of immunity against pertussis after natural infection or vaccination. Pediatr Infect Dis J 24(5 Suppl):S58–S61
27. Witt MA, Katz PH, Witt DJ (2012) Unexpectedly limited durability of immunity following acellular pertussis vaccination in preadolescents in a North American outbreak. Clin Infect Dis: Off Publ Infect Dis Soc Am 54(12):1730–1735. https://doi.org/10.1093/cid/cis287
28. World Health Organization (2018, July 15) WHO UNICEF coverage estimates WHO World Health Organization: immunization, vaccines and biologicals. Vaccine preventable diseases vaccines monitoring system 2018 global summary reference time series: DTP3. http://apps.who.int/immunization_monitoring/globalsummary/timeseries/tswucoveragedtp3.html. Accessed 01 Nov 2018

29. World Health Organization (2018, September 21) WHO World Health Organization: immunization, vaccines and biologicals. Vaccine preventable diseases vaccines monitoring system 2018 global summary reference time series. http://apps.who.int/immunization_monitoring/globalsummary/timeseries/tscoveragebcg.html. Accessed 01 Nov 2018

Zero-Inflated Bayesian Regression Model and a Land Cover Topological Supervised-Classification for Prioritizing Asthma Related Covariables at the State Level

Dewan S. Tahsin, Namit Choudhari, and Benjamin G. Jacob

Abstract Asthma is a common chronic non-communicable condition that has numerous variables, including age, sex, education, income, and race in the United States. Predictive asthma-associated variables were collected from the Centers for Disease Control and Prevention (CDC) Behavioral Risk Factor Surveillance System (BRFSS) database. The county population was obtained from census.gov for Colorado and Kansas. We generated multiple land cover, vegetation, and elevation models employing high-resolution Sentinel 2B-10 m visible and near infra-red data overlaid with georeferenced capture point sentinel sites in ArcGIS Pro. A hierarchical probabilistic Bayesian regression model was constructed in R to determine covariate weights (the intercept value for race in Colorado was 0.8, and in Kansas was 0.7). The assumption was that a non-frequentistic simulation study could generate more robust interval estimates and coverage probabilities using finite samples of asthma-related demographics than the classical statistical methods currently contributed to the literature.

Keywords Asthma · Bayesian regression · Satellite data · Vegetation · Elevation · Land cover · Colorado · Kansas

---

D. S. Tahsin
Department of Epidemiology, College of Public Health, University of South Florida, Tampa, FL, USA
e-mail: dewansaima@usf.edu

N. Choudhari
School of Geosciences, University of South Florida, Tampa, FL, USA

B. G. Jacob (✉)
Department of Global Health, College of Public Health, University of South Florida, Tampa, FL, USA
e-mail: bjacob1@usf.edu

B. Jacob et al. (eds.), *Tech Enabled Global Health Security*, Advanced Sciences and Technologies for Security Applications, https://doi.org/10.1007/978-3-031-86997-6_4

1 Introduction

Asthma is a chronic, non-communicable disease affecting children and adults [50]. Symptoms such as wheezing, shortness of breath, and tightness in the chest are common symptoms of this lung condition, which are usually nocturnal but may also occur early in the morning [12]. According to the CDC [11], the total number of asthmatic patients in the United States was 24,963,874 in 2021. People suffering from asthma experience inflammation and narrowing of the airways, making it more difficult for air to pass through the body during exhalation [33]. Even though asthma is idiopathic, variables such as genetics, occupation, environment, and others have been associated with the onset of this condition [12]. Various forms of asthma can be brought on by many triggers, such as allergens, dust mites, pet fur, and mold spores [10]. Pollen can also cause allergic asthma, while non-allergic asthma can be stimulated by working in an environment with industrial dust. Chemicals can cause occupational asthma, while exercise-induced asthma can be caused by physical activity [32].

The severity of asthma can be estimated by linking certain demographic factors with clinical conditions. A study conducted by Colombo et al. [14] showed that females were more prone to developing asthma and experienced more severe clinical symptoms. Fuseini and Newcomb [20] revealed that females had more asthma symptoms than men during puberty due to the role of sex hormones, even though the mechanisms are unclear. Testosterone can reduce airway inflammation, and ovarian hormones exacerbate the situation [20]. Age is another crucial factor associated with asthma. Maestrelli [31] found that adults with asthma usually do not recover completely, and the disease usually progresses to extreme severity (e.g., asphyxiation, death). Disparities in social, racial, and ethnic groups significantly impact asthma and its prevalence [17, 51]. Race is also considered another vital cofactor, as it has been found that non-Hispanic African Americans have a thirty percent higher chance of developing asthma than non-Hispanic whites [34]. Children of color have a higher rate of mortality from asthma [34]. Cardet et al. [7] showed in their randomized controlled study that people with lower income were also at a higher risk for having severe asthma symptoms regardless of education, race, and medical therapy. A cross-sectional survey of 358 uncontrolled asthmatic patients with access to treatment showed that education level did not seem to be a barrier to asthma treatment and control [19].

Bayesian statistics, also known as 'inverse probability,' is a method that predicts space and time by employing prior distributions [23]. The posterior distribution summarizes the data gathered along with different probabilities of the posterior shaped by the methodology used. Previously, research on pediatric asthma cases was analyzed using a Bayesian paradigm. A partial genome screening analysis was carried out on 1,201 children, where in the estimator determinant dataset, 436 were asthmatic cases, and 765 were controls [46]. Network-based Bayesian multilevel analysis of relevance estimated the posteriors whether a variable was directly essential or not in the asthma model [46]. Another study by Spyriglou et al. (2018) employed the

Bayesian logistic regression approach implemented by the Markov Chain Monte Carlo (MCMC) on 147 children with asthma, showing that the approach resulted in high accuracy considering the demographic, symptomatic factors in predicting asthma. Markovian algorithms are commonly used for complex sample distributions [22]. Shi et al. [38] conducted a Bayesian meta-analysis to demonstrate how physical activity improved asthmatic'adults' quality of life and pulmonary function and the repercussions over time. Bayesian meta-analysis has been employed to compare two drugs that treat asthma, Reslizumab and Benralizumab. The results indicated that Reslizumab was more effective [8].

There are usually many zeros in prognosticative, asthma-related, capture point, sentinel site, diagnostically stratifiable, geosampled, and count data modeling processes. The commonly used models, such as Poisson or geometric distributions, can underestimate the zero-defect probability, making it challenging to identify significant covariate effects to improve the production quality of scalable asthma-related epidemiologic data, capture point, and sentinel site models. This article introduces a flexible class of zero-inflated models, including a non-frequentist, regression model for predicting hierarchical parameters associated with asthma in two states, Kansas and Colorado.

In this experiment, a zero-inflated Bayesian regression estimation method is developed as an alternative to traditionally employed maximum likelihood-based methods in the literature to analyze a multivariate, empirical, geosampled epidemiological dataset containing asthma-related, diagnostic stratified, sociodemographic/economic, racial, and land cover geomorphologic, classified, estimate determinants. Logistic regression models are commonly used to model response variables as dichotomous, categorical variables with several predictors [25]. The contribution of the predictor variable to the response variable in these analyses is expressed through a regression coefficient, Beta. This study estimated Beta using the Bayesian method to compare vulnerability-related asthma model indices between Kansas and Colorado.

Our Bayesian approach utilized a combination of information from the georeferenced, diagnostic stratified, sociodemographic/economic, racial, land cover and another capture point, sentinel site, geosampled, asthma data, and prior information, namely the posterior. We assumed that a zero-inflated Bayesian regression model might be employable in analyzing asthma-related geosampled epidemiologic data to ascertain precise parameter hierarchy. An MCMC with Gibbs sampling was used to obtain the sample from the posterior distribution of the state-level asthma model. We assumed convergent estimates were parsimoniously obtainable, resulting in component scores to comparatively detail asthma-related, capture point, sentinel site, and scalable covariates at the state level.

In statistics, MCMC methods comprise a class of algorithms for sampling from a probability distribution. By constructing a Markov chain with the desired distribution as its equilibrium distribution, one can obtain a sample of the desired distribution by recording states from the chain [21]. The more steps that are included, the more closely the distribution of the sample matches the actual desired distribution. Gibbs samples are beneficial to sampling from multivariate distributions as they reduce the complex problem of sampling from a joint distribution to sampling from the entire conditional (meaning, conditioned on all other variables) distribution of each variable [16]. The problem of sampling from multivariate distributions often arises in Bayesian statistics, where the inference of likely values for a parameter frequently entails sampling not only on that parameter but also on additional parameters required by the statistical model [30].

Previously, studies have not used Bayesian inferentiality to analyze the prior distributions of asthma-related sociodemographic prognosticators using satellite-determined models. Geographical Information Systems Tools were used in a study by Portnov et al. [37] to analyse the medical records of 3922 school children in Northern Israel and showed that even exposure to low concentration airborne particulate matter can be a risk factor for childhood asthma and used a binary logistic regression model for statistical analyses. Spyroglou et al. [39] was able to predict asthma with high accuracy with steadier results in patients, although they did not consider geographical factors like vegetation or elevation while creating their model. A study by Ouédraogo et al. [36] explored the spatial pattern of asthma outcome using a Bayesian spatial regression analyses considering factors like physical environment, social and economical indicators with environmental factors like air pollution and temperature, but did not look at the spatio-temporal models or geographical indices.

The objectives of this research paper are

- Aim 1 Employ satellite models (Land Use Land Cover [LULC], Normalized differential Vegetation Index [NDVI], and Digital Elevation Model (DEM) to determine geolocations of potential georeferencable (GPS indexable) hyperendemicity of asthma.
- Aim 2 Develop a probability of asthma prevalence at the county level by employing a population stratification methodology.
- Aim 3 Generate a zero-inflated probabilistic Bayesian regression model using a Markovian Sampling Distribution to determine the parameter hierarchy. An inferential, zero-inflated Bayesian can determine the causative covariates of high-probability asthmatic cases at the state level.

2 Methods

Study Site

Colorado is a landlocked state located within the continental United States. The state is bordered to the north by Wyoming, northeast by Nebraska, east by Kansas, and southeast by Oklahoma. The'state's boundaries were established in 1861 and measured strictly along the latitude and longitude lines. These lines extended from 37° to 41° North latitude and from 102°02′28" to 109°02′48″ West longitude. Colorado is one of the fastest-growing states, with a population of 5,877,610 residing within its 64 counties [43, 44]. Colorado has a cool and dry climate due to its location. The yearly average temperature is 43.5 degrees Fahrenheit, which is 13.7 degrees cooler than the global average. Precipitation in this state is seventeen inches lower

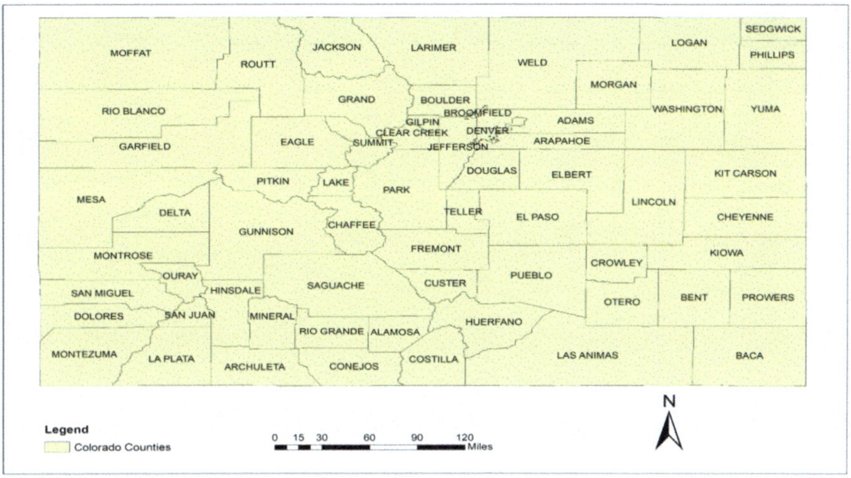

Fig. 1 Map of Colorado State with its counties

than the global mean [15]. Figure 1 shows a map of Colorado state with its counties, created using ArcGIS Pro.

Kansas

The U.S. state of Kansas is surrounded by Nebraska to the north, Oklahoma to the south, Missouri to the east, and Colorado to the west. The longitudinal line extends from 94°35’ to 102°3’ west, and the latitude line extends from 37° N to 40°North. Kansas has a population of 2,940,546 [43, 44] and 105 counties. Kansas has three climates: humid continental, semi-arid steppe, and subtropical. The annual mean temperature is below 52 degrees Fahrenheit, with precipitation ranging from less than 18 inches in the southwest to more than 45 inches in the southeast [28]. Figure 2 shows a map of Kansas state with its counties, created using ArcGIS Pro.

Data Collection

About the Surveillance System

This is a cross-sectional study, the data for the analysis were collected from the Behavioral Risk Factor Surveillance System (BRFSS).

The BRFSS is the leading national system for health-related telephone surveys, established in 1984. This system gathers information regarding health-related risk behaviors, chronic health issues, and the use of preventive services within all 50 states, the District of Columbia, and three U.S. territories. Each year BRFSS collects more than 400,000 adult interviews and it is currently the largest health survey system worldwide [13].

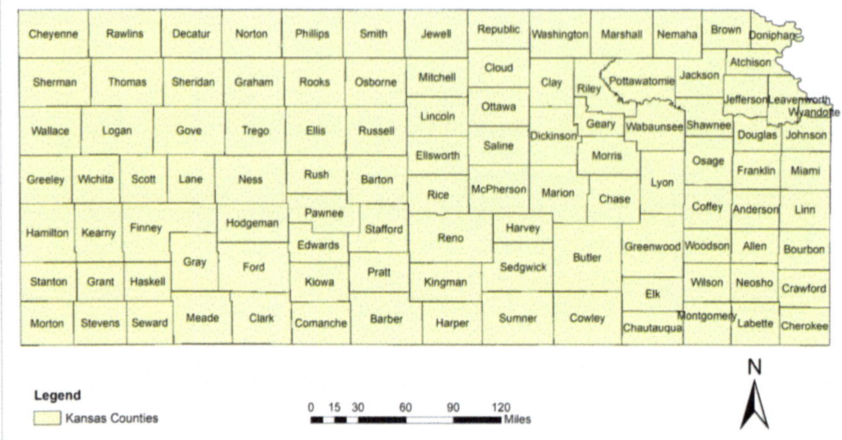

Fig. 2 Map of Kansas State with its counties

Survey Data and Documentation

Data for this experiment was collected from the 2022 BRFSS. The study participants were male and female residents of the U.S. who were above 18 years of age. The recruited population was diverse, including participants representing different races, education levels, employment status, income levels, and presence or absence of asthma (Table 1).

9, 273 individuals from Colorado and 11, 141 from Kansas participated in this study. Exclusions were then made depending on the missing data for some variables. Regressing empirical datasets with missing data can lead to biased results and challenge the validity and reliability of the analysis [24]. County-level data was collected from the United States Census Bureau. This bureau operates under Title 13 and Title 26 of the USA Code, and it acts as a resource for information about the country's population, location, and financial state [45].

Table 1 List of sociodemographic variables

Variable	Description of the measurement unit
Asthma	Yes, no
Gender	Male, Female
Age	18–25, 26–35, 36–45, 46–55, 56–65, 65 or older
Race	White, Black, Asian, American Indian or Alaskan Native, Native Hawaiian, or other Pacific Islander Hispanic, other, multiracial, Hispanic
Education level	Less than high school, high school/ GED, some post-high school, college graduate
Household income	Less than $15,000, $15,000-$24,999, $25,000-$34,999, $35,000-$49,999, $50,000-$99,999, $100,000-$199,999, more than $200,000

Satellite Data Models

Google Earth Pro was used to create the LULC. A high-resolution USGS 1/3 arc-second Digital Elevation Model (DEM) was downloaded from Open Topography [47]. To superimpose all the images, an NDVI map was generated based on the Sentinel 2B-10 m data for the urban areas of Kansas and Colorado.

LULC

High-level thematic resolution Sentinel-2 10-m resolution data was used to create land cover maps, which were mapped using supervised classification in Google Earth Pro. Based on the existing knowledge of the study area's land use/land cover, two LULC categories were classified, viz., urban and vegetation, to analyze the impacts of different land cover patterns and the abundance, occurrence, and distribution of asthma cases. Supervised classification has been extensively used for mapping land use/land cover classes [2]. This high-resolution satellite image avoided misclassifying pixels amongst the other land cover classes by utilizing endmember (sub-pixel) object-based software (ENVI), hence eliminating the noise (geometric, radiometric) and increasing the model output robustness. Figures 3 and 4 demonstrate the LULC of Colorado state and Kansas state, respectively, made by Google Earth Pro.

NDVI

NDVI is one of the most widely used remote sensing indexes [48], which explains the difference between visible (red band) and near-infrared [41]. The red band has

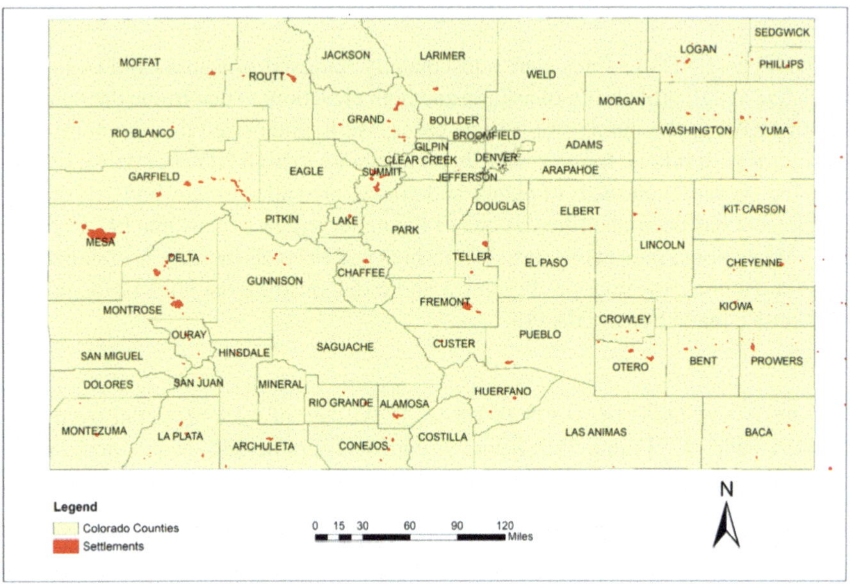

Fig. 3 LULC of Colorado State with its counties

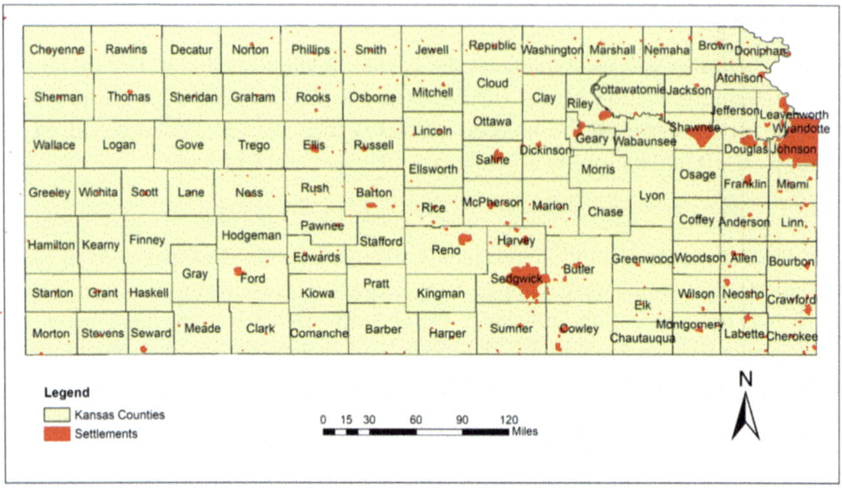

Fig. 4 LULC of Kansas State with its counties

a wavelength of 0.66 μm, and a near-infrared (NIR) band has 0.86 μm. The value of NDVI ranges from positive to negative. Positive values indicate dense vegetation areas; zero or negative values represent sparse vegetation or water/built-up areas. The formula for NDVI is given by Tucker [41] as,

$$\text{NDVI}) = (\rho_{\text{NIR}} - \rho_{\text{red}})/(\rho_{\text{NIR}} + \rho_{\text{red}})$$

The highest NDVI values were subsequently retrieved in Kansas and Colorado, which had a high probability of asthma and high elevation, to determine the relationship of NDVI to the number of asthma cases. We hypothesized that the number of asthma cases should be found primarily on moderate to high NDVI or dense vegetation. The grains, small seeds of plants, and grasses are called pollen, and exposure to this element has been identified as one of the major triggers of asthma [9]. Osborne et al. [35] demonstrated in their article how high pollen levels increase hospitalization rates for asthma exacerbation. Figures 5 and 6 demonstrate the NDVI of Colorado and Kansas, made by ArcGIS Pro.

DEM

High-resolution DEM of ALOS PALSAR DEM data of 12.5-m spatial resolution (Scene 1:Path-152, Frame-550; Scene 2: Path-152, Frame-540; acquisition date of the DEMs: April 26, 2010) was downloaded from the Alaska Satellite Facility [18]. The elevation values were later extracted at each county-level geolocation with a high probability of asthma to analyze the relationship between asthma and LULC changes. A prospective study conducted by Kiechl-Kohlendorfer et al. [29] demonstrated a strong correlation between living in high altitudes and increased risk of hospitalization for asthmatic patients. Remotely sensed georeferenced land cover/

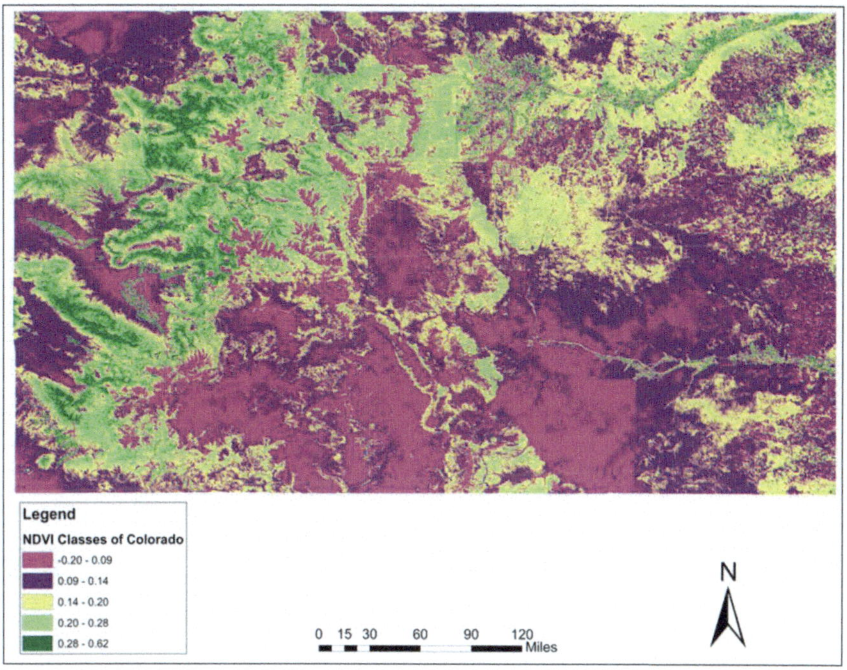

Fig. 5 NDVI of Colorado state

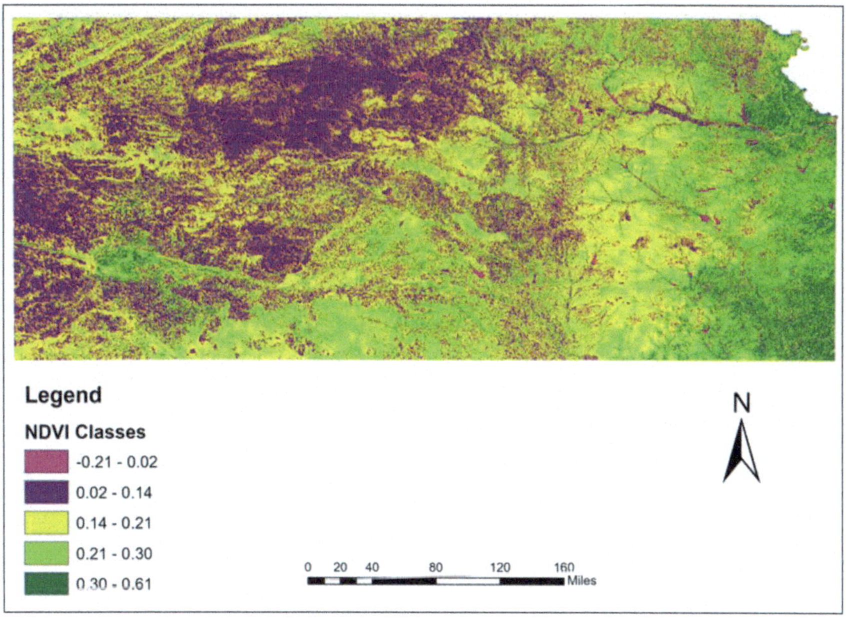

Fig. 6 NDVI of Kansas state

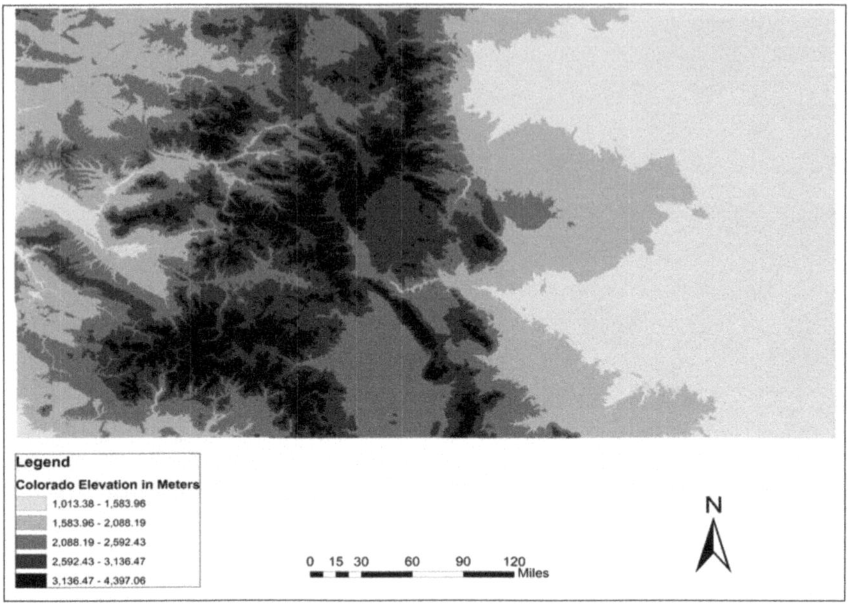

Fig. 7 DEM of Colorado state

non-homogenous, 3-D, explanatory slope coefficients found are in high-elevated areas, especially in dense vegetation LULC [27]. Figures 7 and 8 demonstrate the LULC of Colorado state and Kansas state, respectively, made by ArcGIS Pro.

Bayesian Analysis

The statistical analysis was conducted using R-4.3.1 to carry out the Bayesian inferential analysis to determine the association between asthma and the sociodemographic covariates. We considered a binary regression set-up in which $Y_1, \ldots, Y_n$ were independent Bernoulli random variables such that $\Pr(Y_i = 1|\beta = F(x\ T\ i\ \beta)$, where x_i is a $p \times 1$ vector of known asthma-related covariates associated with Y_i, β is a $p \times 1$ vector of unknown regression coefficients, and $F: R \rightarrow (0, 1)$ was a distribution function. Since we would employ a Bayesian version of the model, a prior distribution for the unknown regression parameter, β, was generated. If $\pi(\beta)$ is the prior density for β, then the posterior density of β given the geosampled asthma data, $y = (y_1, \ldots, y_n)\ T$, was definable as $\pi(\beta|\ y) = \pi(\beta)c(y)\ Y_n i = 1\ F(x\ T\ I\ \beta)y_i 1 - F(x\ T\ i\ \beta)1 - y_i$, where c(y) was the normalizing constant, that is $c(y) := Z\,Rp\ \pi(\beta)\ Y_n i = 1\ F(x\ T\ i\ \beta)\ y_i 1 - F(x\ T\ i\ \beta)1 - y_i d\beta$. In the remainder of this paper, we restrict attention to the regression model with a proper Np(b, B) prior to β. Let X denote the $n \times p$ matrix whose ith row is x T i, and let $R+ = (0, \infty)$.

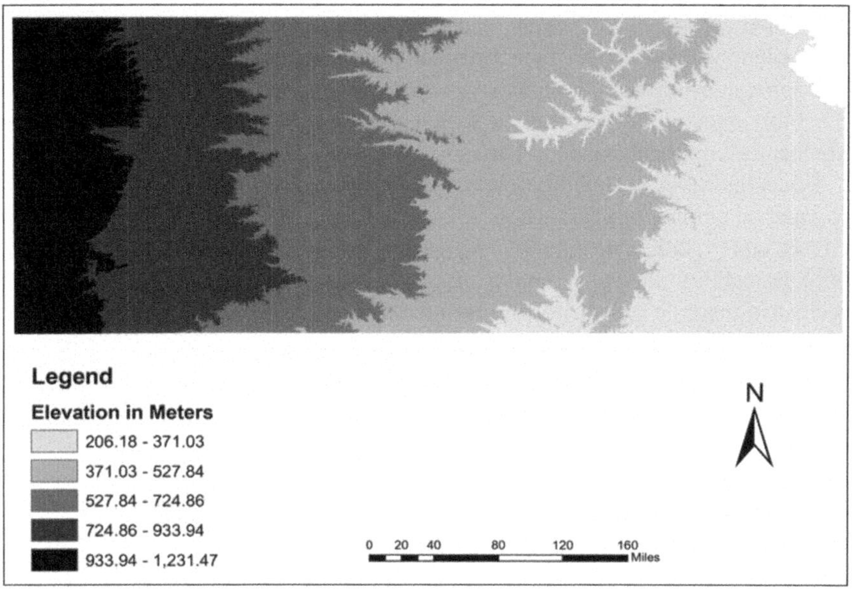

Fig. 8 DEM of Kansas state

We generated a fully inferential Bayesian treatment of the probabilistic matrix factorization model in which model capacity was controlled automatically by model capacity. In doing so, we were able to integrate the empirical georeferenced, geosampled dataset of state-level, asthma-related model parameters, and hyperparameters. We introduced priors to the hyperparameters and maximized the log-posterior of the model over both parameters. The hyperparameters allowed model complexity to be controlled automatically based on the training data (i.e., capture point, sentinel site, scaled-up, asthma-related, county-level, aggregation/non-aggregation-oriented, potential estimator determinants). We employed a joint optimization technique to predictively geolocate county-level, georeferenceable, asthma-related, geo-spatiotemporally estimator determinants, sentinel site, and capture points.

We began with a "prior distribution" based on the relative likelihoods of the geosampled, sociodemographic/economic, land cover, racial stratified, time series, dependent, and estimator determinants. In practice, it is expected to assume a uniform distribution over the appropriate range of values for the prior distribution [16]. Next, we calculated the likelihood of the observed asymptotic distribution as a function of the estimator, determinant, capture point, state-level, geosampled, sentinel site, and discrete integer values by multiplying the likelihood function by the prior distribution. Doing so subsequently normalized the data to obtain a unit probability over all possible stratified, independent, simulated estimator determinants (i.e., posterior distribution). The mode of the distribution approximated probability intervals in the empirical, estimator, determinant, geosampled dataset. In Bayesian inference, a probability interval is a probabilistic region around a posterior movement similar to a frequentist confidence interval [21].

A quantile-based probability interval was calculated in MCMC in R for p(θ|y): > p, interval employing theta, Highest Posterior Density [HPD] = FALSE, MM = FALSE, plot = TRUE statements. The interval function was employed to calculate a 95% probability interval given the marginal posterior samples theta. The function specified a quantile-based probability interval rather than an HPD interval since HPD = FALSE, and multimodality was not considered (i.e., MM = FALSE). The quantile-based 95% probability interval was merely the 2.5% and 97.5% quantiles of the samples for the distribution. Since the distribution was log-normal, it was skewed.

We employed an MCMC algorithm and an autocovariance matrix in R to geospatially quantify stochastic error propagation in Bayesian variables estimated from the asthma covariates geosampled in Kansas and Colorado. In this experiment, all capture point sentinel site-specific characteristics were imposed using the equations:

$$\mu_{\emptyset_1} = \frac{\sum_{j\neq i} \omega_{ij}\emptyset_{ij}}{\sum_{j\neq i} \omega_{ij}} \text{ and } \sigma^2_{\emptyset_1} = \frac{1}{\lambda \sum_{j\neq i} \omega_{ij}} \tag{1}$$

3 Results

In the Bayesian formulation, the specification of the capture point, sentinel site, epidemiologic, asthma-related, state-level, prognosticative model was completed by assigning prior to all georeferenced estimator determinants. We now outline the posterior sampling algorithm for the fixed effects in the zero-inflated Bayesian Inference model. We employed the georeferenced estimator determinants where each x_i for i = 1,..n was assumed to be distributed according to some distribution $p(x_i|\theta)$. Further, θ was an unknown parameter that had to be inferred from the geosampled, state-level, capture point and epidemiologic data. The joint probability of the data was optimally quantifiable by employing the prior distribution that was conditionally independent of the hyperparameters. Bayesian inference determined the posterior distribution of the parameter by employing:

$p(\emptyset) = p(\emptyset) = \prod_{i=1}^{n} p(\emptyset)$; whereby, $p(X|\emptyset, \alpha) = p(X|\emptyset)$ and $p(x_i|\theta, \alpha) = p(x_i|\theta)$ was conditionally independent of the hyperparameters:

$$
\begin{aligned}
p(\theta|X,\alpha) &= \frac{p(\theta,X,\alpha)}{p(X,\alpha)} \\
&= \frac{p(\theta,X,\alpha)}{\int p(\theta,X,\alpha)d\theta} \\
&= \frac{p(\theta,\alpha)p(\theta|\alpha)}{\int p(\theta)p(\alpha)d\theta} \\
&= \frac{\left[\prod_{i=1}^{n} p(x_i|\theta)\right]p(\theta|\alpha)}{\int \left[\prod_{i=1}^{n} p(x_i|\theta\right]p(\theta|\alpha)d\theta}
\end{aligned}
$$

For the fixed regression parameters, a suitable choice was the diffuse prior (i.e., p(γ) const., but a weakly informative Gaussian prior was possible). A second-order Gaussian random walk prior was employed to allow enough flexibility while penalizing abrupt changes in the function. The second-order random walk model is commonly used for smoothing data and quantifying response functions. It is computationally efficient due to the Markov properties of the joint (intrinsic) Gaussian density [21]. An MCMC plot generated that less of the upper tail was includable around the highest posterior density in the quantile-based probability interval. The MCMC algorithm proceeded in two steps:

1. Given current asthma-related parameter values, update the latent at-risk indicators, $w_1, \ldots, w_{n,}$ from their discrete complete conditional distributions.
2. Updating of β_1 using Gibbs sampler.

The MCMC sequence was definable as the reciprocal ratio of the number of MCMC trials needed to achieve the same variance in an estimated quantity as required for independent draws from the target probability distribution in the Bayesian model. Discarding the first set of "burn-in" iterations can ensure that the chain has reached a steady state when estimating Monte Carlo parameters, such as posterior means from sampled asthma covariates after the model had converged, samples from the conditional distribution of the model [21]. The estimation of the mean and the variance for the independent, state-level, scalable asthma parameters were determined empirically by:

$$
n = \left[\sum_{i=-\infty}^{\infty} p(i)\right]^{-1} = \left[1 + 2\sum_{i=1}^{\infty} p(i)\right]^{-1} \tag{2}
$$

After careful inspection of the iterated data, no aberrant values leading to numerical overflow were found.

We checked the statistical efficiency of the MCMC sequence in the zero-inflated, non-frequentist regression model output. Model checking of all data input and compilation was also conducted in R. The number of chains had to be specified before compilation. Syntax checking was employed, which involved highlighting the entire model code and then choosing the sequence model specification. The uncertainty in estimates of quantities derived from an MCMC sequence of random samples was represented by N_k and samples, $\upsilon_{k,}$ which was represented as a probability density function of a scalar quantity υ. The sample mean gave the estimated value of v

$$\hat{\upsilon} = \frac{1}{N_k} \sum_{k=1}^{N_k} \upsilon_k \tag{3}$$

The Monte Carlo error propagation method assumed that the distribution of error variables for each input data layer generated in R from the sampled asthma parameters was known. For each of the data layers, an error surface was simulated by drawing, at random, from an error pool defined by the distribution of the sampled covariates. Error surfaces were added to the input data layers, and the model was run using the resulting data error layers as input. The processes were repeated so that, for each run, a new realization of an error surface was generated for each input data layer. The results of each run were accumulated, and a running mean and standard deviation surface for the output were calculated. This process continued until the running mean stabilized. Since the random error visualizations were positive and negative, the stable running mean was taken as the true model output surface, and the standard deviation surface was used to measure relative error. A simple summary was generated, showing the posterior mean, median, and standard deviation, with a 95% posterior credible interval.

Model estimators were compared using the Deviance Information Criterion (DIC) in R, where $DIC = \underline{D} + P_D$ was the sum of the posterior mean of the deviance (D), a measure of goodness-of-fit, and the effective number of parameters (P_D), a measure of model complexity. A measure of goodness-of-fit based on the DIC values was applied to an R^2DIC, calculated in line with the standard R^2 measure for the asthma state-level model. This was definable as $R^2_{DIC} = 1 - \left(\frac{DIC_k - \underline{D}_{best}}{DIC_{max} - \underline{D}_{best}}\right)$ where DIC_k was the DIC value for the one-fixed parameter and model and was the posterior deviance from the model. Tables 2 and 3 indicate the results of the Bayesian regression of Colorado and Kansas state.

The expected variance in $\hat{\upsilon}$ was the expectation for the ensemble of the sequences generated from the geosampled asthma covariates was expressible as:

$$\begin{aligned}
\sigma^2_{\hat{\upsilon}} &= var(\hat{\upsilon}) = E\{(\hat{\upsilon} - E\{\hat{\upsilon}\})^2\} \\
&= E\left\{\frac{1}{N_k}\sum_j \left(\nu_j - \underline{\nu}\right)\frac{1}{N_k}\sum_k \left(\nu_k - \underline{\nu}\right)\right\} \\
&= \frac{1}{N_k^2}\sum_{jk}\left\{\left(\nu_j - \underline{\nu}\right)\left(\nu_k - \underline{\nu}\right)\right\}
\end{aligned}$$

where

$$\underline{\nu} = E\{\nu\} = E\{\underline{\nu}\} \tag{4}$$

Table 2 Bayesian output of Colorado

Model Info:
Function: stan_glm
Family: Gaussian [Identitty]
Formula: Asthma ~
Algorithm: sampling
Sample: 4000 (posterior sample size)
Prior: see help('prior_summary')
Observations: 2
Predictors: 6

Estimate	*Mean*	*SD**	*10%*	*50%*	*90%*
(Intercept)	381.5	10,379.2	−12,816.4	365.1	13,667.8
Gencases	−0.2	9.7	−12.5	−0.4	12.2
Agecases	−0.1	9.9	−12.5	−0.2	12.6
Educases	−0.9	8.0	−11.1	−0.8	9.5
Racecases	0.7	7.8	−9.6	0.6	10.7
Incomecases	−0.3	10.0	−13.3	−0.3	12.6
sigma	888.1	515.0	407.1	749.6	1542.7
Fit Diagnostics:	*Mean*	*SD**	*10%*	*50%*	*90%*
Mean_PPD	10.5	714.5	−729.9	13.3	757.1
The mean_ppd is the sample average posterior predictive distribution of the outcome variable					
MCMC Diagnostics	*mcse*	*Rhat*	*n_eff*		
(Intercept)	212.9	1.0	2377		
Gencases	0.2	1.0	3101		
Agecases	0.2	1.0	3461		
Educases	0.2	1.0	2371		
Racecases	0.2	1.0	2157		
Incomecases	0.2	1.0	3499		
sigma	11.4	1.0	2046		
Mean_PPD	11.3	1.0	3980		
Log-posterior	0.1	1.0	1497		

[*] Standard Deviation

the autocovariance of the sequence was definable as

$$E\{(\nu_k - \underline{\nu})(\nu_{k+1} - \underline{\nu})\} \tag{5}$$

The normalized autocovariance was

$$p(l) = (\sigma^2)^{-1}E\{(\nu_k - \underline{\nu})(\nu_k - \underline{\nu})\} \tag{6}$$

Table 3 Bayesian output of Kansas

Model Info: Function: stan_glm Family: Gaussian [Identitty] Formula: Asthma, A ~ Algorithm: sampling Sample: 4000 (posterior sample size) Prior: see help('prior_summary') Observations: 2 Predictors: 6					
Estimate	*Mean*	*SD**	*10%*	*50%*	*90%*
(Intercept)	242.3	11,885.6	−14,830.5	−82.4	15,674.1
Gencases	−0.5	9.3	−12.4	−0.4	11.3
Agecases	−0.4	9.3	−12.4	−0.4	11.5
Educases	−0.1	9.5	−12.4	−0.2	12.4
Racecases	0.8	6.6	−7.7	1.0	9.2
Incomecases	−0.3	9.8	−12.6	−0.4	12.4
sigma	1014.9	589.1	464.5	867.5	1741.3
Fit diagnostics:	*Mean*	*SD**	*10%*	*50%*	*90%*
Mean_PPD	−4.5	823.8	−883.6	−7.2	885.0
The mean_ppd is the sample average posterior predictive distribution of the outcome variable					
MCMC diagnostics	*mcse*	*Rhat*	*n_eff*		
(Intercept)	204.5	1.0	3377		
Gencases	0.2	1.0	3512		
Agecases	0.2	1.0	3384		
Educases	0.2	1.0	3502		
Racecases	0.2	1.0	3149		
Incomecases	0.2	1.0	3880		
sigma	13.3	1.0	1954		
Mean_PPD	12.9	1.0	4053		
Log-posterior	0.0	1.0	1550		

* Standard Deviation

where σ^2 was the covariance of ν and *p(l)*, which did not depend on k. The length of the nonzero normalized autocovariance values were:

$$\sigma_{\nu}^{2} = \frac{1}{N_k^2} \sum_{l} \sum_{k=1}^{N_{k-1}} E\{(\nu_k - \underline{\nu})(\nu_{k+1} - \underline{\nu})\}$$
$$= \frac{\sigma^2}{N_k} \sum_{l=-\infty}^{\infty} p(l) \tag{7}$$

The normalized autocovariance was a symmetric function, i.e. $\rho(-l) = \rho(l)$ the sequence sufficiently converged to the target probability density function. The variance of the distribution of the sampled asthma parameters was generated employing

$$\sigma_{\nu}^{2} \approx S^{2} = \frac{1}{N_k - 1}\sum_{j=1}^{N_k}\left(\nu_k - \underline{\nu}\right)^{2}, \tag{8}$$

and the normalized autocovariance was estimated from the sequence using:

$$\rho(l) \approx \frac{1}{S^{2}}\frac{1}{(N_k - l - 1)}\sum_{k=1}^{N_k - 1}(\nu_k - \hat{\nu})(\nu_{k+1} - \hat{\nu}) \tag{9}$$

for lag $1 \geq 0$. The Bayesian regression analyses revealed that the sampled covariate Race was positively associated with asthma levels at both study sites.

Figures 9 and 10 illustrate Colorado and Kansas state showcasing their top 5 counties when race is considered a variable having a significant effect on asthma.

We collapsed the LULC model with NDVI and DEM signatures. We created the mean values where the interaction occurred using the three proxy satellite maps and the Race variable (Table 4).

A probability index was created for asthma for each county in the two states based on the percentage of the asthmatic population. The probability of asthma in Colorado was 0.00017896, and for Kansas, it was 0.00057811 multiplied by each

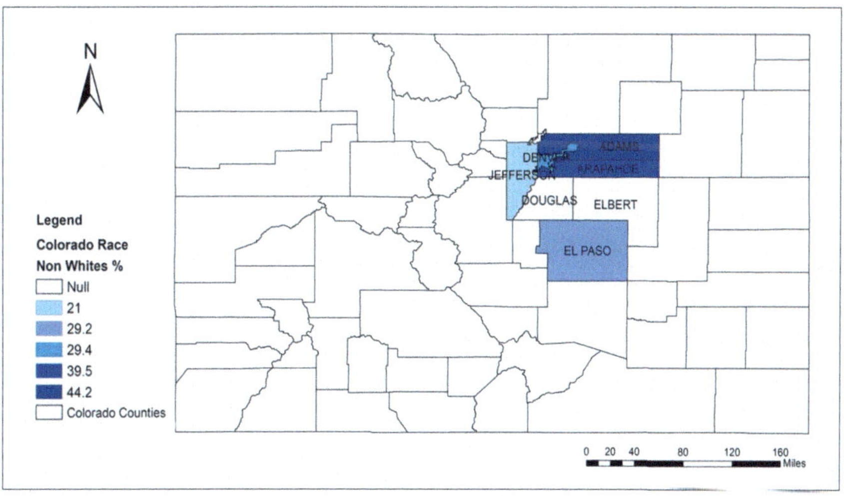

Fig. 9 Map of Colorado state with top 5 counties with asthma according to the asthma probability index

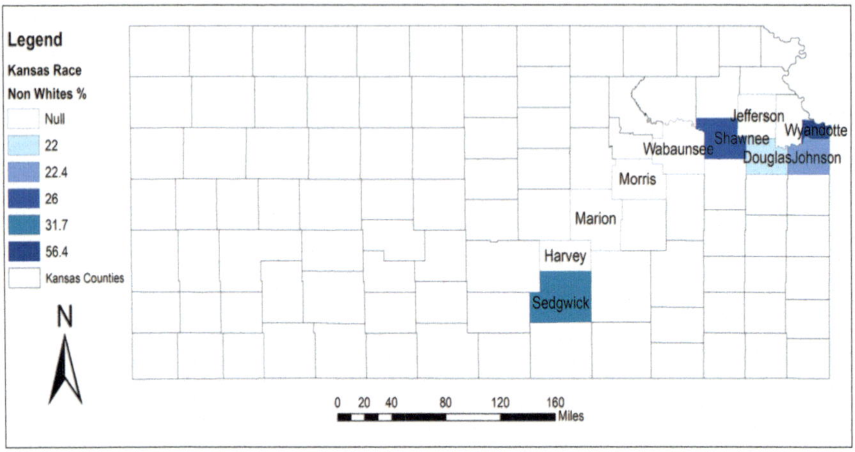

Fig. 10 Map of Kansas state with top 5 counties with asthma according to the asthma probability index

Table 4 Mean values of Kansas and Colorado

Colorado			Kansas		
county	DEM	NDVI	county	DEM	NDVI
El Paso	1971.63	0.138	Johnson	300.27	0.269
Denver	1628.047	0.133	Sedgwick	415	0.193
Arapahoe	1662.003	0.154	Shawnee	307	0.224
Jefferson	2181.165	0.159	Wyandotte	268.9	0.285
Adams	1551.857	0.150	Douglas	295	0.261
Mean	*1798*	*0.146*	*Mean*	*317*	*0.246*

county's population. The top 5 counties with the highest probability of asthma have been listed in Table 5.

Thereafter, we combined the satellite maps to determine discontinuous elevation, levels of vegetation on classified land cover (Figs. 11 and 12).

Table 5 Probability of Asthma at the county level

Colorado			Kansas		
county	Population	Asthma probability	county	Population	Asthma probability
El Paso	740,567	133	Johnson	619,195	358
Denver	713,252	128	Sedgwick	525,525	304
Arapahoe	655,808	118	Shawnee	177,480	103
Jefferson	576,143	103	Wyandotte	165,746	96
Adams	527,575	94	Douglas	119,964	70

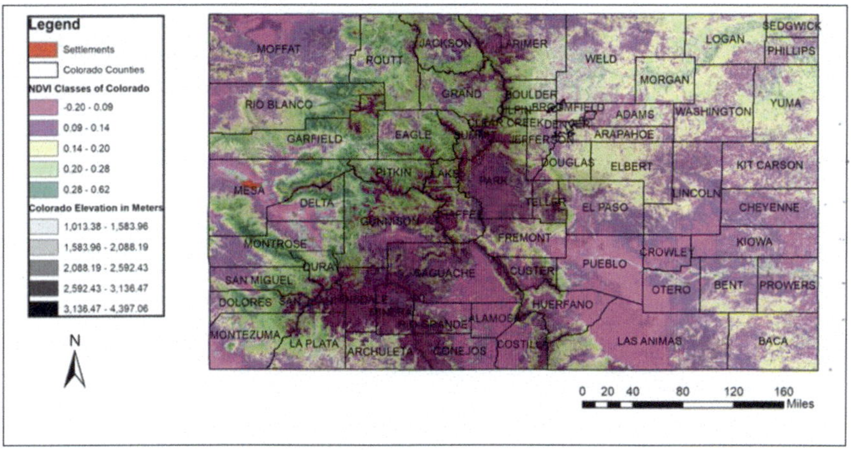

Fig. 11 Collapsed DEM, NDVI, LULC map of Colorado state

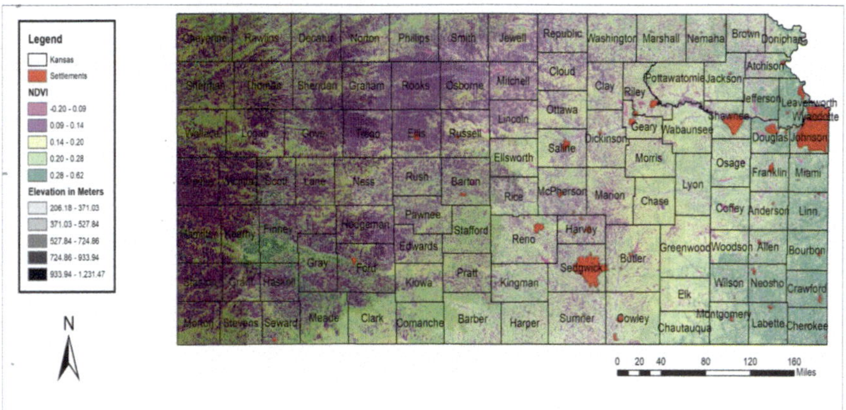

Fig. 12 Collapsed DEM, NDVI, LULC map of Kansas state

According to the results from the Bayesian model, Race was the only covariable associated with asthma. To run the regression model, Race was divided into two categories, Whites and Non-whites (Black, multi-racial, Hispanic). Tables 6 and 7 demonstrate the percentage of asthmatic individuals for each race by separating non-whites into Black, multi-racial, and Hispanic.

We also collapsed gridded, stratified DEM, NDVI, and LULC remotely sensed feature attributes, with the 5 top signatures interpolating county-level georeferenced geolocations (Figs. 13 and 14).

Table 6 Percentage of asthmatic individuals according to race (Bayesian)

	Kansas			Colorado		
Race	Total population	Asthmatic population	Percentage (%)	Total population	Asthmatic population	Percentage (%)
White	6950	759	10.92	9572	958	10
Non-white	2016	221	10.96	1311	156	11.90

Table 7 Percentage of asthmatic individuals according to race

	Colorado			Kansas		
Race	Total population	Asthmatic population	Percentage (%)	Total population	Asthmatic population	Percentage (%)
White	6950	759	10.92	9572	958	10
Black	256	41	16.02	543	58	10.68
Multi-racial	164	23	14.02	242	49	20.25
Hispanic	1555	157	10.01	526	49	9.32

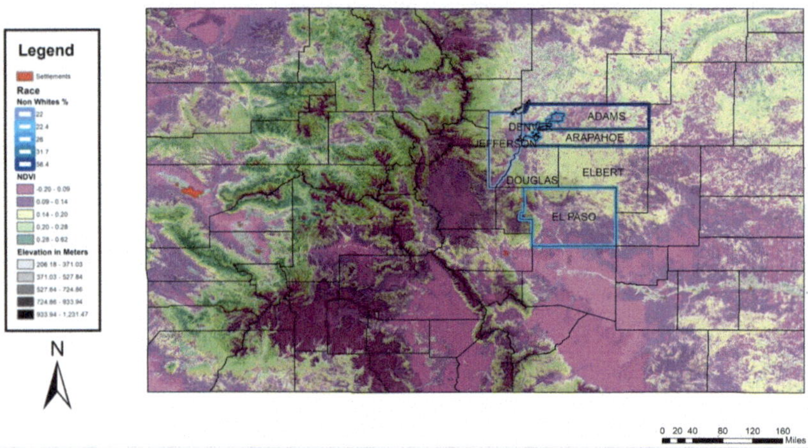

Fig. 13 Final collapsed map of Colorado state

4 Discussion

This functional form makes Bayesian inference for regression intractable [6]. This is because the evidence would require normalizing the product of a prior distribution (e.g., a Gaussian prior on) times the likelihood function in a state-level scalable, capture point, sentinel site asthma time-series model. A similar problem arises for other models with binomial likelihoods parameterized by log odds that allowed the

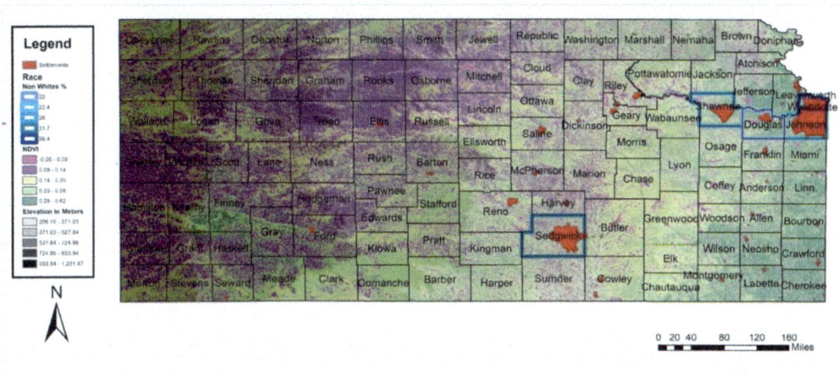

Fig. 14 Final collapsed map of Kansas state

construction of simple Gibbs samplers to generate a hierarchy of asthma-related scalable, state-level forecastable, epidemiologic simulation model covariates.

The utility of the data strategy extends an inferential, Bayesian, regression, asthma state-level model. In doing so, an asthma researcher or epidemiologist can develop useful Gibbs samplers for various Bayesian hierarchical models in which a logit link is employable. These include Bayesian regression with random effects quantified from a negative binomial regression. Of course, given the results herein, it is natural to ask whether these Gibbs samplers are also uniformly ergodic and to what extent methods could be used to model asthma at the state level. We note, however, that the Gibbs sampler is relatively simple for modeling county and state-level sociodemographic parameters of asthma. For example, the Gibbs sampler for the mixed effects model has more than two steps, which is significantly more complex than that of the Polya-Gamma Gibbs sampler, which only requires initialization. Thus, the development of (uniform) minorization conditions for the other Gibbs samplers would likely entail more than just a straightforward extension of the proof for precision modeling sociodemographic parameters of asthma.

The coefficient was treated as a random variable in the zero-inflated Bayesian regression approach. Information from the asthma stratified sociodemographic covariates and landcover epidemiologic, observational data was summarized. Subsequently, prior and data information were combined, producing posterior information. We could draw conclusions about asthma-related parameters based on posterior information. The posterior information iteratively generated simulations of potential, asthma-related, county-level, scalable, epidemiological data using the MCMC sampling techniques. The probability distribution was used to estimate the asthma-related determinants.

Bayesian inference can nonfrequentistically analyze statistical asthma-related models employing prior knowledge about the model or model parameters. The root of such an inference is the Bayes Theorem: $\left\{ \mathrm{P(parameters|\ data)} = \frac{\mathrm{P(data|parameters)} \times \mathrm{P(parameters)}}{p(\mathrm{data})} \propto \text{likelihood} \times \text{prior} \right\}$. For example, suppose we have a geosampled epidemiological empirical state-level dataset of normal observations.

$$\mathrm{X}|\theta \sim N\left(\theta, \sigma^2\right)$$

where sigma is known, and the prior distribution θ is

$$\theta \sim N\left(\mu, \tau^2\right)$$

In this formula, μ & τ, sometimes known as hyperparameters, are also known. If we observe n samples of X, we can obtain the posterior distribution for θ as,

$$\theta|X \sim N\left(\frac{\tau^2}{\frac{\sigma^2}{n}+\tau^2}\hat{X}+\frac{\frac{\sigma^2}{n}}{\frac{\sigma^2}{n}+\tau^2}\mu, \frac{\left(\frac{\sigma^2}{n}\right)\times\tau^2}{\frac{\sigma^2}{n}+\tau^2}\right)$$

According to our Bayesian analysis, Race was the only variable with a positive mean, thus indicating an association with asthma. Structural, social, biological, and behavioral determinants are the complex factors that create racial and ethnic disparities in asthma [3]. Race-based economic disadvantages directly affect the social determinants of health, such as lodging, housing environment, education, employment, and access to healthcare and healthy food [4]. Racism combined with community-level disinvestment in health care leads to subpar health status of the marginalized communities, including American Indians, Black and Hispanic Americans, as well as multi-racial cub groups [4]. Williams et al. [49] revealed in their study that race/ethnicity had a, more significant impact on health than socioeconomic status, even though they were linked. A spatial analysis conducted by Tung et al. [42] suggested that if race, place, and poverty were placed as a directional axis in a three-dimensional matrix, maintaining a healthy lifestyle becomes difficult as there is an expansion of the decision matrix of growing uncertainty (e.g., interventions should prioritize the less represented racial groups to have access to quality health care as well as ensuring proper housing with economic security).

Employing the index probability with the LULC, DEM, and NDVI maps, an epidemiologist or other researcher collaborator would be able to identify potential georeferenceable hot/cold spots of asthma cases for spatially targeting interventions at the state level. Satellite grid-based models can also reveal potential, seasonal, time series georeferenceable evidential prognosticators that can be associated with asthma. These covariates may be scaled up from a geosampled georeferenced capture point sentinel site to the county or state level.

Many prevention tactics can be utilized to prevent asthma. For example, housing with proper ventilation can prevent mold. Mold is a microscopic organism (fungi) that travels through the air. When inhaled by asthmatic individuals, it can trigger the immune system and cause an allergic reaction [1]. Installing air purifiers or filters for optimal air quality inside the house is recommended for asthmatic individuals. Maintaining a record of time and development can also aid in preventing the condition.

Furthermore, asthmatic patients should wear masks during high pollen season to prevent exacerbation [5]. Educating healthcare providers and clinical practitioners can guide patients to prevent and manage existing asthma. Educating the public about the triggers of asthma is essential. Stakeholders must consider this as a multi-level approach to mitigate the impacts of asthma by improving the housing environment, increasing access to healthcare, and educating healthcare workers about the existing racial gap.

Future research efforts for predictive asthma modeling at the county level should utilize time series data. This would create a more robust output since temporality is vital in chronic diseases such as asthma. If time series data is utilized, then there would be a requirement to use a Markovian semiparametric eigen-Bayesian generalized autoregressive conditional heteroscedastic (GARCH) model. (define GARCH). Jacob et al. [26] employed a Markovian semiparametric eigen-Bayesian GARCH model to geo-statistically prognosticatively map the hierarchical diffusion of COVID-19 at the sub-county district level in Uganda. This geospatial temporarily sensitive paradigm can allow parsimonious quantification of stratified covariates associated with hyper/hypo endemicity of asthma. Further, this model can adjust for temporal non-homoscedastic and potentially multicollinearity in a county-level, scalable, capture point, sentinel site asthma-related model.

Estimating semiparametric eigen-Bayesian Markovian algorithmic may also determine the causation of biases due to space and time regression assumptions violations. Eigen-spatial filters may be estimable through stepwise selection procedures [26]. Since the vicissitudes of stepwise selection are attenuated by the fact that all candidate eigenvectors are mutually orthogonal [40]. This may allow diagnosing latent, geo-spatiotemporal, non-Gaussian, zero autocorrelation in sociodemographic, scalable, asthma capture point sentinel sites in semiparametric, eigen-Bayesian, Markovian, eigenvector eigen-geospace. In doing so, any researcher or epidemiologist would be able to account for non-normal parameter uncertainty estimates due to violations of regression assumption (e.g., geo-spatiotemporal non-homoscedasticity, multicollinearity, zero auto-correlation non-Gaussian error variance) as derived from perturbations in model selection. Alternative methods may also be elaborated for estimating algorithmic eigen-spatial filters to address time-series perturbations in Markovian, GARCH, and eigen-Bayesian eigenvector eigen-geospace. This model may allow non-frequentist, adaptive sampling algorithms for optimally heuristically optimizing aggregation/non-aggregation-orients, stratifiable, scalable, capture point, state-level asthma-related sentinel sites.

In conclusion, we employed multiple spatial and non-spatial statistical models to localize areas where hyper/hypo endemic cases of asthma could occur at the state level in Kansas and Colorado. We were able to conduct a proportional population stratification at the county level. We successfully constructed a zero-inflated,, Bayesian, inferential model, which revealed Race as the primary determinant associated with asthma in both states. This is the first contribution to the literature that has revealed an association between hyper/hyperendemicity of asthma at the county level using a supervised LULC, DEM, and NDVI classification. Based on our literature review, this article represents the first research conducted employing satellite signature models and Bayesian analysis for predicting hyper/hyperendemicity of asthma at the county level.

Conflict of Interests No conflict of interest exists, and this paper has not been presented elsewhere.
Correspondence concerning this article should be addressed to Dewan Saima Tahsin, College of Public Health, University of South Florida, 13,201 Bruce B. Downs Blvd, MDC 56, Tampa, FL 33,612.

References

1. American Lung Association (2018) Combating mold and preventing asthma symptoms. https://www.lung.org/blog/mold-and-asthma
2. Ang EML, Arts D, Crawford D, Labatos BV Jr, Ngo KD, Owen JR, Gibbins C, Lechner AM (2021) Socio-environmental land cover time-series analysis of mining landscapes using Google Earth Engine and web-based mapping. Remote Sens Appl Soc Environ 21. https://doi.org/10.1016/j.rsase.2020.100458
3. Asthma and Allergy Foundation of America (2020) Asthma disparities in America. https://aafa.org/asthma-allergy-research/our-research/asthma-disparities-burden-on-minorities/#:~:text=The%20burden%20of%20asthma%20in,asthma%20rates%2C%20deaths%20and%20hospitalizations
4. Beech BM, Ford C, Thorpe RJ Jr, Bruce MA, Norris KC (2021) Poverty, racism, and the public health crisis in America. Frontiers 9. https://doi.org/10.3389/fpubh.2021.699049
5. Bergmann KC, Kugler S, Zuberbier T, Becker S (2021) Face masks suitable for preventing covid-19 and pollen allergy: a study in the exposure chamber. Springer Nature. https://doi.org/10.1007/s40629-021-00180-8
6. Bishop CM (2006) Pattern recognition and machine learning. Inf Sci Stat. https://link.springer.com/book/9780387310732
7. Cardet JC, Louisias M, King TS, Castro M, Codispoti CD, Dunn R, Engle L, Giles BL, Holguin F, Lima JJ, Long D, Lugogo N, Nyenhuis S, Ortega VE, Ramratnam S, Wechsler ME, Israel E, Phipatanakul W (2018) Income is an independent risk factor for worse asthma outcomes. J Allergy Clin Immunol 141(2):754-760.e3. https://doi.org/10.1016/j.jaci.2017.04.036
8. Casale TB, Pacou M, Mesana L, Farge G, Sun SX, Castro M (2019) Reslizumab compared with Benralizumab in patients with eosinophilic asthma: a systematic literature review and network meta-analysis. J Allergy Clin Immunol Practice 7(1). https://doi.org/10.1016/j.jaip.2018.08.036
9. Centers for Disease Control and Prevention (2020) Pollen your health. https://www.cdc.gov/climateandhealth/effects/pollen-health.htm
10. Centers for Disease Control and Prevention (2022a) BRFSS prevalence and trends data. https://nccd.cdc.gov/BRFSSPrevalence/rdPage.aspx?rdReport=DPH_BRFSS.ExploreByTopic&irbLocationType=StatesAndMMSA&islClass=CLASS03&islTopic=TOPIC06&islYear=2022&rdRnd=78395
11. Centers for Disease Control and Prevention (2023a) Most recent national asthma data. https://www.cdc.gov/asthma/most_recent_national_asthma_data.htm
12. Centers for Disease Control and Prevention (2023b) Learn how to control asthma. https://www.cdc.gov/asthma/faqs.htm
13. Centers for Disease Control and Prevention (2014) Behavioral risk factor surveillance system. https://www.cdc.gov/brfss/about/index.htm

14. Colombo D, Zagni E, Ferri F, Canonica GW (2019) Gender differences in asthma perception and its impact on quality of life: a post hoc analysis of the PROXIMA (patient reported outcome and Xolair in the management of asthma) study. Allergy Asthma Clin Immunol 15(65). https://doi.org/10.1186/s13223-019-0380-z
15. Colorado Encyclopedia (n.d.) Weather and climate. https://coloradoencyclopedia.org/things/weather-climate#:~:text=Colorado%20Climate&text=Colorado
16. Cressie N (2015) Statistics for spatial data. Wiley Classics Library. https://www.wiley.com/en-us/exportProduct/pdf/9781119114611
17. Drake KA, Galanter JM, Burchard EG (2008) Race, ethnicity and social class and the complex etiologies of asthma. Pharmacogenomics 9(4). https://doi.org/10.2217/14622416.9.4.453
18. Earthdata (n.d.) Earthdata search. https://search.earthdata.nasa.gov/search
19. Forno E, Celedon J (2014) Asthma and ethnic minorities: socioeconomic status and beyond. Natl Libr Med. https://doi.org/10.1097/aci.0b013e3283292207
20. Fuseini H, Newcomb DC (2017) Mechanisms driving gender differences in asthma. Natl Libr Med 17(3). https://doi.org/10.1007/s11882-017-0686-1
21. Gelman A, Carlin JB, Stern HS, Dunson DB, Vehtari A, Rubin DB (2013) Bayesian data analysis. Texts Stat Sci. https://doi.org/10.1201/b16018
22. Gelman A (2005) Analysis of variance—why it is more important than ever. Ann Stat 33(1). https://doi.org/10.1214/009053604000001048
23. Gelman A, Shalizi CR (2012) Philosophy and the practice of Bayesian Statistics. Br J Math Stat Psychol 66(1). https://doi.org/10.1111/j.2044-8317.2011.02037.x
24. Griffith DA (2003) Gaining understanding through theory and scientific visualization. Spat Autocorrelation Spat Filtering. https://doi.org/10.1007/978-3-540-24806-4
25. Hosmer DW, Lemeshow S (2000). Appl Logistic Regression. https://doi.org/10.1002/0471722146
26. Jacob BG, Izureta R, Bell J, Parikh J, Loum D, Casanova J, Gates T, Murray K, White L, Aceng JR (2023) Approximating non-aymptoticalness, skew heteroscedascity and geo-spatiotemporal multicollinearity in posterior probabilities in Bayesian Eigenvector Eigen-geospace for optimizing hierarchical diffusion-oriented COVID-19 random effect specifications geosampled in Uganda. Am J Math Stat 13(1):1–43. https://doi.org/10.5923/j.ajms.20231301.01
27. Jacob BG, Novak RJ (2014) Integrating a Trimble Recov X 400 Mhz Intel PXA255 Xscale CPU mobile field data collection system using differentially corrected global positioning system technology and real-time bidirectional actionable platform within an ArchGIS cyberenvironemnt for implementing mosquito control. Adv Remote Sens 3(3). https://doi.org/10.4236/ars.2014.33012
28. K-state (n.d.) Kansas climate. https://climate.k-state.edu/basics/
29. Kiechl-Kohlendorfer U, Horak E, Mueller W, Strobl R, Haberland C, Fink FM, Schwaiger M, Gutenberger KH, Reich H, Meraner D, Kiechl S. Arch Disease Childhood. https://doi.org/10.1136/adc.2006.106278
30. Kruschke JK (2014) Doing Bayesian data analysis: a tutorial with R, JAGS, and Stan, 2nd edn. Academic Press, Elsevier. https://nyu-cdsc.github.io/learningr/assets/kruschke_bayesian_in_R.pdf
31. Maestreilli P (2004) Natural history of adult-onset asthma: insights from model of occupational asthma. Natl Libr Med 169(3). https://doi.org/10.1164/rccm.2312012
32. Medline Plus (2021) Asthma symptoms. https://medlineplus.gov/asthma.html
33. National Heart, Lung, and Blood Institute (2022) What is asthma? https://www.nhlbi.nih.gov/health/asthma
34. Office of Minority Health (n.d.) Asthma and African Americans. https://minorityhealth.hhs.gov/omh/browse.aspx?lvl=4&lvlid=15
35. Osborne NJ, Alcock A, Wheeler BW, Hajat S, Sarran C, Clewlow Y, Mennes RN, Hemming D, White M, Vardoulakis S, Fleming LE (2017) Pollen exposure and hospitalization due to asthma exacerbations: daily time series in a European city. Int J Biometeorol 61(10):1837–1848. https://doi.org/10.1007/s00484-017-1369-2

36. Ouédraogo AM, Crighton EJ, Sawada M, To T, Brand K, Lavigne E (2018) Exploration of the spatial patterns and determinants of asthma prevalence and health services use in Ontario using a Bayesian approach. PLoS ONE. https://doi.org/10.1371/journal.pone.0208205
37. Portnov BA, Reiser B, Karkabi K, Cohen-Kastel O, Dubnov J. (2012). High prevalence of asthma in Northern Israel is linked to air pollution by particulate matter: evidence from GIS analysis and Bayesian model averaging. Int J Environ Health Res 22(3):249–269. https://doi.org/10.1080/09603123.2011.634387
38. Shi S, Huang H, Zheng R, Zhang N, Dai W, Liang J, Dai Y (2022) Can long-term regular physical activity improve health outcomes among adults with asthma-evidence from a systematic review and Bayesian meta-analysis. Ann Transl Med 10(6). https://doi.org/10.21037/atm-22-1170
39. Spyroglou II, Spock G, Chatzimichail EA, Rigas A, Paraskakis E (2018) A Bayesian logistic regression approach in asthma persistence prediction. Epidemiol Biostatistics Public Health 15(1). https://doi.org/10.2427/12777
40. Tiefelsdorf M, Griffith D (2017) A Semiparametric filtering of spatial autocorrelation: the eigenvector approach. Sage J 39(5). https://doi.org/10.1068/a37378
41. Tucker CJ (1979) Red and photographic infrared linear combinations for monitoring vegetation. Remote Sens Environ 8(2):127–150. https://doi.org/10.1016/0034-4257(79)90013-0
42. Tung EL, Cagney KAC, Peek ME, Chin MH (2017) Spatial context and health inequity: reconfiguring race, place, and poverty 94:757–763. https://doi.org/10.1007/s11524-017-0210-x
43. U.S. Census Bureau (2023) Colorado—U.S. census bureau quickfacts. https://www.census.gov/quickfacts/fact/table/CO/PST045222
44. U.S. Census Bureau (2023) Kansas—U.S. census bureau quickfacts. https://www.census.gov/quickfacts/fact/table/KS/PST045222
45. U.S. Census Bureau (2019) What we do. The United States census Bureau. https://www.census.gov/about/what.html#par_textimage
46. Ungvari I, Hullam G, Antal P, Kiszel PS, Gezsi A, Hadadi E, Virag V, Hajos G, Millinghoffer A, Nagy A, Kiss A, Semsei AF, Temesi G, Melegh B, Kisfali P, Szell M, Bikov A, Galffy G, Tamasi L, Falus A, Szalai C (2012) Evaluation of a partial genome screening of two asthma susceptibility regions using Bayesian network based Bayesian multilevel analysis of relevance. Plos One 7(3). https://doi.org/10.1371/journal.pone.0033573
47. United States Geological Survey (2021) United States geological survey 3D elevation program 1/3 arc-second digital elevation model. OpenTopography. https://doi.org/10.5069/G98K778D
48. Valdez-Delgado KM, Salazar QG, Moo-Llanes DA, Izcapa-Trevino C, Cruz-Pliego M, Dominguez-Posadas G, Armendariz-Valdez MO, Correa-Morales F, Cisneros-Vazquez LA, Ordonez-Gonzalez JG, Salas IF, Danis-Lozano R (2023) Mapping the urban environments of aedes aegypti using drone technology. Drones 7(9):581. https://doi.org/10.3390/drones7090581
49. Williams DR, Priest N, Anderson NB (2016) Understanding associations among race, socioeconomic status, and health: patterns and prospects. Health Psychol 35(4):407–411. https://doi.org/10.1037/hea0000242
50. World Health Organization (2023) Asthma. https://www.who.int/news-room/fact-sheets/detail/asthma
51. Wypuch-Slusarska A, Krupa-Kotara K, Niewiadomska E (2022) Social inequalities: do they matter in asthma, bronchitis, and respiratory symptoms in children? Natl Libr Med 19(22). https://doi.org/10.3390/ijerph192215366

A Real-Time High-Performance Artificial Intelligent Machine Learned Interactive Mobile iOS App for Optimizing Primary Prevention, Timeliness Diagnosis and Rehabilitation Cardiovascular Emergencies: A Vision for Smart and Connected Health Care

Benjamin Jacob and Joe Bohn

Abstract The speed and accuracy of data transfer in an artificial intelligent [AI] infused machine learned [ML], mobile, interactive, smartphone application (app) can vary significantly depending on network conditions, server response times, optimization, and device performance. Currently, selective search engines embedded in mobile, iOS, health apps merges data based on engineered low-level features and has an order of magnitude of up to 91.3 s per information text in a CPU implementation. Faster R-CNN enables end-to-end detector training on shared convolutional features and shows compelling accuracy and speed. Unlike traditional batch processing, where data is collected and processed periodically in predefined portions, real-time R-CNN integration operates on a near-instantaneous basis. This means that prevention, timeliness diagnosis and rehabilitation of cardiac emergency intelligent smartphone health data captured can be retrieved, transferred, and archived in milliseconds. Currently mobile heart health apps may only employ fixed computational models for various applications due to non-real time data transfer leading to misspecifications in prevention, and rehabilitation of cardiac emergency-related injuries. Non-inclusion of real time vital in cardiovascular signs [e.g., chest pressure, nausea, shortness of breath, vertigo, unilateral facial paralysis etc.] can cause premature death in high-risk cardiac patients. This proposal introduces a Region Proposal Network (RPN) that shares full-image real-time convolutional features within an infused, intelligent, AI-ML detection network in an interactive, continuously self-learning wrist-wearable, smartphone, mobile app for enabling cost-free

B. Jacob (✉)
Center for Global Health and Interdisciplinary Research, College of Public Health, University of South Florida, Tampa, USA
e-mail: bjacob1@usf.edu

J. Bohn
Policy, Practice and Leadership, College of Public Health, University of South Florida, Tampa, USA

B. Jacob et al. (eds.), *Tech Enabled Global Health Security*, Advanced Sciences and Technologies for Security Applications, https://doi.org/10.1007/978-3-031-86997-6_5

region proposals [e.g., instantaneous body physiological responses, changes in symptoms, incentivized protocols for adherence to prescribed medication, diet, physical activity, exercise stress tests and follow-up care, etc.]. A fully merged RPN and Faster R-CNN deep convolutional unified network infused into a mobile iOS app dashboard will be employed to simultaneously train, aggregate and predict object bounds and objectness scores for implementing real-time prevention, timeliness diagnosis and rehabilitation of cardiac emergency protocols whose effective running time for proposals is estimated at 10.3 ms. Advances like R-CNN have reduced the running time of real times detection mobile networks, exposing RPN computation as a bottleneck. However, currently there are no prevention, timeliness diagnosis and rehabilitation of cardiac emergency, heart, health-related, iOS mobile apps in the literature, or on the on-line commercial market that has a real time, R-CNN network infused into a continuously self-learning wrist-wearable for enabling cost-free region proposals which may have multiple applications [e.g., instantaneous detection of quivering or irregular heartbeat, due to arrythmias due to coronary heart disease or cardiomyopathy, creating an interactive lifetime visualization of blood pressure trends, medication compliance, GPS maps of closet locations of defibrillators for public use, cloud storage, security, etc.,]. We introduce a real time RPN, intelligent AI-ML infused interactive mobile app that is trainable end-to-end to generate high-quality region proposals, for real time data retrieval, tracking, transference and archiving of prevention, timeliness diagnosis and rehabilitation of cardiac emergencies within a R-CNN. This context-aware augmented and virtual reality mobile application utilizes location data, object recognition software, and 3D features in a R-CNN/RPN unified real time network to provide state of the art data retrieval, tracking, transference and archivability. The app dashboard enables a unified, deep-learning-based detection system to run at real-time frame rates. The learned RPN improves region proposal quality and thus the overall object detection accuracy.

Keywords Real time · R-CNN prevention · And rehabilitation of cardiac emergency-related · Mobile iOS app

1 Introduction

A scientific statement from the American Heart Association indicates that self-care research and clinical efforts have been hindered by the perceptions of both patients and providers that pharmacological interventions are more effective than lifestyle change, thus warranting researchers to focus on self-management and lifestyle interventions [1]. Prevention, timeliness diagnosis and rehabilitation of cardiac emergencies requires patients to perform daily self-monitoring for changes in weight and symptoms, practice decision making for symptom changes, and maintaining adherence to prescribed medication, diet, physical activity, and follow-up care etc. [2]. Living with the risk of a potential cardiac emergency imposes significant stressors for patients in following daily self-care tasks and lifestyle changes to maintain functional independence and quality of life [3]. Clinical outcomes depend on how well a potential cardiac patient conducts self-care practices at home and seeks early care for symptoms [4]. For self-care of any chronic condition such as prevention, timeliness diagnosis and rehabilitation of cardiac emergencies to be sustained, self-management techniques need to be integrated into one's lifestyle [5].

Given the complexity of prevention, timeliness diagnosis and rehabilitation of cardiac emergencies self-care, assisting patients to manage their own care at home is a key component of prevention and management. Mobile real time, health technology is definable as the "use of smartphones, tablets, and other mobile devices to deliver health care and preventive health services [Global Observatory for eHealth series, Volume 3. Geneva: World Health Organization; 2018-04-19 mHealth: new horizons for health through mobile technologies]. Personal mobile devices are portable and stay with the owner throughout the day which would allow high-risk cardiac patients with constant real time prevention, and rehabilitation of emergency-related data updates.

Currently, mobile phones contain sensors such as accelerometers and cameras that have been leveraged in health care for health education, health management, and medical imaging such as electrocardiogram (EKG), [6]. Among a sample of previous stroke patients, 96% owned a mobile phone and 32% relied on the mobile phone for Internet access, searched health information, and reported moderate self-confidence in using mobile apps [7]. Emerging evidence suggests that mobile technology can serve as a form of support for patients classified as high-risk cardiac patients and may enhance patient-provider collaboration for self-management. According to the Pew Research Center, 40% of American adults use mobile phones, which has more than doubled since 2013. Given the potential growing trend in the use of real time health technology and interactive mobile phone apps, this vision is to determine evidence on the use of real time personal patient data [e.g., blood pressure status, levels of vertigo, syncope or near-syncope confusion, levels of emotional fatigue etc.] in prevention, timeliness diagnosis of cardiac emergencies and rehabilitation self-care.

Recent deep neural network classifiers such as signature maximum likelihood classifiers stochastic satellite signature interpolators, spatial lag and eigenvector spatial filters (ESF) can be deployed in an AI-ML [i.e., random forest (RF) and Support Vector Machine (SVM)] infused an iOS real time interactive, prevention, timeliness diagnosis and rehabilitation of cardiac emergencies mobile app. dashboard in Python. The infusion of the real time AI-ML classifiers into a web-configurable, and customizable mobile, dashboard, smartphone app can provide real-time data detection, retrieval tracking and communication of prevention, timeliness diagnosis and rehabilitation of cardiac emergencies. These classifers can archive and analyze information that promotes the resolution of cardiac-related problems [e.g., atrial fibrillation, or ventricular tachycardia due to side effect of certain medications (such as an EpiPen), low levels of potassium in the blood (hypokalemia) in ischemic and intracerebral hemorrhagic stroke patients etc.]. This functionality may be supplemented by personal data gathered in real-time, such as physiological response data, previously logged messages, and other clinical and nonclinical information. These technological advances can aid in meshing these and other information of prevention, and rehabilitation cardiac emergency-related data [e.g., age, demographics, nature of the symptoms: 'positive' versus 'negative', onset and progression, duration, precipitating factors associated symptoms, for example, headache, loss of awareness, during or after the attacks] in a web-configurable AI-ML, infused, interactive, iOS app available on both wearable and mobile devices. These data can help high risk patients by implementing customizable treatment [i.e., intelligent augmented lifelike avatars for virtual physical examination of previous Heart Failure [HF] patient with cardiac implantable electronic device, analyzing effect of lifestyle-focused text messaging on risk factor modification in patients with coronary heart disease or generating charts and graphs on prehospital medical information system data on prehospital care for real time modeling inequalities in cardiovascular outcomes in minority migrant farm workers, field assessment of cardiac triage for emergency destinations etc.] to determine if there a high a priori probability of a cerebrovascular event. The classifiers can develop a digitized version of a workflow checklist if for the management of acute stroke based on American Heart Association/American Stroke Association and NIHSS guidelines. Regarding rehabilitation therapy, telemedicine demonstrated higher validity and reliability, as well as higher confidence and satisfaction perceived by high-risk cardiac patients [8]. The interactive, infused, AI-ML classifiers in an intelligent, real time, mobile interactive, iOS dashboard can provide customized, prevention, timeliness diagnosis of cardiac emergencies and rehabilitation self-care. [e.g., provide visual, audible and vibratory feedback in case of incorrect movements in order to assist and improve motor rehabilitation, construct a smartphone-based regional convolutional model of care to support patients with cardiac disease transitioning from hospital to the rural communities, continuous on-going, EKG readings from a transient episode of neurological dysfunction caused by a focal brain, spinal cord, or retinal ischemia, without acute infarction], provide accurate differential diagnosis [e.g., thromboembolic stroke in pregnant women or those taking oral contraceptives from hemorrhagic stroke secondary to ruptured aneurysm, or ventricular fibrillation or pulseless ventricular tachycardia as shockable rhythms rather than shocking pulseless electrical activity or asystole, etc.]. For some cardiac injuries the diagnostic challenge is greater, and the 'mimic' rate higher (and more varied), as there is no definitive diagnostic test, hence real time, mobile timeliness diagnosis of cardiac emergencies and rehabilitation data is vital for prevention of death and successful cardiac recovery and outcome. Other cardiac, emergency-related, mobile, smartphone apps employ the features of virtual reality to improve the effectiveness of rehabilitation treatment [9–11] via videos. However, increasing speed and accuracy for interactive training, for example, can allow implementing mobile, smart phone-based, real time, motion-tracking technology with a live cardiac rehabilitation instructor which can improve strength, and balance in cardiac patients. In addition, supervision of patients using exergames (video games with a rehabilitative function for promoting movement aimed at improving strength and coordination of injured limbs) via teleconferencing could improve the effectiveness of exercise and limit errors in research data model outcomes [e.g., regression trends and survival rates in a community-based population to predict the all-cause mortality in coronary artery disease patients with atrial fibrillation]. Evidence from epidemiological investigations, interpopulation studies, secular trends, and community interventions provide a compelling rationale for attacking the cardiovascular epidemic at the community level [12]. The real time iOS mobile dashboard app can have embedded functions which allows a series of questions to assist in deciding the most appropriate workflow for prevention, timeliness diagnosis and rehabilitation of cardiac emergencies. The real time iOS dashboard can be characterized by four functional modules: health reminder, consultation, health information, and patient diary. The dashboard will provide real time medical information and provide rehabilitation exercises. For example, high-risk cardiac patients can scan enter their clinical information to enhance clinical decision for upper limb rehabilitation, post myocardial infraction [MI] event or blood pressure lowering exercise treatment for preventing stroke recurrence by evaluating the effect of immersive virtual reality technology on gait rehabilitation etc.]. Medical staff would be able to access these modules, answer questions, and help with the management of the rehabilitation process remotely.

Our objective in this vision for a smart and connect healthcare is to build and deploy a comprehensive, mobile,, real time iOS app for prevention, timeliness diagnosis and rehabilitation of cardiac emergencies. We will field test the usability and potential efficacy of a real time, R-CNN, RPN mobile, real time, AI-ML, infused, intelligent, iOS app dashboard on high-risk cardiac patients. We will employ the following criteria subsequently to test the real time mobile devise: (1) using a pre-post-test design, (2) providing usability or potential efficacy data, (3) testing interventions using a mobile platform (by itself or as part of a mobile system), (4) including high risk cardiac patients aged 18 years and older representing all minorities and underrepresented groups. We will optimize traditional remote telemonitoring interventions, transitional care, and structured telephone support for heuristically optimizing detection and communication of prevention, timeliness diagnosis and rehabilitation of cardiac emergency-related health data by: 1. examining digital consciousness, games 2. including real time retrievable, transmittable, and archivable, diagnostic, clinical measures such as EKG on mobile phones with self-care measures 3. Generating real time transmittance protocols for non-English and migrant minority populations; and 4. Posting current reminders of prevention, timeliness diagnosis and rehabilitation of cardiac emergency news in real time. We will determine the most effective self-management strategies including complex medication regimen, diet, and exercise recommendations and modification of behavior according to the patient's related symptoms which will be downloaded in real time from the real time app dashboard and statistically analyzed using pre-trained models, if necessitated. Protocol-driven. real time, prevention, timeliness diagnosis and rehabilitation of cardiac emergency management techniques along with education, self-monitoring of symptoms, a flexible diuretic regimen, early care-seeking, prompt health care responses, psychosocial interventions, and professional coordination can be highlighted for successful strategies and for self-management employing the real time dashboard library archives. Less than half a percentage change on behavioral outcomes may result from each self-care component, thus prompting recommendation of interventions with multiple self-care components for a cumulative effect on behavioral outcomes generated from real time, prevention, timeliness diagnosis and rehabilitation of cardiac emergency-related, prognosticative models. Also, novel mobile health technologies will be recommended to serve as conduits for self-management for field validating the real time statistically analyzed high risk, cardiac patient data [e.g., weight management in obese minority stroke survivor, collection of functional outcomes after inpatient rehabilitation, assessment, or cardiac triage for inner city emergency destinations, guidelines for the prevention of fentanyl overdose deaths, from cardiopulmonary arrest: Advanced Cardiac Life Support [ACLS] algorithms i.e., a continuation of resuscitation attempts, for those patients who have been evaluated and found to require advanced resuscitative measures post CPR and defibrillation, timing on clot-busting medication intake etc.]. Given the complexity of self-management, multiple self-care components specific for high risk cardiac patients, the proposed infused, interactive, AI-ML, intelligent, smartphone, mobile, dashboard app will retrieve, transfer and generate prevention, timeliness diagnosis and rehabilitation of cardiac emergency-related data in real time for recommending protocols to achieve desired health outcomes based on early recognition of potential cardiac threat [e.g., decreased blood flow resulting from either embolism into a cerebral supply artery from the heart, or the great proximal vessels, extracranial or intracranial arteries, affected by atherosclerosis, in situ occlusion of small perforating arteries etc.]. For example, communication, concentration, memory, and executive function, as well as spatial awareness, are all cognitive functions which may be impacted by a specific cardiac event [MI, HF, Transient ischemic attack (TIA) etc.], which may manifest as post transient speech disturbance, hyperventilation, altered alcohol intake or missing antiepileptic medication. Peripheral vestibular annoyance, focal seizures, presyncopal symptoms etc. These indicators may be monitored continuously within the wearable, real time, engineered wristwear technology solution. The interactive intelligent mobile devise would have accessory features [e.g., in-ear software for simultaneous interpreting, smart hearing aids, headphones with AI-powered noise cancellation, head-mounted display (HMD) software for augmented reality glasses and virtual reality headsets to create highly immersive high risk cardiac patient user experiences etc.].. The real time app dashboard could also monitor for specific cardiac-related symptoms such as symptoms of a heart attack [e.g. numbness, or tingling in fingers, pain or numbness in lower back or legs, gastrointestinal issues, unexplained weight loss lightheadedness on stand up etc.].

Cardiac events can also trigger mental health issues such as anxiety and depression, as well as movement and coordination problems, paralysis, difficulties swallowing, visual impairment, and urinary incontinence and loss of bowel control. The faster a person is treated after a cardiac emergency the more likely they are to recover from it. While some people recover quickly from cardiac emergencies others may need long-term support. Real time R-CNN, RPN, AI-ML, infused, intelligent, mobile health related apps are available to help aid the recovery process. They can help track appointments and medications, provide language therapy, train the brain, and even lower some risk factors [e.g., rates of asymptomatic carotid artery stenosis, long

Q-T syndrome, where the electrical conduction of the heart is slowed and the heart's natural rhythm is disrupted due to opioid addiction, cardiopulmonary resuscitation (CPR) training videos, for instructing chest compressions (e.g., 100 per minute) etc.,] for future prevention, timeliness diagnosis and rehabilitation of potential emergencies [an aneurysm (enlargement) in the abdominal aorta for deterrence of an ruptured aortic artery, EKG of cardiac arrhythmias associated with transient global amnesia, symptomatic intracerebral hemorrhage etc.]. In this R21, our objective is to develop a workflow for real-time measurable prevention, timeliness diagnosis and rehabilitation of cardiac emergency-related clinical, and other health data in an infused, interactive, AI-ML intelligent, mobile, smartphone app for generation of analysis and feedback with immediate decision-making and clinical and public health action.

We will evaluate the usability and potential efficacy of a real time R-CNN, RPN infused, iOS mobile app for high-risk cardiac patients. We will employ the following criteria to test the real time mobile devise: (1) using a pre-post-test design, (2) provide usability or potential efficacy data, (3) test interventions using a mobile platform (by itself or as part of a mobile system): and, (4) include high risk cardiac patients aged 18 years and older representing all minorities and under-represented groups. The intelligent mobile app will determine the most effective self-management strategies include complex medication regimen, diet, and exercise recommendations and modification of behavior according to cardiac symptoms downloaded in real time from the mobile app and statistically analyzed. Protocol-driven management along with education, self-monitoring of cardiac-related symptoms, a flexible diuretic regimen, early care-seeking, prompt health care responses, psychosocial interventions, and professional coordination will be highlighted for successful strategies and for self-management. Less than half a percentage change on behavioral outcomes may result from each self-care component, thus prompting recommendation of interventions with multiple self-care components for a cumulative effect on behavioral outcomes employing prevention, timeliness diagnosis and rehabilitation of cardiac emergency, real time models generated from the real time iOS dashboard. Also, novel mobile health technologies will be recommended to serve as conduits for self-management for field validating all the high-risk cardiac-related patient data regardless of location or patient status. For example, we will seamlessly integrate clinical medical device software with previous electronic health intervention records archived in the real time dashboard to create a streamlined approach to enhanced remote patient care which can help detect medical contraindications [e.g., the combination of drugs prescribed for HF (e.g., angiotensin converting enzyme inhibitor and beta blocker) with other drugs with blood pressure-lowering effects (e.g., alpha-1 blockers) increasing the risk of hypotension]. Given the complexity of self-management, multiple self-care components specific for high-risk cardiac patients, the prevention, timeliness diagnosis and rehabilitation of emergencies, self-care monitoring interactive, intelligent, mobile app will real time retrieve potential heart-related health data [e.g., loss of balance or coordination, sudden shortness of breath, sudden unilateral weakness, slurred speech, sudden vision loss in one or both eyes or double vision, migraine, seizures syncope etc.]. The app will be able to update newer cardiac medications and generate specific innovative treatment FDA approved protocols [nitric oxide lozenges, videos of robotic-assisted cardiac catheterization or coronary artery bypass surgery, m–glucose cotransporter 2 (SGLT2) inhibitors to reduce the risk of hospitalization for and cardiovascular death among patients with chronic HF and a left ventricular ejection fraction of 40% or less etc.] using a real time, interactive, intelligent, smartphone-assisted, AI-ML infused mobile app.

Specific Aims for a Smart and Connected Healthcare:

Aim 1: **Design, Build and Deploy an interactive intelligent, configurable smartphone mobile application [app] to support prevention, timeliness diagnosis and rehabilitation of cardiac emergency, real time, self-care monitoring**.

Overview: We will design, develop, and validate an infused, real time AI-ML intelligent, smartphone mobile interactive, app using Faster R-CNN and an RPN.

Hypotheses: A Region Proposal Network and a Pre-trained ResNet-50 Tensor Flow object detection real time grid infused into an interactive, intelligent, smartphone infused, AI-ML, iOS dashboard can generate prevention, timeliness diagnosis and rehabilitation of emergency protocols in real time in high-risk cardiac patients.

Experimental Design: We will employ the Unity game engine software and Vuforia 10.22 SDK for creating the dashboard. The Vuforia Area Target Creator application allows generation of an Area Target employing a depth-enabled mobile device, [iPads, and iPhones] [developer Vuforia.com]. Vuforia is an augmented reality software development kit for mobile devices that enables the creation of augmented reality applications in the Unity Hub. We will create a New Project and select the Unity version currently supported by Vuforia Engine. We will download the extension with the installer script from the Developer portal: Vuforia Engine Extension for Unity.

Subsequently, we will add an ARCamera GameObject from the Vuforia Engine menu. We will select the ARCamera and Open Vuforia Configuration. Then we will import the downloaded Unity Asset Package from the menu Assets. Importing the package will allow letting a script make the necessary changes to the Unity project so as to install or upgrade to the latest Vuforia Engine SDK. Subsequently we will select the 3D or 3D (URP) core template. We will navigate to the Vuforia Engine Menu and select Image Target.

Each Vuforia Engine Game Object will be configured in the dashboard. When a real time prevention, timeliness diagnosis and rehabilitation of emergencies, self-care monitoring target is added, it will appear in the Hierarchy. We will select the Image Target GameObject from the Hierarchy and choose from the Type dropdown either.

The Vuforia Web Services (VWS) API can integrate with Vuforia Cloud Services, including the Cloud Recognition service, VuMark Generation API, and the Model Target Web API [13, 14]. We will connect to Cloud Recognition databases on the Vuforia Developer Portal and introduce the Web API into the mobile real time, iOS interactive intelligent app for adding and managing prevention, timeliness diagnosis and rehabilitation of cardiac emergency-related health data. We will employ Cloud Targets Web API and the Cloud Query Web API for information on managing and querying Cloud targets. We will employ the API to upload Image Targets to a Cloud Recognition Database, query Cloud Databases, create Model Targets and generate VuMarks. Our real time app will perform tasks such as authentication, location-aware functions, and provide targeted content and communication for end users [i.e., high -risk cardiac patients]. For Vuforia targets, we will set an Optimize Tracking mode to improve tracking for particular high risk cardiac patient cases [e.g., elevated levels of homocysteine in patients with atherosclerosis or venous thrombosis, usage of thrombolytics for determining risk of thrombo-embolic events in patients with high likelihood of left heart thrombus, such as patients with mitral stenosis, determining side effects of bleeding in a new heart medication in patients with acute pericarditis or subacute bacterial endocarditis, hemostatic defects including those secondary to severe hepatic or renal disease] etc.

In this vision, we introduce an RPN that shares full-image convolutional features with the real time, detection retrieval and transfer network, thus enabling cost-free region proposals using Tensor Flow object detection. The RPN will be trained end-to-end to generate high-quality region proposals, which will be employed by Faster R-CNN for real time detection. Retrieval. tracking, transference and archiving of the prevention, timeliness diagnosis and rehabilitation of cardiac emergency-related health data. To generate the region proposals, we will slide a small network over the convolutional feature output by the last shared convolutional layer in the dashboard. This small network will take as input an n $\times$ n spatial window of the input convolutional prevention, timeliness diagnosis and rehabilitation of cardiac emergency-related health data. Each sliding window will be spatially associated to a lower-dimensional feature 512-d for VGG, with Rectified Linear Unit [ReLU] as in Jacob et al. [13].

We introduce the use of rectified linear units (ReLU) as the classification function in a deep neural network (DNN) in an iOS real time mobile app dashboard. Conventionally, ReLU is used as an activation function in DNNs, with Softmax function as their classification function. However, Jacob et al. [13] employed a classification function by taking the activation of the penultimate layer hn $-$ 1 in the real time neural network. Our assumption is that by multiplying it by stratified, prevention, timeliness diagnosis and rehabilitation of cardiac emergency-related weight parameters θ., we could attain threshold raw scores oi by 0, i.e. f(o) $=$ max(0, oi), where f(o) is the ReLU function. We will provide class predictions y through argmax function, i.e. argmax f x. This feature will be fed into two sibling fully-connected layers—a box-regression layer (reg) and a box-classification layer (cls). We will use n $=$ 3 noting that the effective receptive field in our input georeferenced, prevention, timeliness diagnosis and rehabilitation of cardiac, emergency-related, information data capture points and diagnostic variable(s) will have a maximum threshold of 171 and 228 pixels for VGG. Note that because the mini-network will operate in a sliding-window fashion, the fully-connected layers will be shared across all spatial locations in the neural architecture which will be implemented with an n $\times$ n convolutional layer followed by two sibling 1 $\times$ 1 convolutional layers (for reg and cls, respectively). At each sliding-window location, we will predict multiple region proposals, where the number of maximum proposals for each prevention, timeliness diagnosis and rehabilitation of cardiac emergency-related health data, capture point, sentinel site location and diagnostic variable will be denoted as k. The k proposals will be parameterized relative to k reference boxes. Subsequently, we will unify the RPNs with Faster R-CNN object, real time, detection networks which will employ a scheme that alternates between the RPN tasks for fine-tuning the prevention, timeliness diagnosis and rehabilitation of cardiac emergency-related information while keeping the data fixed in the real time dashboard. A VGG-16 model CNN architecture consisting of 16 layers, including 13 convolutional layers and 3 fully connected layers will be built to achieve strong performance on various computer tasks including real time classification and object recognition of retrievable, transmittable, and or archivable, prevention, timeliness diagnosis and rehabilitation of cardiac emergency-related data capture point and or diagnostic variable(s).

The model's architecture will feature a stack of convolutional layers followed by max-pooling [i.e.., an operation for extracting a small feature map (e.g., $7 \times 7 \times 512$)] layers, with progressively increasing depth. This design will enable the interactive intelligent mobile iOS app to learn intricate hierarchical representations of georeferenceable [i.e. GPS indexable] prevention, timeliness diagnosis and rehabilitation of cardiac emergency-related information, data capture points [cardiovascular fitness center] and diagnostic variables [e.g., CT radiographic angiography and single-photon emission computer topographic values for optimizing imaging modality in assessing atherosclerotic plaque formation as a potential source of retinal hypoperfusion or embolism] which can lead to positive patient outcomes [improving spontaneous limb movements or posturing due to brainstem ischemia or heart valve failure due to aortic stenosis].

In the iOS dashboard Mask R-CNN will extend Faster R-CNN by adding a branch for predicting segmentation masks on each Region of Interest (RoI), in parallel with the existing branch for classification and bounding box regression. The mask branch is a small Fully Convolutional Network [FCN] applied to each RoI for predicting a segmentation mask in a pixel-to pixel manner. Mask R-CNN is simple to implement, and train given the Faster R-CNN framework, which facilitates a wide range of flexible architecture designs. Additionally, the mask branch only adds a small computational overhead, enabling a fast system and rapid experimentation.

The Faster R-CNN detector will consist of two stages. Initially, RPN will propose candidate object bounding boxes. The second stage will extract features using RoIPool from each candidate box and performs classification and bounding-box regression. The features used by both stages will be shared for faster inference. We will also explore another more effective backbone recently proposed by Lin et al. (2021), called a Feature Pyramid Network (FPN). The FPN will uses a top-down architecture with lateral connections to build an in-network feature pyramid from a single-scale input. Faster R-CNN with an FPN backbone will extract RoI features from various levels of the feature pyramid according to their scale. Using a ResNet-FPN backbone for feature extraction with Mask RCNN in the real time dashboard will then provide excellent gains in both accuracy and speed of retrieval, tracking, transferring and archiving prevention, timeliness diagnosis and rehabilitation of cardiac emergency-related data (Fig. 1).

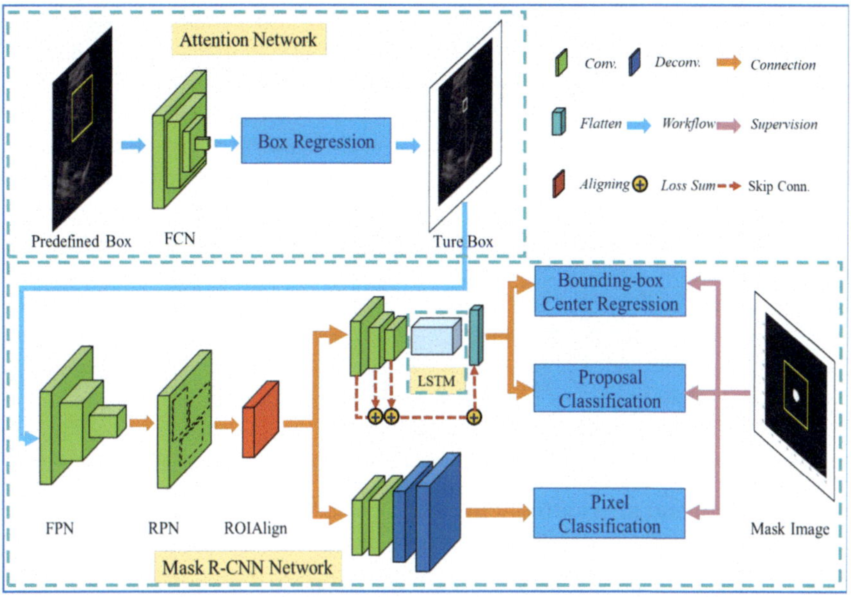

Fig. 1 The proposed deep network. Prevention, timeliness diagnosis and rehabilitation of cardiac emergency-related app neural architecture including FCN and feature pyramid

We will employ the GPU to speed up the computation in the interactive, intelligent, mobile, iOS dashboard for fast, accurate real time, information storage, management and retrieval of the prevention, timeliness diagnosis and rehabilitation of cardiac emergency-related data. In the proposed app, GPU and tensorflow-GPU will be installed in Keras where Mask R-CNN will automatically employ the GPU. Mask R-CNN is a CNN and state-of-the-art in terms of data segmentation and real time data transfer and retrieval is variant of a Deep Neural Network https://viso.ai/deep-learning/mask-r-cnn/. We will leverage cloud-based data storage and processing tools, including Microsoft Azure, AWS, and Google Cloud, to develop a reliable back-end infrastructure for the wearable. Final metrics will be completed based on the Pascal VOC dataset.

Expected Outome. We will formulate satellite real time, parameter estimation as a classification task and use artificial neural networks in the dashboard to efficiently perform robust Bayesian estimation in Python, PyMC3 and ArviZ of the prevention, timeliness diagnosis and rehabilitation of cardiac emergency-related capture point and diagnostic variable selected data. Bayesian eigen-autocorrelation inference will generate and visualize any uncertainty in the posterior distributions. We will show that the network's posterior distribution is centered at the true (unknown) value of a high-risk cardiac patient parameter within the uncertainty-oriented model employing the inverse Fisher information for delineating the ultimate sensitivity limit for the given real time apparatus. The reg layer will have 4 k outputs encoding the coordinates of k boxes, and the cls layer will output 2 k scores that estimate probability of a prevention, timeliness diagnosis and rehabilitation of real time retrieved tracked, transferred and archived cardiac emergency-related information data capture point or diagnostic variable for each proposal generated in the real time dashboard. A simple implementation of a two-stage real time RPN will achieve AR 15 points higher than that of the conventional RPN used in non-real time prevention, timeliness diagnosis and rehabilitation of cardiac emergency-related, health mobile apps surpassing any existing region proposal methods in the literature and in the on line commercial market for real time applications. For the very deep VGG-16 model the real time detection system will have a frame rate of 5fps (including all steps) on a GPU, while achieving state-of-the-art object detection accuracy on PASCAL VOC and MS COCO datasets with only 300 proposals per sampled prevention, timeliness diagnosis and rehabilitation of cardiac emergency-related information data capture point and or diagnostic variable(s). The RPN methods will be implemented on the GPU, in the interactive app dashboard in Python with a runtime of 10.3 ms.

Potential Pitfall and alternative approach: Deep CNNs have achieved remarkable results in image processing tasks. However, their high expression ability risks overfitting. Consequently, data augmentation techniques have been proposed to prevent overfitting while enriching datasets. Recent CNN architectures with more parameters are rendering traditional data augmentation techniques insufficient. Unfortunately, overfitting is a common challenge in interactive, mobile app dashboard models and CNN deep learning projects. It happens when the model learns the training data too well including its noise and outliers. Such a learning may lead to a model that performs well on the training data but badly on new, unseen prevention, timeliness diagnosis and cardiac emergency-related real time detectable, retrievable, and transmittable capture points or diagnostic, health, information-related data. This may be observed when the performance on training data is too low compared to the performance on validation or testing data. Deep learning models, especially CNNs, are particularly susceptible to overfitting due to their capacity for high complexity and their ability to learn detailed patterns in large-scale data. In this vision we will employ several regularization techniques to mitigate overfitting in CNNs to the real time, prevention, timeliness diagnosis and rehabilitation cardiac emergency-related, mobile, interactive app which will include: 1. Dropout method which consists of randomly dropping some neurons during the training process forcing the remaining neurons to learn new features from the input data 2. Batch normalization whereby overfitting is reduced at some extent by normalizing the input layer by adjusting and scaling the activations [this approach may also be used to speed up and stabilize the training process]; and, 3. Pooling Layers will be used to reduce the spatial dimensions of the input prevention, timeliness diagnosis and rehabilitation of the cardiac emergency-related data to provide the real time app dashboard model with an abstracted form of representation, hence reducing the chance of overfitting.

Aim 2. Train and validate an intelligent smartphone app using a GPU cluster with four nodes of GeForce GTX TITAN X, each with 12 GB of memory for generating a real-time, interactive app-generated Faster R-CNN model.

Hypothesis: Faster R-CNN network [i.e., 2 step deep learning object detector] using a pre-trained CNN model can extract and transfer convolutional prevention, timeliness diagnosis and rehabilitation cardiac emergency-related information data capture points and diagnostic variables from the last layer of a trained network in real time.

Experimental Design: We will train the real time sensed prevention, timeliness diagnosis and rehabilitation cardiac emergency-related capture point empirical datasets using the pre-trained CNN model and extract the convolutional feature map from the last layer of the trained network. This step will enable the neural network to understand the key features within the prevention, timeliness diagnosis and rehabilitation of cardiac emergency-related data that separates multiple classes of objects within the high-risk patient data. To train towards feature recognition in real time, the right neural network must be employed and adapted [14]. For this study, we will employ Inception V2 as the base pre-trained convolutional network for extracting the prevention, timeliness diagnosis and rehabilitation of cardiac emergency-related data. Inception V2 is a complex deep learning architecture which we will employ using smart factorization methods to make the convolutions efficient in terms of computational complexity. This architecture will help in achieving the same for the real time detected, retrieved, transmitted and archived prevention, timeliness diagnosis and rehabilitation of cardiac emergency-related problems. After extracting the high-risk cardiac patient data to localize objects of interest we will employ the Faster Region-based CNNs generated in Aim 1 to predefine anchor aspect ratios and scale the prevention, timeliness diagnosis and rehabilitation of cardiac emergency-related capture point and diagnostic variable (s) feature data attribute(s). Anchors, the rectangular boxes that are used to scan objects is vital for real time data transfer which in this proposal will be emplaced during training of the neural network. For the case of our, real time prevention, timeliness diagnosis and rehabilitation of cardiac emergency-related data we will set the base anchor size as (256, 256) pixels, with scaling ratios and aspect ratios as (0.25, 0.5, 0.75, 1.0, 1.50, 2.0, 2.5) and (0.25, 0.5, 0.75, 1.0, 1.50, 2.0, 2.5) respectively. The width and height of each anchor will be set as:

$$\begin{aligned} \text{width anchor} &= \text{scale}\,[\text{i}] * \text{sqrt (aspect ratio}\,[\text{i}]) * \text{base anchor}[0], \\ \text{height anchor} &= \text{scale}[\text{i}] * \text{base anchor}[1]/\ \text{sqrt (aspect ratio}[\text{i}]) \end{aligned}$$

where i is the index of the matrices of scales and aspect ratios. In total, 300 anchors will be generated per capture point or diagnostic variable for training. We will feed our anchors and feature maps to RPN. Here, the task will be to train the network to identify those boxes that were indicative of real time detected, retrieved, tracked transmitted and archived prevention, timeliness diagnosis and rehabilitation of cardiac emergency-related, information data capture points and diagnostic variables. To do so, we will ground truth each box prior to training as a potential prevention, timeliness diagnosis and rehabilitation of real time, cardiac emergency-related information data capture point and diagnostic variable (s) and the classified background. The RPN will consist of small convolutional layers, which will be trained to identify the anchors having the relevant objects of interest (in our case, a potential, prevention, timeliness diagnosis and rehabilitation of cardiac emergency-related data capture point and or diagnostic variable with an "objectness" score and return the ones that most likely contain the information within the dashboard) based on the score along with the center coordinates, width and height of all anchors. We will resize the feature data of the anchors so as to optimize differentiating diagnosis [e.g., such as cerebral amyloid angiopathy, a common cause of lobar cerebral hemorrhage in older patients, which can also present as transient focal neurological episodes, i.e., 'amyloid spells'] learned from the above step into fixed sized feature maps which will serve as input to two branches in the dashboard as: (a) classifier to identify potential prevention, timeliness diagnosis and rehabilitation of cardiac emergency-related data capture points within the anchor; and (b) regression to tighten the anchor for improved accuracy. All steps will be performed in parallel. We will fine-tune the classifier by varying learning rates. We will employ the stochastic gradient descent (SGD) solver for 50,000 iterations with a base learning rate of 2e-5, then another 25,000 iterations by reducing the base learning 2e-6 and the rest 25,000 with 2e-7 for faster convergence. The learning rate for each iteration is shown in Fig. 2.

The real time high-risk cardiac input data will be archived in the real time dashboard memory. We will create classified datasets from them and convert them to tf, tensor objects and tensor slices. The code snippet will embed the prevention, timeliness diagnosis and rehabilitation of cardiac emergency-related data features and labels arrays in a TensorFlow graph as tf. constant () operations in the real time iOS dashboard. During training, one important criterion for correctness is the loss function. Briefly, our loss function in the dashboard will measure the learning ability of a neural network architecture. It will range from 0 to 1 where 0 means perfect learning and 1 means no learning. The goal of training the real time detected, retrieved and or transferred prevention, timeliness diagnosis and rehabilitation cardiac emergency-related data is to minimize the loss of training data. To minimize the loss, we will apply a novel focal loss function which penalizes instances of false negatives which in

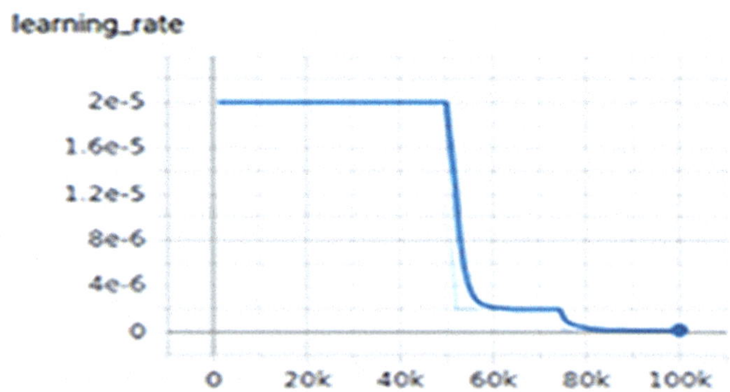

Fig. 2 The learning rate for each iteration in the proposed recovery capital sobriety, interactive AI-ML infused smartphone app in Jacob et al. [13]

our case will be any information data capture point or diagnostic variable classified as background. If the classifier is biased towards the background class this will result in biased learning. (Lin et al. 2017) Focal loss is an improvement over the more standard cross entropy loss, since it operates by lowering the loss of well classified cases and emphasizing the misclassified ones (Lin et al. 2017). We will consider the following equation, $Q = \begin{cases} p & \text{if } y = +1 \\ 1 - p & \text{otherwise} \end{cases}$ Here y will denote the ground-truth of the class (+1 for prevention, timeliness diagnosis and rehabilitation of cardiac emergency-related capture point or diagnostic variable and -1 for background), which will be the model estimate probability for the class label. The Focal loss (FL) for each anchor k will be defined as $FL(k) = -(1 - Q) \log(Q)$. Here, y would denote the ground truth of the class (+1 for a 12-step town hall meeting or cardiology care center and -1 for background), and $p \in [0, 1]$, which would be the app-generated real time dashboard model estimated probability for the class label $y = +1$. In so doing we will be able to determine the focal loss in the cardiac high risk patient data. We will add an additional modulating factor $(1 - Q)^{\lambda}$ to the above loss equation to provide tunability, where $\lambda >= 0$. The focal loss (FL) will be subsequently defined as $FL(Q) = -(1-Q)^{\lambda} log(Q)$ in the neural architecture Here *l* will be a tunable parameter (Figs. 3 and 4).

After applying focal loss, we will determine that overall minimal loss achieved in the iOS, AI-ML infused, real time dashboard based on uncertainty-oriented eigen-Bayesian semi-parametric, non-Gaussian Markovian model. When the loss in training data is similar to the loss in validation data (here the validation data will be also used to train the model), then they should not decrease any further and the process of training will be considered complete. The process of training and validation of the prevention, timeliness diagnosis and rehabilitation cardiac emergency-related prognosticative models will be an iterative process.

It is emphasized here that we will also train the prevention, timeliness diagnosis and rehabilitation of cardiac emergency-related, real time, mobile dashboard app-generated Faster R-CNN model in using the annotated data to detect and locate sources of solutions (meshing of mental health issue data in order to reach a stable

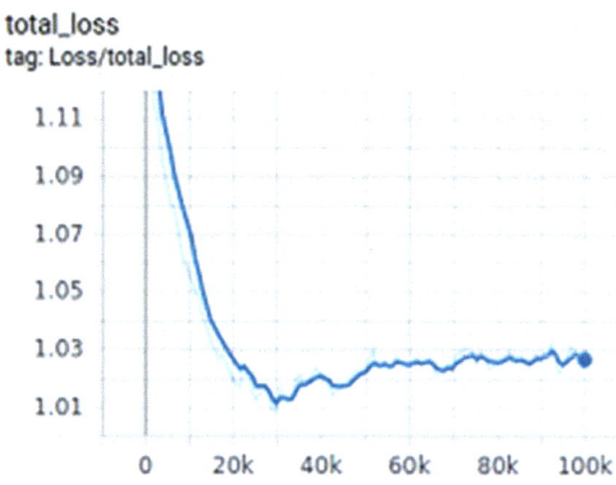

Fig. 3 Total loss graph with focal loss

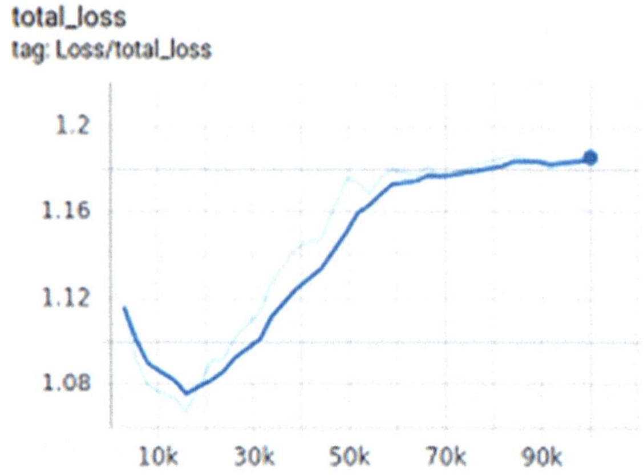

Fig. 4 Total loss graph without focal loss

recovery solution in a high-risk cardiac patient). A testing team will train and validate the smartphone app real time data using a GPU cluster as in Jacob et al. [13]. The cluster will have four nodes of GeForce GTX TITAN X, each with 12 GB of memory. Each high-risk cardiac patient update will be directed into previous models constructed for classification and localization, which will strengthen expected results using the GPU. We will conduct early stopping during iteration. We will monitor the model's performance on validation data during the training process and stopping the training whenever the validation error does not improve anymore. We will also employ noise injection**.** This process consists of adding noise to the inputs or the outputs of hidden layers during the training to make the model more robust and prevent it from a weak generalization. Furthermore, both L1 and L2 will be used to add a penalty to the loss function based on the size of prevention, timeliness diagnosis and rehabilitation of cardiac emergency-related information, real time, data capture point or diagnostic variable predictions weights. More specifically, L1 will encourage the weights to be spare leading to better feature selection. On the other hand, L2 (also called weight decay) will be employed to encourage the weights to be small, preventing them from having too much influence on the data delivery time and hence diagnosis. Data augmentation will also be conducted to artificially increase the size and diversity of the training dataset by applying random transformations like rotation, scaling, flipping, or cropping to the input high risk cardiac patients data from video analog data which may be constrained in poorer networked locations [e.g., migrant farmlands, inner city low income urban residences etc.]

Expected outcome: The default binary cross entropy loss function will increase after 15,000 iterations. The probabilities of detection by the real time mobile app neural network (e.g., the confidence in the forecast dashboard model in making decisions such as whether a high-risk cardiac could benefit from individual treatment or extended group support treatment) will be close to 0.99. Meanwhile the number of anchors containing prevention, timeliness diagnosis and rehabilitation of cardiac emergency-related capture points and diagnostic variables will be lesser in comparison to anchors that contained the background class. In the real time dashboard when a potential prevention, timeliness diagnosis and rehabilitation of cardiac emergency-related information-related data capture point or diagnostic variable is misclassified as background and Q will be small, the modulating factor (1−Q) l, will be close to 1 and this will not affect loss. When Q tends to 1, the modulating factor will be close to 0 and loss for well classified examples will be down weighted. The focal loss (FL) will be definable as $\mathrm{FL}(Q) = -(1-Q)^{\lambda} log(Q)$ in the real time neural app dashboard architecture. The final set of parameters of the real time dashboard will subsequently minimize the data loss criteria. When adopting to Fast R-CNN and Faster R-CNN, RPN will improve the detection mAP by 5.0 points. Computing the True Positives, False Positives, and False Negatives for classification in the validation set will allow for the integration of a final mAP in the iOS mobile app dashboard which will geolocate the area under the precision-recall curve of each real-time detectable. retrievable, transferrable, prevention, timeliness diagnosis and rehabilitation of cardiac emergency-related recovery-oriented information data capture point or diagnostic variable class. The mAP of all app dashboard real time, detected, retrieved, tracked transferred and archived, prevention, timeliness diagnosis and rehabilitation of cardiac emergency-related data in the validation set will be determined to be >0.80.

Potential pitfalls and alternative strategies:

If the pipeline architecture does not perform in real time, a Cascade Region Proposal Network (Cascade RPN) will address the conventional RPN employed to heuristically define the anchors and realign the real time, detected, retrieved, tracked transferred and archived, prevention, timeliness diagnosis and rehabilitation of cardiac emergency-related capture point and diagnostic variable data features to the anchors. First, instead of using multiple anchors with predefined scales and aspect ratios, the Cascade RPN will rely on a single anchor per prevention, timeliness diagnosis and rehabilitation of cardiac emergency-related, real time, mobile dashboard app-generated capture point or diagnostic variable and perform multi-stage refinement. Each stage will subsequently be progressively more stringent in defining positive samples by starting out with an anchor-free metric followed by anchor-based metrics in the ensuing stages. Second, to attain alignment between the features and the anchors throughout the stages, adaptive convolution is proposed that will take the anchors in addition to the real time, prevention, timeliness diagnosis and rehabilitation of cardiac emergency-related data as its input and learns the sampled features guided by the anchors.

Furthermore, if a single false negative exists during testing the mobile dashboard real time detected retrieved and transferred prevention, timeliness diagnosis and rehabilitation of cardiac emergency-related information, data, capture point or diagnostic variable, we will employ a user-understandable metric for revealing the quality of our trained and validated real time neural app network. The metric will be mAP over Intersection over Union (IoU) threshold. The IoU metric will measure the correctness of a given bounding box. It is formally defined as, the area of the intersection of the predicted box and ground truth box which in this proposal will be divided by the area of union of the predicted box and the ground truth box. The mean Average Precision or mAP score will be calculated by taking the mean AP over all classes and/or across all IoU thresholds. Lower IoU values indicate poorer performance. We will subsequently train our Faster R-CNN model using the annotated images to detect and locate sources of prevention, timeliness diagnosis and rehabilitation of cardiac emergency-related data. When the IoU Is 1, then a perfect classification and emplacement of the bounding box occurred. Lower IoU values in an interactive dashboard app indicate poorer performance.

Aim 3 Construct hot/cold spot, eigenvector spatial filtering (ESF) eigenfunction eigen-decomposition models using real time prevention, timeliness diagnosis and rehabilitation of cardiac emergency-related data

Hypothesis A semi-parametric, eigen-Bayesian, non-Gaussian Markovian eigendecomposable eigen-algorithm can tease out any spatial outliers in real time detected retrieved and communicated georeferenced, hot/cold spot prevention, timeliness diagnosis and rehabilitation of cardiac emergency-related data due to violations of regression assumptions in space and time.

Hypothesis Spatial structure can be incorporated into the predictors of prevention, timeliness diagnosis and rehabilitation of cardiac emergency-related information, data, capture point or diagnostic variable employing a generalizable linear model with a combination of the real time prevention, timeliness diagnosis and rehabilitation of cardiac emergency-related eigen-decomposed spatial filter [ESF] eigenvectors generating from a spatial connectivity matrix and for generating precise robust real time data.

We will introduce ESF eigenvectors into the iOS app dashboard employing Python codification in part as a means to circumvent the difficulties pertaining to the estimation of uncertainty-related, hyperendemic/hypoendemic, aggregation/non-aggregation-oriented [i.e., geospatial heteroscedascity and or multicollinearity in the hot and cold spot prevention, timeliness diagnosis and rehabilitation of cardiac emergency-related eigen-decomposed eigenfunction eigenvectors]. ESF methodology has since been extended to address problems of spatial uncertainty in real time traffic pollution network autocorrelation (Jing 2024), curb cut mapping geospatial artificial intelligence [geo-AI], infused machine learning [ML], prognosticative, recovery capital sobriety mobile app development (Jacob et al. 13,) and others.

Experimental design: After we have built an interactive, real time, iOS app dashboard for self-time measuring prevention, timeliness diagnosis and rehabilitation of cardiac emergency-related data, the AI-ML components [e.g., Support Vector Machine[SVM] Random Forest (RF)] will train the Faster R-CNN using the, real-time, retrieved, self-measured, patient information. The technical design effort for such an app would be to address the diffusion of real-time, retrieved prevention and rehabilitation information about potential georeferenced sub-meter resolution [e., WorldView 4 visible and near infra-red [NIR] 0.3 m spatial resolution hot spots [, e.g., cluster of rehabilitation fitness centers, cardio-healthy restaurants, detox facilities,] and cold spots (e.g., satellite mapped urban parks where opiate usage and drug transaction occurs, bars, vape and cigar shops etc.) based on county-level, field-verified, scalable geolocations in a designated county geographical area with massive data dissemination on live streaming platforms platforms such as X (formerly known as Twitter), Instagram, YouTube, Reddit, and Gab. Integration middleware or platforms play a crucial role in real-time data integration by providing the infrastructure and tools necessary to connect different data sources, transform data formats, and route data. Delays in processing and delivering real-time data can result in poor user experiences [13]. Hence, we will implement efficient data pipelines and utilize technologies like data streaming and caching to minimize latency. Since the pipeline for a real time, prevention, timeliness diagnosis and rehabilitation model might aggregate data from files in a distributed file system, we will apply random perturbations to each capture point or diagnostic variable and merge randomly selected cardiac-related data or images into a batch for training. The pipeline for a text model might involve extracting symbols from raw text data, converting them to embedding identifiers with a lookup table, and batching together sequences of different lengths. The tf.data API will introduce tf.data. Dataset abstraction in the real time dashboard will be represented as a sequence of elements, in which each element consists of one or more prevention, timeliness diagnosis and rehabilitation, cardiac emergency related. high-risk model components. For example, in a hot /cold spot image pipeline, an element might be a single training example, with a pair of tensor components representing the image or a information bit and its label. Our data source library will construct a dataset from real time retrieved and transmitted high-risk cardiac data which we will subsequently store in memory or in one or more files in the dashboard. We will employ a data transformation to construct datasets from one or more tf.data. and import tensorflow as tf. Furthermore, we will analyze engagement and interest in the real-time monitoring of scalable, signature-retrieved, real time, patient input the prevention, timeliness diagnosis and rehabilitation of cardiac emergency-related data and provide a differential assessment of the evolution of discourse regarding a county-wide scale for each platform and its users. Scalability of real time mobile devices is of paramount concern, as real time smartphone apps must continue to perform optimally as data sources and user bases grow. One key objective in this vision will be to promote patient sponsor, cardiac recovery specialist, and other collaborator relationships on ChatGPT 'space by using the app real time dashboard, for example, to determine possible consequences of health-related threats (e.g., relapse-induced overdoses, deaths due to delays in door-to-needle time for acute ischemic stroke or lack of coronary calcium scans) in high risk cardiac patients. Features like live chats, collaborative document editing, or multiplayer gaming will be employed to leverage the prevention, timeliness diagnosis and rehabilitation of cardiac emergency-related data to facilitate dynamic and interactive user experiences to facilitate real-time interactions within the iOS app dashboard to create a sense of connection and engagement. We will develop innovative wearable device application solutions and perform reliable integrations with multiple data layer APIs to ensure full functionality. In addition, the impact that this tool will have on the empirical research and policy development by disseminating information through the app 's preprint servers can allow for rapid media coverage and a consequent impact on implementing real time prevention, timeliness diagnosis and rehabilitation of cardiac emergency-related choices. We will fit information spreading with potential, real-time-generated dashboard models. The unique competitive advantage of our app dashboard is it will employ a real-time detection system neural architecture with single scale features for optimal retrieval and transfer which will be infused into an iOS app where a pyramidal feature hierarchy can be computed by a ConvNet. Aside from being capable of representing higher-level semantics in real-time, ConvNets are robust to variance in scale and can facilitate information recognition from features computed on a single input scale in an eigenfunction eigen-spatial filter algorithm [13]. By naturally leveraging the pyramidal shape of a ConvNet's feature hierarchy, a feature pyramid will have strong semantics at all scales which can prevent high risk cardiac-related injuries due to non-detection of large vessel occlusion strokes, for example, or non-employment of accurate prehospital scales using eigenvector maps generated in a real time dashboard. These convolution layers will extract feature attributes (, past ultrasound evaluation of the carotid arteries that screens for buildup of fatty plaques, tests performed using a 6-lead EKG designed to identify Atrial Fibrillation. etc.) from sample datasets by performing convolution operations utilizing filters on real-time input dashboard data to generate pre-trained research models employing mobile real-time tracking of high-risk cardiac patients and instant, secure inter-team communication to initiate and sustain prevention, timeliness diagnosis and rehabilitation of cardiac emergencies at the county level. All task analyses will occur within the interactive, web-based iOS dashboard in real-time. The dashboard will be integrated with handheld devices to enhance capabilities for video footage. The process flow will consist of data processing, multiscale image segmentation, feature extraction, construction of ensemble AI-ML algorithms, and accuracy assessment. [a pulse waveform graph, Standup test to measure fatigue and fitness. recording of beats-per-minute etc.] We will use 70% of the data for training and 30% for testing. Cluster images will be classified using manual methods and with SVM/RF classifiers (among other ML classifiers) and rule-based classifiers which will be used for segmentation and classification. We will be comprehensive in coverage of both analysis tasks and AI /ML training algorithms. To compare the performances of the classifications for testing forecasted georeferenced hot/cold spot images. We will use overall accuracy, the Kappa coefficient, quantity and allocation disagreement.

A Bayesian maximum likelihood signature interpolation algorithm in Python will run the captured signatures over the entire county employing commercial high-resolution satellite data (0.3 m spatial resolution Worldview 2 visible and NIR band data) to identify previously unknown geolocations of prevention, timeliness diagnosis and rehabilitation of cardiac emergency- related capture point and diagnostic variable data. The model will assume that at least some of the spatial variation observed can be modeled by random processes with autocorrelation and models [GPS data transfer to notify people in the surrounding area who can respond quickly and put their life-saving skills to use before EMS arrives. guides responders to directions to the nearest automated external defibrillator].

External validation of models will assess how well the models perform on independent ground truth data, and performance metrics will include sensitivity, specificity, AUC, and kappa. The minimum sample size required to estimate sensitivity of the eigen-spatial filter eigen-algorithms in detecting retrieving and transferring prevention, timeliness diagnosis and rehabilitation of cardiac emergency- related capture point data will be based on Buderer's formula assuming an alpha of 0.05, desired precision of 0.05, and $z1 - \alpha\ 2/2 = 3.84$.

Expected outcomes Assuming an expected sensitivity of the georeferenced hot and cold spot models in the capture point or diagnostic variable interpolated epicenters of 92%, based on our previous literature [13, 14], the minimum number of potential data capture point or diagnostic variables required for ground truthing should be 324. The tf.data API will make it possible to handle large amounts of high-risk cardiac data, read from different data formats, and perform complex transformations. Employing only a limited number of calibration measurements, our real time, iOS, AI-ML intelligent, infused, interactive. dashboard app will outperform every standard calibration method in the literature or on the on-line commercial market. We will be able to identify non-zero [i.e., non-chaos], geo-spatiotemporal autocorrelation coefficients throughout the intervention study site based on the aggregated/non-aggregated detected, retrieved, tracked, transferred and archived real time prevention, timeliness diagnosis and rehabilitation of cardiac emergency-health related data. Zero autocorrelation is geographic chaos (Griffith 2003). We will leverage web-native languages such as JavaScript, HTLMl5 and CSS to create a intuitive non-intrusive real time wearable devise. We will also create highly immersive applications for augmented reality glasses, like the Google Lens, and virtual reality headsets, such as Oculus or HTC VIVE headset. We will successfully generate a python code using GeoPandas and LibPySAL for the eigenfunction eigenvector, hot/cold spot, cardiac maps.

Pitfalls and alternative strategies:

The Python package has an implementation but has a distinct disadvantage. The implementation converts the 3-channel, satellite, pixel image into channel greyscale image before clustering. Losing the 3-channel signature pixel information may result in degraded clustering that is unable to capture the granularity of groups sufficient to identify where prevention, timeliness diagnosis and rehabilitation of cardiac emergency-related variable/geolocations occur. Alternatives to the ISODATA algorithm will be investigated. Self-Organizing Maps (SOMs) is an unsupervised machine learning clustering algorithm in ArcGIS Pro which preserves geometric relationships between clusters {www.erisri.com}. Orfeo-Toolbox also provides a similarly capable implementation of K-Means clustering. The 0.3 m per pixel WorldView satellite imagery collected over the intervention study site will be processed through usage of unsupervised clustering algorithms. We will compare these unsupervised clustering techniques for application in the remote sensed prevention, timeliness diagnosis and rehabilitation of cardiac emergency- capture point data. Each alternative algorithm will be traded to determine best performance. In addition, each algorithm parameter will be tuned, and these will all be investigated and traded as well.

References

1. Siegel B, Moser DK, Buck HG, Dickson VV, Dunbar SB, Lee CS, Lennie TA, Lindenfeld J, Mitchell JE, Treat-Jacobson DJ, Webber DE (2017) American heart association council on cardiovascular and stroke nursing. Council on peripheral vascular disease. Council on quality of care and outcomes research self-care for the prevention and management of cardiovascular disease and stroke: a scientific statement for healthcare professionals from the American heart association. J Am Heart Assoc 6(9):1–27
2. Moser DK, Dickson V, Jaarsma T, Lee C, Stromberg A, Riegel B (2012) Role of self-care in the patient with heart failure. Curr Cardiol Rep 14(3):265–275
3. Harkness K, Spaling M, Currie K, Strachan P, Clark A (2015) A systematic review of patient heart failure self-care strategies. J Cardiovasc Nurs 30(2):121–135
4. Coorey GM, Neubeck L, Mulley J, Redfern J (2018) Effectiveness, acceptability and usefulness of mobile applications for cardiovascular disease self-management: systematic review with meta-synthesis of quantitative and qualitative data. Eur J Prev Cardiol 25:505–521
5. Arnett D, Blumenthal R, Albert M et al (2019) ACC/AHA guideline on the primary prevention of cardiovascular disease: a report of the American College of Cardiology/American heart association task force on clinical practice guidelines. Circulation 140:e596–e646
6. Lee S, Squillace G, Smeets C, Vandecasteele M, Grieten L, de Francisco R, Van Hoof C (2015) Congestive heart failure patient monitoring using wearable Bio-impedance sensor technology. Engineering in Medicine and Biology Society
7. Remo EFJ, Mack L, San Juan AB, Elliott AF (2016) Features and usability assessment of a patient-centered mobile application (HeartMapp) for self-management of heart failure. Appl Nurs Res 32:156–216
8. Li L, Huang J, Wu J, Jiang C, Chen S, Xie G et al (2020) A mobile health app for the collection of functional outcomes after inpatient stroke rehabilitation: pilot randomized controlled trial. JMIR Mhealth Uhealth
9. Cai H, Lin T, Chen L, Weng H, Zhu R, Chen Y et al (2021) Evaluating the effect of immersive virtual reality technology on gait rehabilitation in stroke patients: a study protocol for a randomized controlled trial. Trials 22:91
10. Lee K (2019) Speed-interactive pedaling training using smartphone virtual reality application for stroke patients: single-blinded, randomized clinical trial. Brain Sci 9:295
11. Choi YH, Paik NJ (2018) Mobile game-based virtual reality program for upper extremity stroke rehabilitation. J Vis Exp 133:56241
12. Blackburn H (1997) Epidemiological basis of a community strategy for the prevention of cardiopulmonary diseases. Ann Epidemiol S7:S8–S13
13. Jacob (2024) Closing the gap on addiction recovery engagement with an AI-infused convolutional neural network technology application—a design vision. Benjamin Jacob and Heather McDonald and Joe Bohn, American Journal of Neural Networks and Applications. https://api.semanticscholar.org/CorpusID:268476703
14. Jacob BG, Loum D, Kaddumukasa M, Kamgno J, Djeunga HN et al (2021) Geospatial artificial intelligence infused into a smartphone drone application for implementing 'seek and destroy' in Uganda. Am J Entomol 5(4):92–109

Intentionally Violating the Univariate Assumption of Homogeneity of Variance Covariance Matrices Employing Non-infinitesimal Default Axiomatic Kolmogorovian Probabilities: The Sub-multiplicativity of the Frobenius Norm Quasi-Newtonian Algorithms and Semi-parametric Markovian Non-gaussian Eigen-Bayesianism for Resolving Tail-Weight Non-trivial Deviant Trajectories in Space Time and Geography for Robustifying Forecasting Hyperendemic Aggregation-Oriented Georeferenceable TB Estimator Determinants in Florida, USA

Benjamin G. Jacob, Ismael Hoare, Namit Choudhari, Sasha Mosich, Aarya Satardekar Heather McDonald, Anusha Parajuli, Spuritha Bhandaru, Alexander Gambrell, and Anthony Masys

Abstract The capability to optimally regressively forecast pulmonary tuberculosis (TB) in Florida is demonstrated. We develop analytical approaches for accurately assessing exogenous forces of infection while accounting for individual, county-level characteristics that may affect susceptibility to disease and co-factors that may affect transmissibility. An empirical georeferenced epidemiological dataset

B. G. Jacob (✉) · I. Hoare · S. Mosich · A. S. H. McDonald · A. Parajuli · S. Bhandaru · A. Gambrell · A. Masys
Global Infectious Diseases, College of Public Health, University of South Florida, Tampa, FL, USA
e-mail: bjacob1@usf.edu

N. Choudhari
School of Geosciences, University of South Florida, Tampa, FL, USA

B. Jacob et al. (eds.), *Tech Enabled Global Health Security*, Advanced Sciences and Technologies for Security Applications, https://doi.org/10.1007/978-3-031-86997-6_6

of geographical, spatiotemporal (henceforth, geo-spatiotemporal), diagnostic, stratified, descriptive, TB-related, explanatory, discrete integer values was acquired from the Center of Disease Control (CDC). A non-homogeneous, gamma-distributed, mean, negative, binomial regression was constructed to treat Poisson overdispersion in the determinants in scipy.stats.poisson. A Newton's method was employed to find zeroes of a function g of the stratified CDC variables which was provided by: $x_{n+1} = x_n - \left[J_g(x_n)\right]^{-1} g(x_n)$ in scipy.optimize.newton while $\left[J_g(x_n)\right]^{-1}$ evaluated the model outcome for (x_n). Newton's method found zeroes of a function g in the stratified TB data as generated by: $x_{n+1} = x_n - \left[J_g(x_n)\right]^{-1} g(x_n)$ where $\left[J_g(x_n)\right]^{-1}$ was the left inverse of the matrix $J_g(x_n)g$ which was evaluated for x_n. Broyden's Method solved for F(x) = 0. We computed the whole Jacobian only at the first iteration and conducted rank-one update at the other iterations. The secant method was found to be a non-robust estimator. Subsequently, we introduced a novel non-linear Kalman filter that employed semi-parametrization gradients for optimization of the CDC data. The semi-parametric approximation was based on a joint Gaussian assumption, which in this experiment was equivalent to minimizing an approximation to the Kullback-Leible divergence. Since residual, zero, autocorrelation coefficients may have provoked non-Gaussianism in the model output due to violations of regression assumptions [ie., multicollinearity] in space, time, and geography, we employed Markovian, semi-parametric, eigen-Bayesianism and a conditional heteroscedastic [GARCH] paradigm in PyMC3 for robustification of random geographical chaos in the prognosticated, TB-related, eigen-spatial, filtered estimated determinants. In the non-frequentistic geo-spatiotemporal, residual analysis we specified a prior uncertainty about the model parameters. The forecasts revealed the temporal error variance in the residuals. We generated a diagnostic summary report which revealed levels of heteroscedasticity [i.e., unequal scattered distribution of residuals], and /or inflated standard errors in the regressors in eigenvector eigen-Bayesian eigen-geospace. The fit of the final model revealed that the adjusted estimator determinant 'Personal Income < 20,000' was statistically important to the county-level, capture point, sentinel site scalable geolocations (adjusted improvement χ^2 was—1.299). Diagnostically stratifiable. non-Gaussian, TB-related, temporality, multicollinear, and or zero autocorrelated model heteroscedastic biasness due to violations of regression assumptions may be detectable and treatable in eigen-Bayesian, eigenvector eigen-geospace. Our final model revealed that from 2025 to 2030, Duval, Orange, and Broward counties would require immediate intervention to prevent TB transmission. The model also revealed that from 2035 to 2050, Hillsborough and Palm Beach counties could become hyperendemic without the implementation of control strategies. All non-Gaussian semiparametric, Markovian, space time and geography compatibility tests in this experiment were performed in Python 3.8, however, the eigen-spatial filter algorithms are executable on MacOS 11.3 and Linux Ubuntu Server 20.04 LTS environments.

Keywords Tuberculosis · Poisson · Negative binomial · Eigen-Bayesian · Markovian semi-parametric · Eigenvector · Eigen-geospace · Florida

1 Introduction

Precision forecast maps targeting and prioritizing transmission-oriented, diagnostically stratified, potential estimator determinants associated with a subcounty, or district-level, scalable, transmission-related, aggregation/non-aggregation-oriented i.e., hot/cold spot requires disturbance-free regressors. (e.g., non-Gaussianity of the error variance) for non-asymptotically optimally reflecting the hierarchical, diffusion-related dissemination of the sampled determinants. Statistical error or uncertainty is the amount by which an observation differs from its expected value (Freedman 2008), the latter being based on the entire population from which the statistical unit was chosen randomly. The expected value, being the mean of the entire population, may be typically unobservable in an empirical, asymptotical, vulnerability-oriented, geo-spatiotemporal dependent, scalable, county or district-level, prognosticative model, and hence the statistical error may not be observable. A residual (or fitting deviation) is an observable estimate of the unobservable statistical error, which can be embedded in., geo-spatiotemporal, non-Gaussian, deviant trajectories rendered from empirically regressed epidemiological datasets of georeferenced, stratifiable, county or district-level, scalable, TB, parameterizable sampled, vulnerability-oriented, estimated determinants.

Jacob et al. [44] approximated asymptoticalness, skew heteroscedasticity and geo-spatiotemporal multicollinearity in posterior probabilities in eigen-Bayesian eigenvector eigen-geospace for heuristically optimizing hierarchical diffusion-oriented, COVID-19, random effect [RE]specifications sampled in Uganda. The authors present two space–time model specifications, one based upon the generalized linear mixed model (GLMM), and the other upon Moran eigenvector space–time filters for the identification of eigen-optimization algorithms to fit a COVID-19, prognosticative, district-level, regression capture point, model to a training dataset of multicollinear, asymptotical, skew heteroscedastic, sentinel site estimator determinants, and other non-Gaussian non-normalities due to violations of regression assumptions in space time and geography. They did so to learn more about how regression functions could characterize spilled-over, hierarchical diffusion of the viral infection in Uganda at the sub-county district-level. Their objective was to prioritize and target hyper/hypoendemic, transmission-oriented, explanatory clinical, sociodemographic and environmental variables. A Moran eigen-spatial filtering technique was employed, which performed an eigenfunction, second order, eigen-spatial filter eigendecomposition of the REs in varying, temporally dependent, georeferenced, potentially clinical, environmental, and socio-economic, hyper/hypoendemic, diagnostically stratified, transmission-oriented, aggregation/non-aggregation-oriented, estimator determinants which rendered (SSRE) and spatially unstructured (SURE) components. The RE model incorporated eigen-decomposed eigenvectors derived from a geographic connectivity matrix to account for SSRE and SURE in standardized z scores stratified by multi-month, viral, infection yield due to spill-over, hierarchical diffusion of the virus at the sub-county, district-level. The authors calculated the conditional probabilities and derived the conditional distribution functions for the regressed diagnostic determinants, including the probability density function, the cumulative density function, and the quantile function. A Poisson random variable mean response specification was written as follows: $LN(\mu_{it}) = \alpha + \beta_{\text{day}} LN(\mu_{\text{day}}) + \sum_{k=1}^{K} e_{sitk}\beta_{sk} + \sum_{h=1}^{H} e_{\text{Hith}}\beta_{\text{Hh}} + RE_i + LN(P_i)$ where e_{sitk} and e_{Hith} respectively in numpy.random.poisson was the ith elements of the K < NT and H < NT selected eigenvectors. E_{stk} and E_{Hth} were extractable from the doubly centered space–time SWMs $(\mathbf{I}_T \otimes \mathbf{C}_3 + \mathbf{C}_T \otimes \mathbf{I}_n)$ and $(\mathbf{I}_T \otimes \mathbf{C}_{\text{H}} + \mathbf{C}_T \otimes \mathbf{I}_n)$ The expectation attached to the equation, i.e., RE ≡ SURE was satisfiable, with both having trivial SSRE components. In the semiparametric eigen-Bayesian Markovian context, the SSRE component was parsimoniously modeled with a conditional autoregressive specification which captured residual, geographically chaotic, non-Gaussian, non-homoscedastic, non-asymptotical, multicollinear infected, observational evidential prognosticators in the georeferenced, diagnostic stratified, non-zero autocorrelated COVID-19, sub-county, hot/cold spot clustering covariate propensities. The model's variance implied a substantial variability in the prevalence of COVID-19 across districts due to the hierarchical diffusion of the virus. The conclusions derived from the experiment revealed that site-specific, semi-parametric, eigen-decomposable, eigen-spatial filtered georeferenced, signature, epidemiological scalable variables are useful in revealing the influence of heterogeneity of variances due to violations of regression assumption in space time and geography and hence would be more accurate in optimally forecasting stratified, COVID-19, hyper/hypoendemic, sub-county, scalable, transmission-oriented, district-level, locations capture point, sentinel site, georeferenced covariates compared with a global model in which the non-homogenous erroneous estimators and their non-evidential uncertainty-oriented probabilities do not vary across eigenvector eigen-Bayesian eigen-geospace. Although hierarchical diffusion can quantify regression error in time space, and geography, semi-parametric, Markovian, eigen-Bayesianism has never been employed for identifying and treating, geo-spatiotemporal, geosampled estimator determinant deviant trajectories in TB data in space, time and geography.

In this experiment, we provide an intuitive geometric interpretation of the covariance matrix for quantitating a dataset of county-level TB clinical, sociodemiographic and environmental, stratified time series, Center of Disease Control, [CDC] archived prognosticative, estimator determinant, county-level space–time, and geography uncertainty model. Most textbooks and contributions in the literature explain the shape of data based on the concept of covariance matrices. Instead, we take a backward approach and explain the concept of covariance matrices based on the shape of the data for provoking estimator determinant non-Gaussian, non-homogeneity due to violations of regression assumptions in space–time and geography. In a previous article (see Jacob et al. [42]), we discussed the concept of variance and provided a derivation and proof of the well-known formula to estimate the sample variance in an empirical, epidemiologic, georeferenceable, prognosticative, district-level, diagnostically stratifiable, scalable multi-drug-resistant tuberculosis [MDR-TB], multivariate, estimator determinant prognosticative model. In [42], we employed a second order eigenfunction eigen-decomposition algorithm associated with a Moran's coefficient to investigate scalable, district-level non-linearity in an empirical epidemiological dataset of georeferenced, MDR-TB, semi-parameterizable dataset of estimator determinants geosampled in San Juan de Lurigancho (SJL) Lima, Peru. Our technique attempted to remove the inherent latent autocorrelation in an eigen-autocorrelation forecast model by introducing appropriate eigen-orthogonalizable surrogate variants. We also constructed a robust, generalizable, Bayesian Poissonian hierarchical, model to generate unbiased uncertainty-oriented regression estimator determinants for qualitatively assessing resistance to four commonly used drugs in TB treatment: isoniazid, rifampin, ethambutol, and streptomycin. Initially, data on the residential addresses of individual patients with smear-positive MDR-TB were geocoded in ArcGIS Pro. Next, the geosampled data were matched automatically and interactively within the cartographic database. The MDR-TB explanatory data feature attributes were then calculated and digitally overlaid onto sub-meter resolution satellite data within a 1 km buffer of 31 georeferenced health centers using a 10 m^2 grid-based algorithm in ArcGIS. Global autocorrelation statistics were then generated by decomposing the sampled data into positive and negative eigen-spatial filter eigenvectors using a second order, eigen-decomposition eigen-algorithm. Bayesian Poisson projections were then rendered employing normal priors for each of the geosampled prognosticators. A Residual Moran's coefficient (MC) minimization criterion was subsequently applied to the clinical coefficients generated from the second order eigenfunction eigen-decomposition eigen-algorithm to detect any unaccounted latent non-Gaussian zero-autocorrelation error in the estimator determinants. The model accounted for approximately 14% of the pseudo-replicated erroneous information in space, time, and geography and exhibited positive residual non-zero. Gaussian autocorrelation. The experiment proved an explanatory Bayesian count variable Poissonian specification with a conditionally autoregressive prior and a residual Moran's coefficient minimization criterion could quantitate erroneous random non-Gaussian heterogeneity in stratified, MDR-TB, resistant strain, leptokurtotic data distributions for heuristically optimizing georeferenceable, regression-based treatment protocols. [e.g., differentiating patient database 4-month TB treatment regimen (high-dose daily rifapentine with moxifloxacin, isoniazid, and pyrazinamide) and 6-month RIPE (rifapentine, isoniazid, pyrazinamide, and ethambutol) regimen. In so doing, georeferenceable, clinical stratified, time series covariates may be prioritized at the county or district level and treated for identifying the spatial distribution of high- risk populations.

In this experiment, we show that the standard deviation, as the square root of the variance, provides a measure of how much scalable, geo-sampled, county-level, capture point, sentinel site, uncertainty estimates occur in eigenvector eigen-Bayesian, eigen-geospace due to violations of regression assumptions [geo-spatiotemporal, skew, multicollinearity non-homoscedasticity, zero autocorrelation etc.]. Furthermore, we tease out the outlier and over-dispersed data in an empirical epidemiological dataset of county-level, TB-specified georeferenced, estimator determinant, county-level, regression-related, forecasted, model residuals employing semiparametric, non-frequentist, Markovian, eigen-Bayesian, conditional, autoregressive, eigenvector second order non-heteroscedastic eigen-algorithms.

We employed the following formula to compute population variance Var(X) $= \Sigma(X_i - X)^2/N = \Sigma x_i^2/n$ where each element of the vector was a scalar random in the empirical, time series, dependent, semiparametric, Markovian, eigen-Bayesian, conditional, autoregressive, county-level, TB-, vulnerability-oriented, regression model framework. Each georeferenceable, diagnostically scalable capture point, sentinel site, county-level, time series dependent, estimator determinant was representable as a finite number of observed, empirical, non-homogenous, discrete, integer values. Thereafter, the empirical sampled dataset of georeferenced, eigen-spatial filtered estimator determinants was employed to specify a joint probability distribution. In the study of probability, given at least two clinical, sociodemographic or environmental, diagnostic stratified, random, regressively forecastable explanatory variables X, Y, …, that is definable on a probability space, the joint probability distribution for X, Y, … would be a probability distribution that renders the probability that each of X, Y, … falls in any particular range, or discrete set of integer values specifiable for that specific variable in an epidemiological TB-related, diagnostic regression model (e.g., a bivariate distribution based on two random, scaled-up, capture point, eigenizable, estimator determinant, capture point, sentinel site, county-level, aggregation/non-aggregtaion-oriented, stratifiable evidential, uncertainty-oriented, prognosticators) [see [42]].

In the context of TB-related, regression analysis, several methods are applicable to control for the statistical effects of dependencies amongst erroneous evidential forecastable observations due to violation of space, time, and geography regression assumptions. Maximum likelihood or Bayesian approaches may account for some dependencies in a prognosticative, epidemiological, TB, diagnostic, stratified, regression-related framework, whereas recent eigen-spatial filtering eigen-Bayesian Markovian approaches [44] focus on semi-parametrically removing inconspicuous latent non-Gaussian, propagation, zero autocorrelation coefficients due to violations of regression assumptions in space, time and geography. In this experiment, we propose an eigen-algorithmic, semiparametric, non-frequentistic, Markovian, conditional, autoregressive, prognosticative, eigen-Bayesian, uncertainty-oriented, forecast modeling approach that allows epidemiologists, TB researchers clinicians and others to deal explicitly with (a) lagged autoregressive models, and, (b) simultaneous autoregressive, georeferenceable, uncertainty-related, stratifiable, estimator determinant, scalable, evidential paradigms. As in one eigen-algorithmic, eigen-spatial filtering approach, a specific subset of eigen-orthogonalizable, eigenfunction eigenvectors rendered from a transformed spatial link matrix are employed to capture dependencies amongst the perturbations of a georeferenced, capture point, sentinel site, scaled-up, county-level, diagnostic stratified, geosampled, estimator determinant, potential, aggregation/non-aggregation-oriented, hyper/hypoendemic regression-related, non-frequentistic, uncertainty-oriented TB model. However, we assumed the optimal subset in the proposed filtering model would be more accurately identifiable by an objective function that minimizes latent residual autocorrelation rather than maximizes a model fit. We assumed that the proposed objective function may have the advantage that it leads to a robustification of smaller subset of selected, diagnostically stratifiable, uncertainty-oriented, TB-related, scalable, estimator determinants in eigen-algorithmic, non-frequentistic, semiparametric, Markovian, eigen-Bayesian, conditional, autoregressive, eigenvector eigen-geospace due to violations of regression assumptions in space, time and geography.

In semiparametric, eigen-Bayesian, Markovian, conditional, autoregressive statistics, the Wishart distribution may be the conjugate prior of the inverse covariance-matrix of a multivariate-normal random-vector. In eigen-Bayesian semiparametric, Markovian probability theory (see [44]), if the posterior distributions p(θ|x) are in the same family as the prior probability distribution p(θ), the conjugate distributions, and the prior could be a conjugate prior for the likelihood function. For example, the Gaussian family is conjugate to itself (or self-conjugate) with respect to a Gaussian likelihood function: if the likelihood function is Gaussian, choosing a Gaussian prior over the mean in any empirical, epidemiological, prognosticative, georeferenceable, aggregation/non-aggregation-oriented, diagnostic, stratifiable, hyper/hypoendemic, TB-related, geo-spatiotemporal prognosticative, county or district-level, scalable, signature model. Such a paradigm could theoretically heuristically, optimize, targeting and prioritizing, georeferenceable, capture point, sentinel site, county-level, hot/cold spot geolocations and their respective, eigenizable, scalable, county or district-level, geosampled, estimator determinants. This prior may thus ensure that the posterior distribution is also Gaussian in a eigen-algorithmic, semiparametric, eigen-Bayesian, non-frequentistic, Markovian, conditional, autoregressive, diagnostically stratifiable, epidemiologic, capture point, sentinel site, prognosticative, TB-related, model. We assumed that a non-Gaussian distribution could be a conjugate prior to the likelihood (that is opposed to Gaussianism in these autoregressive prognosticative paradigms as suggested by the contributions in the literature) [e.g., Griffith 2003]. In eigenvector eigen-Bayesian probability theory, if the posterior distribution p is in the same probability distribution family as the prior probability distribution p(θ), the prior and posterior would be the conjugate distributions, and the prior would be a conjugate prior for the likelihood function p [44]. The concept and the term “eigen-Bayes conjugate prior” are introduced by Jacob et al. [44]. This uncertainty-oriented model is based on the Bayesian decision theory of Raiffa and Schlaifer (1961). Jacob et al. [44] employ eigen-spatial filter concepts discovered independently by Griffith et al. (2003).

We will now expand on the eigen-Bayesian, eigen-algorithmic, on-frequentistic, semiparametric, Markovian, conditional, autoregressive, and statistical probability of Jacob et al. [44]. The eigen-Bayesian view is based on a probability that is assignable to a hypothesis, whereas, under frequentist inference, a hypothesis is typically tested without being assigned an error probability. Semiparametric, eigen-Bayesian non-frequentistism, belongs to the category of eigen-algorithmic, semiparametric, Markovian, conditional, autoregressive, statistical evidential probabilities which may be usable to evaluate the probability of a hypothesis based on an uncertainty approximation in space, time and geography. The eigen-Bayesian probabilist specifies a prior probability similar to a Bayesian inferential hierarchical generalizable prognosticative model. This, in turn, is then updated to a posterior probability in the light of new relevant evidential data (e.g., georeferenceable, aggregation/

non-aggregation-oriented, scaled-up, sentinel site, TB-related, non-Gaussian, hyper/hypoendemic diagnostically stratifiable, epidemiological capture point sentinel site forecasted county-level, scalable capture point locations). The non-frequentistic, semiparametric, eigen-Bayesian, Markovian, conditional, autoregressive, statistical interpretation of probability can be interpreted as an extension of propositional logic that enables reasoning with hypotheses; that is, with propositions whose absolute truth or falsity is unknown.

Semi-parametric, eigen-Bayesian Markovian inference refers to statistical inference where reasoning uncertainty is quantified employing probability. In the classical frequentist inference model, parameters and hypotheses are fixed. Probabilities are not assigned to parameters or hypotheses in frequentist inference. Statistical probabilistic models specify a set of statistical assumptions and processes that represent how the sample data are constructable robustly and parsimoniously in space, time and geography. These uncertainty-oriented prognosticative, scalable, vulnerability-oriented models can specify several parameters that may be modifiable during an interpolation process [i.e., levels of propagation-oriented, non-Gaussianism due to violations of regression assumptions in space, time, and geography]. For example, an epidemiological regression-related, TB-related, diagnostically stratifiable, county-level, scalable, prognosticative signature, capture point, sentinel site model can be represented as samples from a Bernoulli distribution, which models two possible outcomes [transmission or no transmission].

In semiparametric, Markovian, eigen-Bayesian, conditional, autoregressive, non-frequentistism, the Wishart distribution may be the conjugate prior of the inverse covariance-matrix of a multivariate-normal random-vector of eigenized TB-related epidemiologic prognosticators,. Hence, the probability distribution may originate from a family of probability distributions definable over symmetric, nonnegative-definite, random matrices, i.e., matrix-valued random variables. The multivariate normal distribution, Gaussian distribution, or joint normal distribution is a generalization of the one-dimensional [1-d] (univariate) normal distribution to higher dimensions Downey (2021).

In probability theory and statistics, the multivariate normal distribution, multivariate Gaussian distribution, or joint normal distribution is a generalization of the 1-d (univariate) normal distribution to higher dimensions Hosmer and Lemeshew (2000). One definition is that a random vector is said to be k-variate normally distributed if every linearizable combination of its k components has a univariate normal distribution. Its importance derives mainly from the multivariate central limit theorem. The central limit theorem states that the sum of independently distributed random variables with finite variances will tend to a normal distribution as the number of variables grows Bolstad et al. (2016).

The uncertainty distribution in an empirical, epidemiologic, geo-spatiotemporal, georefenceable, county-level, semiparametric, TB-related, eigen-Bayesian, Markovian, conditional, autoregressive, inferential, prognosticative, regression model framework may lead to the hypergeometric Wishart generator distribution with the

Wishart generator and the Wishart as special cases. Important statistical characteristics may be iteratively derivable for regressing a county-level, sentinel site, georeferenceable, potential hyper/hypoendemic, diagnostically stratifiable, TB-related, clinical, sociodemiographic and environmental, empirical dataset of aggregation/non-aggregation-oriented, geo-spatiotemporal prognosticatable, eigenizable, iteratively, interpolatable capture point, sentinel site, estimator determinant. The significance of this generator distribution is further demonstrated by assuming a special case as a prior for the underlying matrix, variate, normal model in Python.

Jacob et al. [44] employed an advanced spatial autocorrelation measure in a Bayesian hierarchical non-time series dependent framework that incorporated multivariate spatial weights. Currently, the ways in which uncertainty-oriented spatial units interact in space, time and geography [e.g., Gaussian error variance] may be quantified. The formula below captures spatial interdependence.

$$EDE = \frac{1}{n}\sum_{i=1}^{n}\sum_{j=1}^{n} W_{ij}\left(f_i - f_j\right)^2 / \sum_{i=1}^{n}\sum_{j=1}^{n} W_{ij}$$

- (n) is the number of observations,
- (f_i) and (f_j) are the values of the variable of interest for the (i)th and (j)th observations, respectively,
- (W_{ij}) is the weight assigned to the pair of observations (i) and (j).

The Python code is similar to Moran's *I*, just using the inequalipy library:

```
import numpy as np
import inequalipy as ineq
# Assuming your_data_array is a 1-D numpy array
your_data_array = np.array([...]) # replace with your data
# Assuming epsilon and kappa are the inequality aversion
parameters
epsilon = ... # replace with your value
kappa = ... # replace with your value
# Calculate the Kolm-Pollak EDE
kp_ede = ineq.kolmpollak.ede(your_data_array, epsilon, kappa)
# Calculate the Kolm-Pollak Index
kp_index = ineq.kolmpollak.index(your_data_array, epsilon,
kappa)
print(f'Kolm-Pollak EDE: {kp_ede:.4f}')
print(f'Kolm-Pollak Index: {kp_index:.4f}')
```

In this experiment, we employed a methodology in Python for creating new matrix variate distributions in an epidemiologic, georeferenceable, capture point, sentinel site, scaled-up, potential, hyper/hypoendemic, aggregation/non-aggregation-oriented, eigen-decomposable, epidemiological, TB-related, prognosticative, hierarchical model for every county in Florida. We employed a "Gaussian" prior for iterating the geosampled, TB-related, stratified, georeferenced, county-level, capture point, sentinel site scalable model, eigenizable, estimator determinants using it as a springboard for the creation of a newly developed model which acted as a prior for the covariance matrix of the matrix variate, non-frequentistic, semiparametric, eigen-algorithmic, eigen-Bayesian, Markovian, conditional, autoregressive, non-Gaussian eigenvector model.

Non-frequentistic, eigen-algorithmic, epidemiological, uncertainty-oriented, prognosticative modeling signature is a form of statistical reasoning [44], Griffth (2003). Here it was involved in calculating and updating probabilities to make robust predictions of georeferenceable, capture point, sentinel site, stratifiable, TB-related, estimator determinant, scalable, hot/cold spot, epidemiological stratified forecasts as regressively derived from an empirical, geosampled, dataset of stratified, georeferenceable, clinical, sociodemographic and, or environmental, estimator determinants geo-spatiotemporally associated to hyper/hypoendemic, aggregation/non-aggregation-oriented, transmission-related, scalable, capture point, sentinel site, county-level epidemiological foci.

Amongst our research testing hypotheses for our non-frequentistic, semiparametric, eigen-Bayesian, Markovian, conditional, autoregressive, sentinel site, diagnostic, stratifiable, estimator determinant, uncertainty-oriented model was a random matrix Y of dimension $p \times r$ which had a matrix normal distribution. It was constructable from georeferenceable, capture point, geosampled, scalable, sentinel site, diagnostically stratifiable, eigen-orthogonalizable, estimator determinant where M, Ω, and Φ if $f(Y \mid M, \Omega, \Phi) = (2\pi) - rp^2|\Phi| - p^2|\Omega| - r^2 \exp\{-12\,\mathrm{tr}(\Omega - 1(Y - M)\Phi - 1(Y - M)T)$. In the model, M was a matrix of dimension $p \times r$ representing the expected values of Y, Ω, a covariance matrix of dimension $p \times p$ containing the variances and covariances of the p variables, [i.e., aggregation/non-aggregation-oriented, geosampled, eigenizable estimator determinants]. This experiment also included determining if Φ was a covariance matrix of dimension $r \times r$ containing the variances and covariances computed between the scalable, georeferenceable, epidemiologic, stratified, potential, hyper/hypoendemic, aggregation/non-aggregation-oriented, capture points (with respect to all the Floridian, county-level, sentinel site, geo-spatiotemporal geosampled geolocations). We assumed that this non-frequentistic, eigen-algorithmic, eigen-Bayesian, Markovian, conditional, autoregressive, prognosticative, epidemiological, semiparametric model output would be able to optimally, heuristically, optimize, geo-spatiotemporal, target and prioritization of, the regressively forecastable, capture point, sentinel sites and their stratifiable, empirical, geosampled, hot/cold spot, estimator determinants for every county in Florida.

In probability theory, a log-normal distribution is a continuous probability distribution of a random predictive explanatory variable whose logarithm is normally distributed [38]. Hence, we assumed if an empirical, geo-spatiotemporal, geosampled, scalable, capture point, county-level, georeferenceable, sentinel site, diagnostically stratifiable, county-level, TB-related, vulnerability-oriented, prognosticative model random variable is log-normally distributed, then the residual model, potential, aggregation/non-aggregation-oriented, hyper/hypoendemic epidemiological forecasts, [i.e, simulated, targeted and prioritized, residualclinical evidential observational descriptors] should have a non-normal distribution. Likewise, if the optimally scaled-up, regressed, county-level, geosampled, diagnostic, stratified, capture point, sentinel site, observational capture point, evidential prognosticators have a non-normal distribution, then according to Jacob et al. [44] the heuristically optimizable, explanators [e.g., clinical hierarchical diffusion of a TB-related, streptomycin -resistant, strain stratified regressor), or an eigen-spatial filterable, zero autocorrelatable, estimator determinant of an eigen-orthogonalized Euclidean, distance measurement (e.g., a rural clinic geolocation to a centroid of a scalable, capture point, sentinel site, county-level, georeferenced, farmer's market,), or sociodemiographic, explanative estimator determinant (e.g., urban, residential, highly, TB infected, internally, displaced, migrant worker camp) should be parsimoniously robustly derivable from a gridded, panchromatic, sub-meter, resolution, 3-dimensional[d] digital, elevation model (DEM). This capture point, sentinel site, signature paradigm may have a geo-spatiotemporal log-normal distribution. A DEM is a representation of the bare ground (bare earth) topographic surface of the Earth, excluding trees, buildings, and any other surface objects [www.esri.com]. DEMs are created from a variety of sources. USGS DEMs used to be derived primarily from topographic maps [www.usgs.gov].

In Jacob et al. [44] a product moment correlation coefficient (i.e., the Moran's coefficient) was employed to quantitate local spatial variation in multiple, georeferenced, time series, dependent, empirical, capture point, prognosticative, scaled-up, sentinel site, district-level, geosampled, diagnostic, stratified, MDR-TB, explanatory variables geosampled in San Juan de Lurigancho, Lima, Peru. Initially, QuickBird 0.61 m, spatial, resolution data, encompassing visible bands and the near infra-red (NIR) bands, were selected to map images of land use land cover (LULC) and 3-D, DEM, stratified, geomorphological, explanatory, topological estimator determinant, feature attributes of the georeferenced, epidemiologic, intervention study site. Data of residential addresses of individual patients with smear-positive MDR-TB were geocoded, prevalence rates calculated, and subsequently digitally overlaid onto the satellite data (visible and NIR, waveband, QuickBird, 0.61 m, spatial, resolution data) within a 2 km buffer of 31 georeferenced, health centers, using a 10 m^2 grid-based matrix. Geographical Information System (GIS)-gridded measurements of each health center. This data was generated based on preliminary base maps of the georeferenced, capture point, sentinel site, empirical, epidemiologic, clinical, environmental and sociodemographic data, capture points aggregated to block groups and census tracts within each buffered scaled-up, district-level area. A model of the epidemiological intervention study site was constructed based on the 3-D DEM to determine multivariate, diagnostically stratifiable, LULC, classifiable, time series dependent, geomorphological, terrain-related, regressively forecastable covariates (e.g., 3-D. catchment, watershed variable, slope coefficients) geo-spatiotemporally associated with the georeferenced, geosampled, grid-stratified, MDR-TB, potential hyper/hypoendemic, aggregation/non-aggregation-oriented, eigenized estimator determinants. Pearson's correlation was employed to evaluate the linear relationship between the DEM, derived, catchment watershed 3-D slope coefficients, and the epidemiologic geosampled data. A SAS/GIS bridge software module was employed to calculate univariate statistics and to perform linear and non-linear regression employing the prognosticated evidential, stratified estimator determinants. The estimates generated from an autocorrelation analysis were subsequently eigendecomposed into eigen-orthogonalized bases employing a negative binomial regression with a non-homogeneous, gamma, distributed mean. Results of the 3-D, DEM, geomorphological, eco-hydrological, terrain-related analyses indicated a statistically non-significant, linear relationship between georeferenced health centers and the geosampled, eigenized, estimator determinant, Capture point elevation'. The data exhibited positive, residual, geo-spatiotemporal autocorrelation (PSA) and the eigendecomposition of Moran's coefficient into uncorrelated, grid-stratifiable, LULC pattern components revealed global spatial heterogeneities necessary to capture latent, non-zero, eigen-autocorrelation in the stratified, vulnerability-oriented, semi-parameterized MDR-TB, time series, dependent, probabilistic, vulnerability-oriented model, eigen-spatial filtered, eigen-orthogonalized, hyper/hypoendemic, aggregation/non-aggregation-oriented, estimator determinants. Zero autocorrelation cannot reveal data trends as it only reflects geographic chaos Griffith (2003).

In this experiment, regression analysis in Python was undertaken with the aim of identifying whether clinical explanators, in combination with sociodemographic and income factors, are independently significant predictors of TB incidence at the county level in Florida. Statistical significance was defined as an obtained alpha level of 0.05 or lower. Before regression analysis, a correlation analysis procedure was conducted to test multicollinearity so as not to impair the projected models.

We also attempted to solve general nonlinear unconstrained optimization problems iteratively for minimizing the epidemiological, TB, county-level model. A Gaussian function that satisfies signature interpolation conditions was imposed. Our assumption was that these conditions could provide a model that behaves like the objective function about an eigen-autocorrelated, georeferenced, geo-spatiotemporal, geosampled, potential, hyper/hypoendemic, aggregation/ aggregation-oriented, geospatial cluster [i.e., potential, capture point, sentinel site, county-level, scalable, hot spot], estimator determinant, dataset of TB-related, signature, diagnostic, stratifiable clinical, sociodemographic and or environmental, model iterates. The model functions in this experiment involved second-order derivatives. The fundamental idea behind our use of quasi-Newton methods was to maintain an approximation to the Hessian matrix since the techniques were used to find either zeroes or local maxima and minima of functions as an alternative to Newton's method. This we e used if the Jacobian or Hessian was deemed unavailable or wastoo expensive to compute at every iteration. The "full" Newton's method requires the Jacobian to search for zeros, or the Hessian for finding extrema, or it needs to be approximated it with a positive-definite matrix Hazewinkle (2002), which may be updatable, we assumed, from iteration to iteration using information computed from previous steps that reduced to Newton's method [hence i.e.,quasi-Newtonian].

One of the goals of this experiment was to describe, implement, and test uncertainty-oriented, simulation estimation trends in a uniform, systematic, and consistent way. Hence, in the first part of the paper, we derive regression and classical quasi-Newton methods, discuss their relative benefits, and show how to implement them when scaling up, georeferenced, capture point, sentinel site, diagnostically stratifiable, county-level, eigen-spatial filtered, geosampled, TB-related, prognosticative, vulnerability-oriented model estimator determinants. In the second part, we investigate more recent probabilistic, error, variation, quantification signature models, explain their motivation and theory, and analyze their performance for optimally heuristically optimizing, geolocating georeferenceable, capture point, sentinel site, TB-related, epidemiological stratified, county-level, hyper/hypoendemic, aggregation/non-aggregation-oriented, time series, dependent, regression, model, eigen-orthogonalizable, eigen-spatial filter, diagnostically, stratifiable, estimator determinants.

Our assumption was that a robust version of a Principal component analysis (PCA) may be expressable as the maximum likelihood [ML] solution of a scalable, non-zero, autocorrelatable, 'eigen-Bayesianizable', eigen-algorithmic, semiparametric, Markovian, non-frequentistic, TB-related, diagnostically, straifiable, georeferenceable, prognosticative, model. Principal component analysis is a popular technique for analyzing large datasets containing a high number of dimensions/features per sampled, observation, increasing the interpretability of data while preserving the maximum amount of information and enabling the visualization of multidimensional data [38]. This reformulation, we assumed, would determine the optimal number of principal components retainable in a prognosticative empirical, capture point, sentinel site, eigen-Bayesian, semiparametric, Markovian eigenvector, county-level epidemiological model for optimally heuristically optimizing, stimulating georeferenceable, capture point, sentinel site, county-level, scalable, hyper/hypoendemic, aggregation/non-aggregation-oriented TB stratifiable, geolocations and their, geosampled, hot/cold spot, estimator determinants i.e., causation covariates predictively.

In this experiment, we constructed an autoregressive, prognosticative, generalizable, eigen- TB-related, semiparametric, non-frequentist, eigen-Bayesian, Markovian, conditional, autoregressive, statistical, hierarchical, georeferenceable model in Python, to prognosticate the true dimensionality of an empirical, geosampled dataset of georeferenced, CDC archived, diagnostically stratified, estimator determinant, sentinel site, county-level, epidemiologic, data capture points. Our assumption was that our experimental results would reveal that the proposed scalable, eigen-filterable, prognosticative model estimator determinants could achieve simultaneous optimal dimensionality selection and accurate principal components recovery for optimally optimizing, geolocating scaled-up, diagnostically stratifiable, TB-related, county-level, hyper/hypoendemic, aggregation/non-aggregation-oriented, georeferenceable geolocations and their statically significant estimator determinants based on the random matrix theory.

Random matrix theory studies properties of $N \times n$ matrices A chosen from some distribution on the set of all matrices. As dimensions N and n grow to infinity, one observes that the spectrum of A tends to stabilize. This is manifested in several limit laws, which may be regarded as random matrix versions of the central limit theorem. Among them is Wigner's semicircle law for the eigenvalues of symmetric Gaussian matrices, the circular law for Gaussian matrices, the Marchenko-Pastur law for Wishart matrices $W = A * A$ where A is a Gaussian matrix, the Bai-Yin and Tracy-Widom laws for the extreme eigenvalues of Wishart matrices W.

The ongoing debate about the concept of statistical significance and the reliability of null hypothesis significance tests and p-values has brought the advent of various eigen-Bayesian semiparametric, Markovian hypothesis tests for detection of violations of regression assumptions in space time and geography h. However, careful calibration of the prior parameters may be necessary for the type I error rates or power of these alternatives to be any better than other, scalable, capture point, sentinel site, geo-spatiotemporal, regression uncertainty-oriented, forecast models contributed in the literature. Also, the availability of various tests for the same statistical problem leads to the question of which test to choose based on which criteria in eigenvector, eigen Bayesian, eigen-geospace.

We also constructed a generalized autoregressive conditional heteroscedasticity (GARCH) model. This model created by Bollerslev (1986) is an important type of time series model for optimally rectifying temporal heteroscedasticity. It explicitly models a time-varying conditional variance as a linear function of past squared residuals and of its past values. Here a non-frequentistic, semiparametric, eigen-Bayesian, epidemiological, GARCH model was constructed for elucidatively rendering, unbiased, georeferenceable, hyper/hypoendemic, aggregation/non-aggregation-oriented, transmission-related, stratifiable, explanatorial, hot/cold spot and their capture point, sentinel site, TB-related, time series, estimator determinants using a Markov chain Monte Carlo [MCMC] eigen-algorithmic sampling distribution. In statistics, MCMC methods are a class of algorithms for sampling from a probability distribution based on constructing a Markov chain with the desired equilibrium distribution Gelman et al. (2005).

We assumed an eigen-spatial filter, MCMC simulation procedure may be suitable for fitting scalable, hierarchical, diffusion-related, inferential, times series, dependent, georeferenced, capture point, sentinel site, county-level, TB-related, semiparametric, epidemiological, hot/cold spots and their diagnostically stratifiable, georeferenceable heteroscedastic, multicollinear and or zero autocorrelated hyper/hypoendemic, aggregation/non-aggregation-oriented, stratified, eigen-orthogonalizable estimator determinants. In probability theory and statistics, a Markov chain is a stochastic process that satisfies the Markovian property (usually characterized as "memorylessness") [www.mathworld.wolfram.com]. We assumed an empirical, geosampled epidemiological dataset of georeferenceable, geo-spatiotemporal, geosampled, non-zero, eigen autocorrelated, county-level, TB-related, epidemiological model stratifiable forecasts [i.e., hyper/hypoendemic, transmission-related, aggregation/non-aggregation-oriented, eigen- decomposable, geo-spatiotemporal, non-heteroscedastic and/or non-multicollinear, estimator determinants] could account for any asymptotical, multivariate bias in a simulated, empirical, prognosticated model output. In so doing, we assumed we would be able to quantitate model errors in eigen-algorithmic, non-frequentistic, Markovian, semiparametric, eigen-Bayesian, conditional, autoregressive, eigenvector, eigen-Bayesian eigen-geospace. We also assumed that an empirical, eigen-algorithmic non-homogenous, eigen-spatial, filtered, epidemiological, eigen-autocorrelated, geosampled dataset of transmission-oriented, time series dependent, hyper/hypoendemic, aggregation/non-aggregation-oriented, clinical, sociodemiographic and environmental, geo-spatiotemporal, stratified, estimator determinants, could optimally forecast and epidemiologically delineate potential, interventional, scalable, study site, georeferenceable, hot/cold spot geolocations (e.g., capture point, transmission-related, sentinel sites) at the county-level in Florida. The fundamental issues in a scalable, hierarchical, potential, eigen-orthogonalizable, georeferenceable, TB-related, epidemiological empirical, capture point, sentinel site, estimator determinant, vulnerability-oriented analysis is the violations of regression assumptions geo-spatiotemporally which can lead to numerous problems in an eigen-algorithmic, capture point, sentinel site, prognosticative vulnerability analysis including bias, distortion and outright errors in the conclusions.

We assumed a conjugate prior within the Gamma distribution may be initially employable for drawing non-frequentistic, statistical inference on a negative binomial corrected, Poissonian distribution derived from an empirical, multivariate, geosampled, potential, time series dependent, hyper/hypoendemic, transmission-related, georeferenceable, diagnostically stratifiable, TB-related, estimator determinant, capture point, sentinel site, county-level, heteroscedastic and multicollinear, estimator determinant dataset. In semi-parametric, Markovian, eigen-Bayesian, conditional, autoregressive, heteroscedastic, multicollinear, and or zero autocorrelated model inference, the prior distribution of a parameter and the likelihood of the observed data are combinable to quantitate the posterior distribution of the parameter optimally [44]. If the prior and the posterior belong to the same parametric family in an eigendecomposed, eigen-spatial filtered, explanatorial, geo-spatiotemporal, estimator determinant, capture point, conditional autoregressive model, the scalable sentinel site, diagnostically stratifiable residuals will optimally heuristically, optimize, targeting and prioritizing, georeferenceable, aggregation/non-aggregation-oriented, potential, scalable, hot/cold spots and their evidential observational prognosticators since the prior would be be conjugate for the likelihood in non-frequentistic, semiparametric, eigen-algorithmic, Markvian, eigen-Bayesian, eigenvector eigen-geospace.

We assumed that the estimation of an empirical epidemiological dataset of MCMC simulated, post-treated, georeferenced, GARCH, time series, eigen-spatial filters may also determine causation of biasness in prognosticative, capture point, sentinel site, TB-related, signature, hot/cold spot, epidemiological county-level eigen-Bayesian, semi-parametric, Markovian, eigen-algorithmic eigen-orthogonalized prognosticative models. According to Jacob et al. [44] these eigen-spatial filters may be estimated by means of a stepwise selection procedure in space, time, and geography. Since the vicissitudes of stepwise selection are attenuated by the fact that all candidate, exploratory TB-related eigenvectors are mutually eigen-orthogonal, this may allow diagnosing latent, non-Gaussian, zero autocorrelation in semiparametric, Markovian, eigen-Bayesian, eigenvector, eigen-Bayesian eigen-geospace. In so doing, any researcher would be able to account for uncertainty-oriented non-Gaussian estimates in space, time, and geography as derived from disturbances in model selection due to violations of regression assumption [e.g., non-homoscedastic, geo-spatiotemporal, non-Gaussian error variance, geographical chaotic zero autocorrelation etc.,]. We assumed alternative methods may also be elaboratable for estimating eigen-spatial filters to address perturbations in the non-frequentistic, Markovian, semiparametric, eigen-Bayesianizable, GARCH-oriented, eigenvector eigen-geospace employing adaptive, sampling signature, geosampled eigen-algorithms for optimally heuristically optimizing, aggregation/non-aggregation-oriented, diagnostically stratifiable, scalable, georeferenceable, capture point, scalable, sentinel sites.

We considered a Poissonian that was based on the log-normal, heterogeneity of an epidemiologic, explanatory, TB-related, probabilistic, uncertainty-oriented, regression estimation model, which we constructed from the stratified, archived, CDC clinical, sociodemiographic and environmental, county-level, sentinel site, geosampled, data, capture points in every county in Florida. We developed an uncertainty estimation method for the empirical count-related, regression model, prognosticative, estimator determinants for optimally optimizing, geosampled, hyper/hypoendemic, stratifiable, potentially generalizable, estimator determinants. We also developed a series expansion approach in which dependence between regressable count variables was introduced by means of unobserved, stochastically varying components embedded in the stratified, hyper/hypoendemic, aggregation/non-aggregation-oriented, empirical model geosampled archived CDC estimator determinant dataset. Extensions that accommodate excess zeros, censored data and multivariate generalizations were analyzed.

We employed a second-order, eigenfunction eigen-decomposition, eigen-algorithm associated with a Moran's coefficient to investigate non-Gaussism due to violations of regression assumptions in space time and geography in the empirical, georeferenced, multivariate, capture point, scalable, sentinel site, county-level, archived, CDC geosampled, estimator determinants eigen-algorithmically. In statistics, Moran's *I* is a measure of eigen-autocorrelation Moran (1950). Geo-spatiotemporal eigen-autocorrelation is characterizable by a correlation in a signal among nearby locations in space (Griffth 2003). The second-order eigenfunction eigen-decomposition technique attempted to remove the inherent, latent, zero, autocorrelation, and other propagation bias due to violations of regression assumptions in specie time and geography in the capture point, sentinel site, uncertainty-oriented, simulated GARCH, empirical model georeferenced dataset of geosampled, hyper/hypoendemic, aggregation/non-aggregtaion-oriented, county-level geolocations and their estimator determinants by introducing appropriate empirical, scaled-up, prognosticative, semi-parametric, Markovian, TB, stratified surrogate variants.

Therefore the objectives in this experiment were to: (1) construct multivariate, county-level empirical, eigen-Bayesianizable, semi-parametric, eigen-algorithmic, scalable, capture point, sentinel site, georeferenceable, exploratory, geo-spatiotemporal, weighted, regression equations for non-asymptotically unbiasing, county-level, diagnostic, stratifiable, observational, evidential estimator determinants associated to hyper/hypoendemic, aggregation/non-aggregation, transmission-related, capture point, epidemiologic, sentinel site foci; (2) generate an autoregressive-oriented, probabilistic, uncertainty-oriented error matrix employing the non-frequentistic estimator determinants; (3) employ quasi-Newton second-order optimization methods to approximate the inverse of the Hessian matrix to either find zeroes,or local maxima and minima of functions; (5) estimate the Bayes process to determine significant departures from variable Gaussianism; (6) test Wishart and other priors within an inferential, eigen-algorithmic, hierarchical, semiparametric, prognosticative, eigen-Bayesian model framework; (7) generate eigen-spatial filters of residual, time series dependent, diagnostic statistics to identify estimator, determinant, multicollinearity and non-Gaussian zero autocorrelation in eigen-Bayesized, no semiparametric, Markovian, conditional, autoregressive, eigenvector eigen-geospace.

Finally, we geo-spatiotemporally quantitate residual temporal heteroscedasticity in the epidemiological, geosampled, CDC archived, TB-related estimator determents using GARCH model post-interpolation. We assumed that we would be able to optimally heuristically optimize, targeting and prioritizing, georeferenceable, scaled-up, capture point, sentinel site, county-level, potential aggregation/non-aggregation-oriented, hyper/hypoendemic, hot/cold spot, transmission sites in Florida geo-spatiotemporally. This experiment constitutes the first attempt to extend conventional, time series, dependent, empirical, epidemiologic prognosticative, diagnostically stratifiable, vulnerability-oriented, georeferencable, TB-related, signature, capture point, sentinel site, scalable, regressively forecastable, county-level, regression model techniques [e.g., employing a heterogenous covariance matrix for constructing a semiparametric, 'eigen-Bayesianizable,' Markovian, eigen-algorithmic, eigen-spatial filterable, GARCH-related, approximation eigenvector formulation]. Our objective was to contribute to literature a methodology for optimally, heuristically optimizing geo-spatiotemporally targeting and prioritizing georeferenceable, hyper/hypoendemic, aggregation/non-aggregation-related capture point, sentinel site, scalable, county-level, hot/cold spots and their regressively forecastable capture point, sentinel site, county-level, scalable, estimator determinants while robustifying biasness due to violations of regression assumption in space time and geography in non-frequentistic, semiparametric, eigen-Bayesian, eigen-algorithmic, Markovian, conditional, autoregressive, eigenvector eigen-geospace.

2 Material and Methods

2.1 Study Site

Florida is a state located in the southeastern region of the United States. The state is bordered to the west by the Gulf of Mexico, to the north by Alabama and Georgia, to the east by the Atlantic Ocean, and to the south by the Straits of Florida and Cuba. Florida is the 22nd largest state, the 3rd most populous, and the 8th most densely populated in the United States. Jacksonville is the most populous municipality in Florida and the largest city by area in the contiguous United States. The Miami metropolitan area is Florida's most populous urban area. The city of Tallahassee is the state capital. There is a peninsula between the Gulf of Mexico, the Atlantic Ocean, and the Straits of Florida, which has the longest coastline in the contiguous United States, approximately 1,350 miles (2,170 km). Florida is the only state that borders both the Gulf of Mexico and the Atlantic Ocean. Much of the state is at or near sea level and is characterized by sedimentary soil. The climate varies from subtropical in the north to tropical in the south.

2.2 Regression Analyses

The relationship between the archived, stratified, CDC county-level, TB-related, epidemiologic, data, capture point, sentinel sites and each individual, geosampled, prognosticative, potentially eigenizable, clinical, sociodemographic, and environmental, stratified, estimator determinant was investigated by single variable regression analysis in Python.

We imported libraries and pandas as pd import matplotlib.pyplot as plt from sklearn.linear_model. Subsequently we imported PoissonRegressor from sklearn.pipeline and then imported Pipeline from sklearn import metrics. The package scikit-learn is a widely used Python library for machine learning, built on top of NumPy and some other packages which provides the means for preprocessing data, reducing dimensionality, implementing regression, classifying, clustering, and more [www.python.org]. Below is the Python code we employed to calculate the probability mass function (PMF) of a Poisson distribution in the epidemiological county-level, TB, stratified, estimator determinant, scalable prognosticative model using the scipy.stats module:

```
Import Numpy as Np
Import Matplotlib. Pyplot as Plt
From Scipy.Stats Import Poisson
# Parameters for the Poisson Distribution
Lambda_parameter = 3 # Mean Arrival Rate
# Generate Some X Values
X_values = np.Arange(0, 10)
# Calculate the PMF for Each X Value
```

```
Pmf_values = poisson.Pmf(X_values, Lambda_parameter)
# Plotting the PMF
Plt.Bar(X_values, Pmf_values, Color = 'skyblue')
Plt.Title('Poisson Distribution PMF')
Plt.Xlabel('Number of County-Level Covariates')
Plt.Ylabel('Probability')
Plt.Show()
```

We developed an analytical approach to estimate exogenous forces of the infection, while accounting for individual-level, geosampled, estimator determinant characteristics that may have affected susceptibility to TB disease, and co-factors that may have affected transmissibility. Since the epidemiological county-level, TB prevalence data were derived from binomial fractions, a regression model was employed, as is standard practice for the analysis of sampled, infectious, disease data [e.g., [44], Jacob et al. 2021, Griffth 2006). A Poisson probability regression analyses was employed to infer the relationship between the diagnostic, potentially eigen-spatially filterable, eigen-algorithmic, capture point, sentinel site, count data, and the archived empirical, stratified, county-level, sampled characteristics (i.e., independent variable) (Table 1).

The regression analyses assumed independent counts (i.e., Ni), taken at multiple, county-level, georeferenced zip-code geolocations [i.e., $i = 1, 2, \ldots, n$]. The stratified, clinical, environmental and sociodemographic, estimator, determinant counts were described by a set of variables denoted by matrix $\mathbf{X}_i$, where a $1 \times p$ was a vector of the discrete, integer values [i.e., covariate weightage] for delineating the scaled-up, transmission-oriented, county-level, capture point, sentinel sites i. The expected value of these data was given by $\mu_i(\mathbf{X}_i) = n_i(\mathbf{X}_i)\exp(\mathbf{X}_i\beta)$ where β was the vector of non-redundant sampled estimator determinants in the probabilistic model were the Poisson rates given by $\lambda_i(\mathbf{X}_i) = \mu_i(\mathbf{X}_i)/n_i(\mathbf{X}_i)$. The rates parameter $\lambda_i(\mathbf{X}_i)$ was both the mean and the variance of the Poisson distribution for each stratified, estimator determinant i. The dependent variable was county-level, capture point, sentinel site, TB prevalence. The Poisson regression model assumed that the stratified prognosticators were equally dispersed, that is, that the conditional variance equaled the

Table 1 Parameterizable determinant samples within stratified geolocations of endemic TB-related in the county study sites as entered in Python

Variable	Description	Units
GCP	Ground control points	Decimal-degrees
DISCAP	Distance from county-level capture point	Meters
EDUC	El	Meters
RCE	Race	Percentage
GEN	GENDER	Percentage
AGE	AGE	Percentage
PPOS	Previous positive cases	Numeric value
ECON	Income	Percentage
TEMP	Daily temperature	Rate in celsius

condition mean. Partial correlations were definable after introducing the concept of conditional distributions.

We initially restricted ourselves to just the conditional distributions rendered from the stratified, Poissonian, county-level, regressed estimator determinant normalized distributions. We noted an $n \times 1$ random vector **Z** which we partitioned into two random vectors, **X** and **Y**, where **X** was an $n_1 \times 1$ vector and **Y** was an $n_2 \times 1$ vector in the equation $Z = (XY)$. The conditional distribution properties of the scaled-up, clinical, environmental and sociodemographic, asymptotical regression coefficients were then definable.

We partitioned the mean vector and covariance matrix in a corresponding manner. That is, $\mu = (\mu 1\, \mu 2)$ and $\Sigma = \left(\sum_{11} \sum_{21} \sum_{12} \sum_{22}\right)$. In so doing, μ_1 the means for the scalable, diagnostic, stratified, prognosticative, potentially TB-related, eigenizable, estimator determinants in the archived CDC dataset x_1, and $\sum_{11}$ were optimally quantifiable along with the variances and covariances for set x_1. The matrix $\sum_{12}$ provided the covariances between the variables in set x_1 and set x_2 as did matrix $\sum_{21}$ Any distribution for a subset of variables from multivariate normal conditional on known values for another subset of diagnostic variables has a multivariate normal distribution (Draper and smith 2019).

We noted that the conditional distribution of x_1 in the epidemiologic, stratified, county-level, TB estimator determinant, prognosticative vulnerability model was given the known values for x_2 which was illustratable as multivariate normal with a mean, vector covariance matirx $=$ $\mu 1 + \sum_{12} \sum -122(x_2 - \mu 2) \sum_{11} - \sum_{12} \sum -122 \sum_{21}$. The procedure employed machine leaning estimation to find the stratified regression coefficients. The data was then log-transformed before analyses to normalize the distribution and minimize standard error.

There was considerable overdispersion in the capture point, sentinel site, Poissonian probabilistic, model epidemiological forecasts. An assumption that must be fulfilled for a Poisson distribution is the mean value of data equals to the variance value (or so- called equidispersion) (Haight 1967). If the variance value is greater than the mean value, it is called overdispersion [38]. Overdispersion can be detected by dividing the residual deviance by the df (Draper and Smith 1985). If this quotient is much greater than one, the negative binomial distribution should be used.

Epidemiologists, clinicans and TB researchers have suggested various statistical procedures to estimate capture point, sentinel site, county, or district-level, scalable, geosampled, parameterizable estimator determinants employing frequency models. In particular, the Poisson regression model, with Poisson error structure, has been employed in recent years (e.g., Ghazvini et al. 2019). However, it has been recognized that stratifiable, geosampled, count, or frequency-oriented, epidemiologic, scalable, diagnostically stratified, regression, TB-related, empirical data often display overdispersion, i.e., the presence of greater variability. Commonly, the cause of overdispersion in a Poissonian, regression, TB-related, prognosticative, explanatory, count variable model is an excess zero probability on the response variable (i.e., statistical dispersion) than would be expected.

Subsequently, a Poissonian mixture model with a negative binomial distribution was employed where the mean of the Poisson distribution was itself a random variable drawn from the Gamma distribution; thereby, introducing an additional free parameter into the empirical, epidemiologic, capture point, sentinel site, county-level, TB-related, regression model, probability distribution. Gamma distribution is a continuous probability distribution which occurs naturally in the processes where the waiting times between events are relevant (Draper and Smith 1985). The family of negative binomial distributions was a two-parameter family which included several parameterizations for treating the over-dispersed, county-level, estimator determinant, geosampled epidemiologic data.

The Poisson distribution had one free parameter and which did not allow for the variance to be adjusted independently of the mean. Hence, a parameterization technique was employed in PyPI. such that any two, geosampled, stratified, georeferenced, capture point, sentinel site, county-level, TB scaled-up, transmission-related, evidential prognosticators p and r were quantifiable employing $0 < p < 1$ and $r > 0$. Under this parameterization, the PMF of the any residual asymptotical prognosticative variables with a NegBin $(r,\ p)$ distribution took the following form: for $k = f(k;\ r,\ p) = \binom{k+r-1}{k} \cdot p^r \cdot (1-p)^k 0,\ 1,\ 2,$ where $\binom{k+r-1}{k} = \frac{\Gamma(k+r)}{k!\Gamma(r)} = (-1)^k - \binom{-r}{k}$ and $\Gamma(r) = (r-1)!$

We employed an alternative parameterization for quantitating the stratified, sampled, epidemiologic, capture point, sentinel site data employing the mean $\lambda : \lambda = r \cdot \left(p^{-1} - 1\right) p = \frac{r}{r+\lambda}$ In so doing, the mass function became: $g(k) = \frac{\lambda^k}{k!} \cdot \frac{\Gamma(r+k)}{\Gamma(r)\,(r+\lambda)^k} \cdot \frac{1}{\left(1+\frac{\lambda}{r}\right)^r}$, where λ and r were the parameters. Under this parameterization, we were able to generate: $_r\underrightarrow{\lim}_\infty g(k) = \frac{\lambda^k}{k!} \cdot 1 \cdot \frac{1}{\exp(\lambda)}$ in the TB model outcome which we noted resembled the mass function of a Poisson-distributed random variable quantified with a Poissonian rate [i.e., λ.] In other words, the alternative negative binomial distribution rendered from the regressed, county-level, capture point, diagnostically stratified, TB, specified, potentially eigenizable, estimator determinants generated from the CDC archived dataset converged to the Poisson distribution, and r controlled the deviation from the Poisson. This made the negative binomial model suitable as a robust alternative to the regression-based, Poissonian, county variable, prognosticative model framework for optimally conducting probabilistic, forecast-related epidemiologic modeling on the sentinel site, transmission-oriented, stratified, clinical, environmental and sociodemographic, empirical, evidential observational variables.

The negative binomial distribution of the regressed estimator determinants arose as a continuous mixture of Poisson distributions where the mixing distribution of the Poisson rate was a Gamma distribution. The Gamma–Poisson distribution is a two-stage model for the distribution of a discrete variable (Draper and Smith 1985) [e.g., the counts of TB related cases in a county zip code] We assumed to fit a Gamma–Poissonian model to the stratified, estimator determinant, capture point, sentinel site, county-level epidemiologic data there would be a requirement to differentiate between variation at two levels. Essentially, we drew lambda from a Gamma distribution and then employed lambda in a Poissonian model. How well we were able to make this distinction was influenced by the general level of variation; the ratio between variation at the two stages and the sample size of the archived, CDC, estimator determinant model dataset. The Gamma distribution was a very flexible distribution which took on varying shapes during the regression TB model construction. It can be expected that for small sample sizes the parameters of the Gamma distribution may be be subject to poor accuracy (Hogg and Craig 1978).

We investigated the prognosticative, Gamma–Poissonian, epidemiological, TB-related, county-level, vulnerability model employing a simulation study. The objective was to illustrate the potential problems in employing the flexibility of the Gamma distribution to handle over-dispersed, capture points. We provided our own formulation of the Gamma–Poissonian model and employed statistical testing to determine if the empirical epidemiologic data was randomly distributable employing a scaled-up, Gamma–Poissonian, capture point, sentinel site, vulnerability-oriented, county-level prognosticative model.

Mathematically, the computations for constructing our capture point, sentinel site model were quite straight forward in PyPI.; $f(k; r, p) = \int \infty 0 \, f \, Poisson \, (\lambda)(k) \cdot f \, Gamma \, (r, 1-pp)(\lambda) d\lambda = \int \infty 0 \lambda kk!e-\lambda \cdot \lambda r-1e-\lambda(1-p)/p(p1-p)r\Gamma(r)d\lambda = \int 0\infty$. We collected the relevant terms for the model $(1-p)rp-rk!\Gamma(r) \int_{\infty} 0\lambda_r+k-1e-\lambda/pd\lambda$ which in the residuals was reducable to $(1-p)rp-rk!\Gamma(r)pr+k\Gamma(r+k)$ and subsequently simplifiable to $= \Gamma(r+k)k!\Gamma(r)pk(1-p)r = \Gamma$. Gamma is a sort of generalizable factorial, so the normalizing coefficient is a binomial coefficient (Draper and Smith 1985).

We started with the expression for the negative binomial in scipy.stats.nbinomand which made the following replacement $\alpha = r$, $\beta = 1-p$. We generated the following expression: $P(k) = \beta\alpha\Gamma(\alpha)\Gamma(k+\alpha)k!(1-\beta)k$. We also conducted stirling on both the denominator and numerator of the second term which rendered $P(k) = \beta\alpha\Gamma(\alpha)(k+\alpha-1)k+\alpha-1ekek+\alpha-1kk(1-\beta)k$. We split the term $(k+\alpha-1)k+\alpha-1 = (k+\alpha-1)k(k+\alpha-1)\alpha-1(-1)$. Next, we simplified the first term as $(k+\alpha-1)k = kk(1+\alpha-1k)k$. The second term was approximately $k\alpha-1$ as $k \gg \alpha-1$ existed in the, epidemiological, signature model output. Subsequently, the ek and k cancelled out and we had $P(k) = \beta\alpha\Gamma(\alpha)k\alpha-1e\alpha-1(1+\alpha-1k)k(1-\beta)k$. Since $\lim n \to \infty(1+xn)n = ex\lim$ [based on Draper and Smith (1985)], we employed this equation on the two last terms of the expression in the capture point, sentinel site, county-level, estimator determinant, vulnerability model to obtain $e\alpha-1e-\beta k$, which yielded the Gamma expression. In mathematics, the gamma function (represented by Γ, the capital letter gamma from the Greek alphabet) is one commonly used extension of the factorial function to complex numbers [38]. The gamma function is defined for all complex numbers except the non-positive integers (Draper and Smith 1985).

We defined the non-homogeneous gamma distributed mean, negative binomial model which here was P(k) = Γ(k + r)k!Γ(r)(1 − p)rpk employing the capture point, sentinel site, stratified, scaled-up, hypo/hyperendemic parameters r, p and the Gamma distribution which was definable employing $P(x) = \beta\alpha\Gamma(\alpha)x\alpha - 1e - \beta x)$ with parameters α, β. We assumed $p \rightarrow 1$, therefore $\langle k\rangle = pr1 - p \rightarrow \infty$ which meant that the sampled, clinical, sociodemographic and environmental, stratified, estimator, determinant, discrete, integer values of k for which P was non-negligible was very high. We verified this by simulating some county-level, random, capture point, sentinel site, scalable, estimator determinant, discrete integer values from the negative binomial and the mixture and the compared frequencies.

In probability theory and statistics, the negative binomial distribution is a discrete probability distribution that models the number of failures in a sequence of independently and identically disturbed (i.d.d.) Bernoulli trials before a specified (non-random) number of successes (DeGroot 1986). The mass function in our county-level, negative, binomial model distribution was f (x) = P (X = x) = (x − 1 r − 1)(1 − p) x − r p r. This output revealed that for any stratified, randomized, regressed, explanatory variable X was a valid PMF. The negative binomial distribution also arose as a continuous mixture of Poisson distributions (i.e., a compound probability distribution) where the mixing distribution of the Poisson rate in the model was a Gamma distribution. As such, we were able to determine and view the, non-homogeneity, in the county-level, negative binomial, model, Poissonian(λ) distribution, where λ was itself a random variable distributed as a Gamma distribution with shape r and scale $\theta = (1 - p)/p$, or correspondingly rate $\beta = p/(1 - p)$. To display the intuition behind these equations we considered two independent Poissonian processes, "Success" = i.e., no transmission-related variable and "Failure" i.e., transmission-related variable, with p and 1 − p. Together, the Success and Failure processes were equivalent to a single Poissonian process of intensity 1, where an occurrence of the process in the TB-related, epidemiologic model forecast was a success if a corresponding independent probability p; otherwise, it was a failure.

The Success count followed a Poisson distribution with mean pT, where T was the waiting time for r occurrences in a Poissonian process of intensity 1 − p, i.e., T was Gamma-distributed with shape parameter r and intensity 1 − p. Thus, the capture point, scaled-up, sentinel site, county-level, negative binomial distribution was equivalent to a Poissonian distribution with mean pT, where the random variate T was Gamma-distributed with shape parameter r and intensity (1 − p). This was true as $\lambda = pT$ was Gamma-distributed with shape parameter r and intensity (1 − p)/p in the stratified, capture point, sentinel site, prognosticated model output. The following formal derivation (which did not depend on r being a, stratified TB-related, estimator determinant, geosampled discrete, integer value) confirmed our intuition.

$$f(k) = \int_0^\infty Poisson(k|\lambda) \cdot Gamma(\lambda|r,\ (1-p)/p)\, d\lambda$$

$$= \int_0^\infty \frac{\lambda^k}{k!} \exp\,(-\lambda) \cdot \frac{\lambda^{r-1} \exp\,(-\lambda p/\,(1-p))}{\Gamma(r)\,((1-p)/\,p)^r} d\lambda$$

$$= \frac{1}{k!\Gamma(r)} p^r \frac{1}{(1-p)^r} \int_0^\infty \lambda^{(r+k)-1} \exp\left(-\lambda/(1-p)\right) d\lambda$$
$$= \frac{1}{k!\Gamma(r)} p^r \frac{1}{(1-p)^r} (1-p)^{r+k} \Gamma(r+k)$$
$$= \frac{\Gamma(r+k)}{k!\Gamma(r)} p^r (1-p)^k.$$

2.3 Spatial Analyses

Initially, a misspecification perspective for the potential, eigenizable, geo-spatiotemporal, scalable, capture point, sentinel site, county-level prognosticative model was generated in Python Spatial Analysis Library (PySAL) assuming that the sampled, georeferenced, potential, hyper/hypoendemic, aggregation/non-aggregation-oriented, transmission-related, risk-based, TB-related parameter fit was $y = X\beta + \varepsilon^*$ (i.e., a standard regression equation). The primary function of the model generation was for identifying the autocorrelated disturbances ε^* in the stratified, clinical, environmental and sociodemographic, georeferenced, potential, county level, hyper/hypoendemic, aggregation/non-agrgegation-oriented, capture point, sentinel site estimator determinants. The stratified latent coefficients of the determinants were decomposed into a white-noise component, ε, and a set of unspecified and mis-specified model outputs that had the structure $y = XB + \underbrace{E\gamma + \varepsilon}_{=\varepsilon^*}$. White noise is a univariate or multivariate discrete-time stochastic process, whose terms are i.i.d. with a zero mean [69]. We employed PySAL to generate these two types of similarity measures.

In spatial autocorrelation analysis, the spatial weights are used to formalize the notion of spatial similarity (Griffith 2003). Here we used the queen contiguity:

```
wq = lp.weights.Queen.from_dataframe(df)
wq.transform = 'r'
```

So the spatial weight between neighborhoods i and j indicated if the two explanatory, stratified, clinical, environmental and sociodemographic, potential, georeferenced, hyper/hypoendemic, aggregation/non-agrgegation-oriented, capture point, sentinel site, estimator determinants were neighbors (i.e., geographically similar). We also measured attribute similarity to pair up with this concept of spatial similarity. The spatial lag was derived using y = df['median_pri'].

```
ylag = lp.weights.lag_spatial(wq, y)
ylag
import mapclassify as mc
ylagq5 = mc.Quantiles(ylag, k=5)
f, ax = plt.subplots(1, figsize=(9, 9))
df.assign(cl=ylagq5.yb).plot(column='cl', categorical=True, \
```

```
k=5, cmap='GnBu', linewidth=0.1, ax=ax, \
edgecolor='white', legend=True)
f. ax.set_axis_off()
plt.title("Spatial Lag Median Price (Quintiles)")
plt.show()
```

The quantile stratified, clinical, environmental and sociodemiographic, georeferenced, capture point, sentinel site, estimator determinant map for the spatial lag tended to enhance the impression of value similarity in space. [i.e., a local smoother].

However, we still had the challenge of visually associating the value of the estimator determinants in a neighborhood with the value of the spatial lag of values for the focal unit. The latter is a weighted, stratified, TB-related, clinical, environmental and sociodemographic, georeferenced, hyper/hypoendemic, aggregation/non-agrgegation-oriented, geo-spatiotemporal, geosampled, capture point, sentinel site, estimator determinants in the focal unit's neighborhood. To complement the precise visualization of these associations we employed formal statistical measures of spatial autocorrelation. We began with a simple case where the variable under consideration was binary. This was useful to unpack the logic of spatial autocorrelation tests for epidemiological, prognosticative county-level TB modeling So even though our attribute estimator determinants were continuously valued we converted them to a binary case to illustrate the key concepts:

```
y.me yb = y > y.median()
sum(yb)dian()
```

We had multiple county-level neighborhoods listed with the stratified, clinical, environmental and sociodemographic, potential, georeferenceable, hyper/hypoendemic, aggregation/non-agrgegation-oriented, capture point, sentinel site, TB-relatable county-level, time series, dependent, estimator determinants above the median and the remainder below the median.

```
yb = y > y.median()
labels = ["0 Low", "1 High"]
yb = [labels[i] for i in 1*yb]
df['yb'] = yb
```

The spatial distribution of the binary variable immediately arose questions about the juxtaposition of the "transmission" and "no transmission" county-level areas which we adjusted for in the epidemiological, TB model using:

```
fig, ax = plt.subplots(figsize=(12,10), subplot_
kw={'aspect':'equal'})
df.plot(column='yb', cmap='binary', edgecolor='grey',
legend=True, ax=ax)
```

A join existed for each geosampled neighbor pair of stratified georeferenced, explanatorial, TB observations, and the joins were reflected in our binary spatial weights object. We then uses the esda package from PySAL to carry out join count analysis:

```
import esda
yb = 1 * (y > y.median()) # convert back to binary
wq = lp.weights.Queen.from_dataframe(df)
wq.transform = 'b'
np.random.seed(12345)
jc = esda.join_counts.Join_Counts(yb, wq)
```

The resulting stratified, georeferenced, TB-related, clinical, environmental and sociodemographic, hyper/hypoendemic, aggregation/non-agrgegation-oriented, capture point, sentinel site, estimator determinant observed counts for the TB county-level, transmission quantified based on the unique number of joins in the spatial weights object. Thereafter PySAL employed random spatial permutations of the observed attribute values to generate a realization under the null of complete spatial randomness. This was repeated a large number of times (999 default) to construct a reference distribution to evaluate the statistical significance of our observed county-level TB counts. The average number of potential, hyperendemic aggregation-oriented polygon joined from the synthetic realizations was labelled as clinical, environmental and sociodemographic, georeferenced, hyper/hypoendemic, aggregation/non-agrgegation-oriented, capture point, sentinel site, estimator determination in import seaborn as sbn.

```
sbn.kdeplot(jc.sim_bb, shade=True)
plt.vlines(jc.bb, 0, .12, color='r')
plt.vlines(jc.mean_bb, 0,.12)
plt.xlabel('BB Counts')inanount.
```

A pseudo p-value summarized the regressed county-level TB counts Since this was below conventional significance levels, we rejected the null of complete spatial randomness in favor of spatial autocorrelation for the geosampled data. First, we transformed our weights to be row-standardized, from a current binary state: wq.transform = 'r' and y = df['median_pri'].

Moran's I i a test for global autocorrelation was employed:

```
np.random.seed(12345)
mi = esda.moran.Moran(y, wq)
mi.I
```

The value for the statistic needed to be interpreted against a reference distribution under the null of CSR. PySAL subsequently we imported seaborn as sbn.

```
sbn.kdeplot(mi.sim, shade=True)
plt.vlines(mi.I, 0, 1, color='r')
plt.vlines(mi.EI, 0,1)
plt.xlabel("Moran's I")loyed a similar approach as we saw in
the join count analysis: random spatial permutations.
```

Here our observed TB geosampled value s were in the upper tail, and thus statistically significant: mi.p_sim. In addition to the Global autocorrelation statistics, PySAL has many local autocorrelation statistics [wwwpython.org]. We computed a local Moran statistic using the np.random.seed(12,345).

```
import esda
wq.transform = 'r'
lag_price = lp.weights.lag_spatial(wq, df['median_pri'])
wq
price = df['median_pri']
b, a = np.polyfit( TB county covariates lag1)
f, ax = plt.subplots(1, figsize=(9, 9))
b, a = np.polyfit(price, lag_, 1)
f, ax = plt.subplots(1, figsize=(9, 9))
plt.plot(price, lag_covariate '.',)
# dashed vert at mean of the covariates
plt.vlines(covariate.mean(), lag_covariate.min(), lag_
covariate.max(), linestyle='--')
# dashed horizontal at mean of lagged price
plt.hlines(lag_covariate.mean(), covaraite.min(),cobariate
.max(), linestyle='--')
# red line of best fit using global I as slope
plt.plot(covariate, a + b*covariate, 'r')
plt.title('Moran Scatterplot')
plt.ylabel('Spatial Lag of covariate')
plt.xlabel(' covariate')
plt.show()
```

Now, instead of a single I statistic, we had an array of local I statistics, stored in the. Is attribute, and p-values from the simulation were in p_sim which we determined using li = esda.moran.Moran_Local(y, wq) li.q. We again tested for local clustering using permutations, but here we employed conditional random permutations (different distributions for each focal location): (li.p_sim < 0.05).sum()We distinguished the specific type of local spatial association reflected in the four quadrants of the Moran Scatterplot.

The PySAL model output created multiple, customized, topographic, capture point, sentinel site, signature, georeferenced, county-level, grid-stratifiable LULC maps. The misspecification term was $S(T) = S(t)$ exp. Quantification of the interpretive land cover patterns was renderable from the distribution of the regressed, time-series, dependent, transmission-related, aggregation/non-aggregation-oriented, georeferenced, TB-related, hyper/hypoendemic, diagnostic, stratified estimator determinants which in this experiment was a requirement to describe independent key dimensions of the underlying uncertainty processes in the empirical, geosampled, scaled-up, sentinel site, county-level, georeferenced, CDC, county-level, epidemiologic, data capture points. We were able to define a pattern in the misspecification term. Python provided an efficient interactive tool for organizing and analyzing the stratified, aggregation/non-aggregation-oriented, epidemiological, hot/cold spot, time series, dependent, clinical, environmental and sociodemiographic, sentinel site, county-level, scaled-up, data, capture points that was georeferenced geo-spatiotemporally.

Next, an autoregressive, hyper/hypoendemic, epidemiologic prognosticative model was generated employing a non- stratified explanatory variable **Y**, as a function of a nearby regressor **Y** in the hot/cold spot, scaled-up model. A capture point, sentinel site, county-level, stratified, clinical, environmental or sociodemographic, potentially residually, eigenizable, geo-spatiotemporally specifiable, geosampled, TB, estimator determinant indicator covariate I (i.e., an autoregressive response) and the residual of **Y** were treated as a function of a nearby geosampled **Y** residual (i.e., an SAR or spatial error specification). For autoregressive, quantitative, time-series, dependent, prognosticative, vulnerability-oriented, hot/cold spot, epidemiologic modeling, the SAR model furnishes an alternative specification that frequently is written in terms of matrix **W** (Griffith 2005). Here, the spatial covariance of the potentially, eigenizable, hyper/hypoendemic, aggregation/non-agrgegation-oriented, empirical, stratified, scalable, georeferenceable epidemiologic, data was a function of the matrix $\left(\mathbf{I}-\rho\mathbf{C}\mathbf{D}^{-1}\right)\left(\mathbf{I}-\rho\mathbf{D}^{-1}\mathbf{C}\right) = \left(\mathbf{I}-\rho\mathbf{W}^{T}\right)(\mathbf{I}-\rho\mathbf{W})$, where T denoted the matrix transpose. The resulting matrix was symmetric and was considered a second-order specification as it included the product of two spatial structure matrices (i.e., $\mathbf{W}^{T}\mathbf{W}$) This matrix restricted, positive, discrete, integer values of the autoregressive parameter to the more intuitively interpretable range of $0 \leq \hat{\rho} \leq 1$. A Python code was generated for an AR model which was trained using statsmodels.tsa.ar_model as:

```
import matplotlib. pyplot as plt
# Load AutoReg class from statsmodels.tsa.ar_model modul
from statsmodels.tsa.ar_model import AutoReg
# Load and plot the time-series data
url='https://raw.githubusercontent.com/jenfly/opsd/.csv'
df = pd.read_csv(url,sep=",")
df['Consumption'].plot()
```

Euclidean distances between the georeferenced capture point, sentinel site, scaled-up, county-level, stratified, potential, georeferenced, hyper/hypoendemic, aggregation/non-aggregation-oriented, estimator determinants were definable in terms of an n-by-n geographic weights matrix, **C**, whose c_{ij} values were, 1 if the georeferenceable county-level, geolocations i and j were deemed nearby, and 0 otherwise. Adjusting this matrix by dividing each row entry by its row sum subsequently rendered **C1**, where **1** was an n-by-**1** vector of ones which converted the regression-based matrix to matrix **W**. The resulting AR model specification with no geosampled, scaled-up, capture point, sentinel site, estimator determinants (i.e., the pure autoregression specification) subsequently took on the following form: $\mathbf{Y} = \mu(1-\rho)\mathbf{1} + \rho\mathbf{W}\mathbf{Y} + \varepsilon$, where μ was the scalar conditional mean of **Y**, and ε was an n-by-1 error vector whose parameters were statistically i.i.d. "normalized" random variates. Spatial autoregressive models are fit using datasets that contain observations on geographical areas or on any units with a spatial representation [26]. Approximate standard errors for the stratifiable, eigenizable, county-level, specifiable, forecast-oriented, vulnerability, estimator determinant signature model was computable as the square roots of the diagonal elements of the estimated covariance matrix. The covariance matrix for analyzing the estimator determinants was spatially expressable employing $E\left[(\mathbf{Y}-\mu\mathbf{1})'(\mathbf{Y}-\mu\mathbf{1})\right] = \Sigma = \left[\left(\mathbf{I}-\rho\mathbf{W}'\right)(\mathbf{I}-\rho\mathbf{W})\right]^{-1}\sigma^{2}$ where $E(\bullet)$ designated the calculus of expectations, **I** was the n-by-n identity matrix denoting the matrix transpose operation and σ^2 was the non-asymptotical error variance [i.e., the variance of how far the non-homogenous, hyper/hypoendemic, aggregation/non-aggregation-oriented, prognosticated, TB-related, stratified, estimator determinants were spread out, geographically].

Subsequently, we employed a Hessian matrix for geo-spatiotemporally quantitating the empirical, county-level, geosampled dataset of georeferenced, stratified, scaled-up, hot/cold spot, capture point, sentinel site, evidential observational prognosticators.

Below isthe code for the Hessian we employed:

```
def hessianComp ( func, x0, epsilon=1.e-5):
f1 = scipy.optimize.approx_fprime( x0, func, epsilon=epsilon)
# Allocate space for the hessian
n = x0.shape[0]
hessian = np.zeros ( ( n, n ) )
# The next loop fill in the matrix
xx = x0
for j in range( n ):
xx0 = xx[j] # Store old value
xx[j] = xx0 + epsilon # Perturb with finite difference
# Recalculate the partial derivatives for this new point
f2 = scipy.optimize.approx_fprime( xx, func, epsilon=epsilon)
hessian[:, j] = (f2 - f1)/epsilon # scale...
xx[j] = xx0 # Restore initial value of x0
return hessian
```

As in Newton's method, we employed a second order approximation to find the minimum of a function $f(x)$ in the estimator determinant sampled dataset. We employed a Taylor series of $f(x)$ to generate $f(x_k + \Delta x) \approx f(x_k) + \nabla f(x_k)^T \Delta x + \frac{1}{2}\Delta x^T B \Delta x$ where (∇f) the gradient was s an approximation to the Hessian matrix. In mathematics, the Hessian matrix or Hessian is a square matrix of second-order partial derivatives of a function (Binmore and Davies 2007). Here, this matrix described the local curvature of a function of the scaled up, capture point, sentinel site, stratified, TB-related, geosampled, potentially, eigenizable estimator determinants. Given the function $f(x_1, x_2, \ldots, x_n)$, if all second partial derivatives of f existed and were continuous over the domain of the function, then the Hessian matrix of f in the prognosticative, epidemiologic, estimator, determinant, vulnerability-oriented, geo-spatiotemporal, county-level, TB model was parsimoniously derivable from $H(f)_{ij}(\mathbf{x}) = D_i D_j f(\mathbf{x})$, where $x = (x_1, x_2, \ldots, x_n)$ and D_i were the differentiation operator with respect to the i-th argument. The matrix rendered was $H(f) = \begin{bmatrix} \frac{\partial^2 f}{\partial x_1^2} & \frac{\partial^2 f}{\partial x_1 \partial x_2} & \cdots & \frac{\partial^2 f}{\partial x_1 \partial x_n} \\ \frac{\partial^2 f}{\partial x_2 \partial x_1} & \frac{\partial^2 f}{\partial x_2^2} & \cdots & \frac{\partial^2 f}{\partial x_2 \partial x_n} \\ \vdots & \vdots & \ddots & \vdots \\ \frac{\partial^2 f}{\partial x_n \partial x_1} & \frac{\partial^2 f}{\partial x_n \partial x_2} & \cdots & \frac{\partial^2 f}{\partial x_n^2} \end{bmatrix}$. We noted the determinant of the time series, dependent, sensitive matrix in our scaled-up, capture point, sentinel site, prognosticate epidemiologic, model was Hessian.

We then calculated the first derivative with numpy using the numpy.gradient() function.

```
data = np.random.rand(30,50,40,20)
first_derivative = np.gradient(TB county data)
```

We employed a 2D function to determine the 'flattest' area below a threshold in the TB prognosticative uncertainty model. The second derivatives were given by the Hessian matrix. Here a Python code was written as:

```
def hessian(x):
"""
Calculate the hessian matrix with finite differences
Parameters:
- x : ndarray
Returns:
an array of shape (x.dim, x.ndim) + x.shape
where the array[i, j, ...] corresponds to the second
derivative x_ij
"""
x_grad = np.gradient(x)
hessian = np.empty((x.ndim, x.ndim) + x.shape, dtype=x.dtype)
for k, grad_k in enumerate(x_grad):
# iterate over dimensions
# apply gradient again to every component of the first
derivative.
tmp_grad = np.gradient(grad_k)
for l, grad_kl in enumerate(tmp_grad):
hessian[k, l, :, :] = grad_kl
return hessian
x = np.random.randn(100, 100, 100)
hessian(x)
```

We were also interested in the magnitude of the second derivatives, so we derived the Laplace operator which was implemented by scipy.ndimage.filters.laplace. In mathematics, the Laplace operator or Laplacian is a differential operator given by the divergence of the gradient of a scalar function on Euclidean space [26]. In this experiment this operator was delineated as the trace (sum of diagonal elements) of the Hessian matrix. We revealed the difference in county results between the derivatives which in this experiment was coded as:

```
second_derivative_abs = np.abs(laplace(tb county data)) and
the minimum of the following:            second_derivative_abs =
np.zeros(data.shape) hess = hessian(data)
# based on the function description; would [-1] be more
appropriate?
for i in hess[0]: # calculate a norm
for j in i[0]:
second_derivative_abs += j*j
```

A derivative between the county-level Hessian matrices was generated for quantifying gradient in the, county-level, TB, epidemiological, stratified, estimator determinant model. The generator function for the scalable, prognosticative, county-level, epidemiological, TB-related, archived CDC, sample data was (1 − np.exp (−10*xi**2 − yi**2))/100.0 with xi, yi which was generated using np.meshgrid.

Hessian matrices are employable in large-scale optimization problems within Newton-type methods for epidemiological modelling because they are the coefficient of the quadratic term of a local Taylor expansion of a function (Jacob et al. 2018). In mathematics, the Taylor series or Taylor expansion of a function is an infinite sum of terms that are expressed in terms of the function's derivatives at a single point (Banner and Adrian 2007). The Taylor series of a real or complex-valued function $f(x)$, that is infinitely differentiable at a real or complex number a, is the power series (Kolk and Duistermaat 2010). The general form of the Taylor series in our epidemiological, signature county-level, scalable, capture point, sentinel site, estimator determinant, signature, county-level, epidemiological TB model was $f(x + a) = f(x) + af'(x) + (a^{2/2})f''(x) + \ldots + (a^{n/n!})d^n f(x)/dx^n + \ldots$ For the case of $f(x) = x^{(1/2)}$, for resolving a quadratic for the model the first two derivatives were $f(x) : f'(x) = (1/2)x^{(-1/2)} f''(x) = -(1/4)x^{(-3/2)}$. Thus, in the multivariate, prognosticative risk model, $y = \frac{1}{2}f\left(\mathbf{x}^T + \Delta\mathbf{x}\right) \approx f(\mathbf{x}) + J(\mathbf{x})\Delta\mathbf{x} + \Delta\mathbf{x}, H(\mathbf{x})\Delta\mathbf{x}$ where J was the Jacobian matrix was a vector (i.e., the gradient) for the scalar-valued functions. The full Hessian matrix can be difficult to compute in practice; in such situations, quasi-Newton algorithms have been developed that use approximations to the Hessian (Nocedal and Stephen 2000).

We were able to visualize the Hessian matrix as a gradient when it was applied a second time for each component of the first gradient calculated from the matrix Below is a python implementation defining the gradient in the county-level, epidemiological, TB, diagnostic, stratified, estimator determinant, prognosticative model using the Hessian matrix:

```
import numpy as np
#Gradient Function
def gradient_f(x, f):
assert (x.shape[0] >= x.shape[1]), "the vector should be a
column vector"
x = x.astype(float)
N = x.shape[0]
gradient = []
for i in range(N):
eps = abs(x[i]) * np.finfo(np.float32).eps
xx0 = 1. * x[i]
f0 = f(x)
x[i] = x[i] + eps
f1 = f(x)
gradient.append(np.asscalar(np.array([f1 - f0]))/eps)
x[i] = xx0
return np.array(gradient).reshape(x.shape)
#Hessian Matrix
def hessian (x, the_func):
N = x.shape[0]
hessian = np.zeros((N,N))
gd_0 = gradient_f( x, the_func)
eps = np.linalg.norm(gd_0) * np.finfo(np.float32).eps
for i in range(N):
xx0 = 1.*x[i]
x[i] = xx0 + eps
```

```
gd_1 = gradient_f(x, the_func)
hessian[:,i] = ((gd_1 - gd_0)/eps).reshape(x.shape[0])
x[i] =xx0
return hessian
```

Quasi-Newton methods are methods used to either find zeroes or local maxima and minima of functions, as an alternative to Newton's method (Broyden 1972). We employed quasi-Newton methods which are second-order optimization algorithms to approximate the inverse of the Hessian matrix for the scaled up, georeferenced, epidemiological, stratified, TB-related, county-level, estimator determinant, forecast model using the gradient, meaning that the Hessian and its inverse did not need to be calculated precisely for each step of the algorithm. Using scipy.optimize.newton we generated:

```
import numpy as np
import matplotlib.pyplot as plt
from scipy import optimize
def f(x): return (x**3 - 1) # only one real root at x = 1
We employed the secant method:
root = optimize.newton(f, 1.5)
root
1.0000000000000016
root = optimize.newton(f, 1.5, fprime2=lambda x: 6 * x)
root
1.0000000000000016
  Subsequently fprime  was provided, We used the Newton-Raphson
method.
root = optimize.newton(f, 1.5, fprime=lambda x: 3 * x**2)
roo1.0
Both fprime2 and fprime were provided. We used Halley's
method:
root = optimize.newton(f, 1.5, fprime=lambda x: 3 * x**2,
fprime2=lambda x: 6 * x)
root1.0
```

In numerical analysis, Halley's method is a root-finding algorithm used for functions of one real variable with a continuous second derivative [38]. We found roots for a set of related starting TB sampled diagnostic stratified values and/or function parameters. We provided them as an array of inputs:

```
f = lambda x, a: x**3 - a
fder = lambda x, a: 3 * x**2
rng = np.random.default_rng()
x = rng.standard_normal(100)
a = np.arange(-50, 50)
vec_res = optimize.newton(f, x, fprime=fder, args=(a, ),
maxiter=200)
```

The above was the equivalent of solving for each sampled county-level, geosampled, explanatory discrete integer value in (x, a) separately in a for-loop, just faster:

```
loop_res = [optimize.newton(f, x0, fprime=fder, args=(a0,),
maxiter=200)
for x0, a0 in zip(x, a)]
np.allclose(vec_res, loop_res)
True
```

We plotted the results found for all the geosampled county-level values:

```
analytical_result = np.sign(a) * np.abs(a)**(1/3)
fig, ax = plt.subplots()
ax.plot(a, analytical_result, 'o')
ax.plot(a, vec_res, '.')
ax.set_xlabel('$a$')
ax.set_ylabel('$x$ where $f(x, a)=0$')
plt.show()
```

In this experiment. the "full" Newton's method required the Jacobian to search for zeros, and the Hessian for finding extrema in the scaled-up, capture point, sentinel site, empirical, epidemiologic, estimator, determinant TB dataset. The quasi-Newton methods are among the most widely used methods for nonlinear optimization in regression-related, geosampled, vulnerability-oriented prognosticative epidemiological modeling (Nocedal and Stephen 1999). Subsequently, the Taylor series of a function $f(x)$ about a scaled-up, capture point, county-level, stratified clinical, sociodemographic or environmental, hyper/hypoendemic, aggregation/non-aggregation-oriented, TB-related, epidemiologic, sentinel site capture point a up to order n was found using Series[f, {x, a, n}]. The n th term of a Taylor series of a function f was computable in the Wolfram Language employing SeriesCoefficient [f, {x, a, n}] which was quantifiable by the inverse Z-transform $a_n = Z^{-1}\left[\frac{1}{z-a}\right](n)$. In mathematics and signal processing, the Z-transform converts a discrete-time signal, which is a sequence of real or complex numbers, into a complex frequency domain representation (Lynn 1986). We considered this product as a discrete-time equivalent of the Laplace transform.

In mathematics, the Laplace transform, is an integral transform that converts a function of a real variable (usually t, in the time domain) to a function of a complex variable s (in the complex frequency domain, also known as s-domain, or s-plane) (Hazewinkle 2000). We assumed the transform had many applications for optimally heuristically, optimizing, stratifiable, hyper/hypo-endemic, aggregation/non-aggregation-oriented, georeferenceable, county-level, TB-related, epidemiologic, scaled-up, sentinel site capture points as it is a tool for solving differential equations. In particular, it transforms ordinary differential equations into algebraic equations and convolution into multiplication.

We then coded the Taylor Series by writing out each county-level prognosticated model term individually. We combined these terms in a line of Python code to estimate e2. The code below calculated the sum of the first five terms of the Taylor Series expansion of ex*ex*, where x = 2. Note the math module needed to be imported before math. factorial() could be employed.

```
import math
x = 2
e_to_2 = x**0/math.factorial(0) + x**1/math.factorial(1)
+ x**2/math.factorial(2) + x**3/math.factorial(3) + x**4/
math.factorial(4)
print(e_to_2)
```

Our Taylor Series approximation of e2 w was calculable as 9.0. We compared our Taylor Series approximation to Python's math.exp() function. Python's math.exp() function raises to any power [www.python.org]. In our case, we wanted to use math.exp(2) as we wanted to calculate e2. For the diagnostically stratifiable, county-level, scalable, epidemiological, TB, prognosticative county-level model our Taylor Series approximation 9.0. was not that far off the calculated value 9.368764… using Python's exp() function. We employed a loop to calculate the Taylor Series Instead of coding each term in he epidemiological, county-level model individually, We generated a syntax for coding a loop in Python using the range() where the function was:

```
for <var> in range(<num>):
<code>
```

where < var> was any georeferenced, clinical, socioeconomic or environmental, stratified, variable name and <num > was any geosampled, discrete integer value that determined how many times the < code > in the loop ran. We recreated our approximation of e2 with 11 terms using a *for loop*. Note we needed to set the variable from_2 to 0 before the loop started. The mathematical operator + = in the line e_to_2 + = x**i/math.factorial(i) was equivalent to e_to_2 = e_to_2 + x**i/ math.factorial(i).The re sult 9.0 was the same as the result we calculated when we wrote out each model term of the Taylor Series individually.

The final Taylor series of the geosampled, epidemiological, county-level, stratified model functions here $\frac{1}{1-x} = \frac{1}{1-a} + \frac{x-a}{(1-a)^2} + \frac{(x-a)^2}{(1-a)^3} + \ldots \cos x = \cos a - \sin a(x - a) - \frac{1}{2}\cos a(x \quad a)^2 + \frac{1}{6}\sin a(x - a)^3 + \ldots, e^x = e^a\left[1 + (x-a) + \frac{1}{2}(x-a)^2 + \frac{1}{6}(x-a)^3 + \ldots\right], \ln x = \ln a + \frac{x-a}{a} - \frac{(x-a)^2}{2a^2} + \frac{(x-a)^3}{3a^3} - \ldots$ To derive the Taylor series of a function $f(x)$, we used the integral of the $(n+1)$ derivative $f^{(n+1)}$ of $f(x)$ from the sentinel site, geosampled, county-level, capture point x_0 to an arbitrary, georeferenced, hyper/hypoendemicTB-related, scaled-up, stratified, county-level, transmission-related, sentinel site geolocation, x given by $\int_{x_0}^{x} f^{(n+1)}(x)dx = \left[f^{(n)}(x)\right]_{x_0}^{x} = f^{(n)}(x) - f^{(n)}(x_0)$, where $f^{(n)}(x_0)$ was the n th derivative of $f(x)$ evaluated at x_0. Hence it was simply a constant. By integrating the diagnostic stratified, vulnerability-oriented, nant, prognosticative vulnerability-oriented county model a second time we achieved $\int_{x_0}^{x}\left[\int_{x_0}^{x} f^{(n+1)}(x)dx\right]dx$:

$$\begin{aligned}
&= \int_{x_0}^{x}\left[f^{(n)}(x) - f^{(n)}(x_0)\right]dx \\
&= \left[f^{(n-1)}(x)\right]_{x_0}^{x} - (x - x_0)f^{(n)}(x_0) \\
&= f^{(n-1)}(x) - f^{(n-1)}(x_0) - (x - x_0)f^{(n)}(x_0),
\end{aligned}$$

which was obtainable when $f^{(k)}(x_0)$ was a constant. The r in the prognosticative, epidemiologic model output revealed $\int_{x_0}^{x}\int_{x_0}^{x}\int_{x_0}^{x} f^{(n+1)}(x)(dx)^3 = f^{(n-2)}(x) - f^{(n-2)}(x_0) - (x - x_0)f^{(n-1)}(x_0) - \frac{(x-x_0)^2}{2!}f^{(n)}(x_0)$, which went up to $n+1$ integrations then rendered $\frac{\int\cdots\int_{x_0}^{x} f^{n+1}(x)(dx)^{n+1} = f(x) - f(x_0) - (x-x_0)f'(x_0)}{n'} - \frac{(x-x_0)^2}{2!}f''(x_0)$ Rearranging this equation generated the one-dimensional [1 – d] Taylor series $f(x) = f(x_0) + (x - x_0)f'(x_0) + \frac{(x-x_0)^2}{2!}f''(x_0) + \ldots + \frac{(x-x_0)^n}{n!}f^{(n)}(x_0) + R_n$. $\sum_{=k=0}^{n} \frac{(x-x_0)^k f^{(k)}(x_0)}{k!} + R_n$. Here, R_n was a remainder term. (i.e., the Lagrange remainder was $R_n = \underbrace{\int\cdots\int_{x_0}^{x}}_{n+1} f^{(n+1)}(x)(dx)^{n+1}$.

Rewriting the repeated integral rendered $R_n = \int_{x_0}^{x} f^{(n+1)}(t)\frac{(x-t)^n}{n!}dt$ Here $g(x)$ was the signature, epidemiological model output revealed $\int_{x_0}^{x} g(x)dx = (x - x_0)g(x^*)$ when $x^* \in [x_0, x]$.The Mean Value Theorem states if f is continuous on the closed interval [a,b] and differentiable on open interval (a,b), then exists a point in the interval (a,b) such that f'(c) is equal to the function's average rate of change over [a,b] (Jaynes and Edwin 2003).

Here integrating $N + 1$ rendered $R_n = \frac{(x-x_0)^{n+1}}{(n+1)!}f^{(n+1)}(x^*)$ Note the the Lagrange remainder R_n was definable in the Taylor series as a complex stratified, scaled-up, TB-related capture point, sentinel site estimator determinant based on the Cauchy integral formula;

$f(z) = \frac{1}{2\pi i}\int_C \frac{f(z')dz'}{z'-z} = \frac{1}{2\pi i}\int_C \frac{f(z')dz'}{(z'-z_0)-(z-z_0)} = \frac{1}{2\pi i}\int_C \frac{f(z')dz'}{(z'-z_0)\left(1-\frac{z-z_0}{z'-z_0}\right)}$. In the model C was $\frac{|z-z_0|}{|z'-z_0|} < 1$ so, hence it followed that $f(z) = \frac{1}{2\pi i}\int_C \sum_{n=0}^{\infty} \frac{(z-z_0)^n f(z')dz'}{(z'-z_0)^{n+1}} = \frac{1}{2\pi i}\sum_{n=0}^{\infty}(z - z_0)^n \int_C \frac{f(z')dz'}{(z'-z_0)^{n+1}}$ in the Cauchy integral $f(z) = \sum_{n=0}^{\infty}(z - z_0)^n \frac{f^{(n)}(z_0)}{n!}$.

A Taylor series of the stratified, capture point, sentinel site, estimator determinants as $f(x, y)$ hich was then generated by:

$$\begin{aligned}
f(x + \Delta x, y + \Delta y) = &f(x, y) + \left[f_x(x, y)\Delta x + f_y(x, y)\Delta y\right] \\
&+ \frac{1}{2!}\left[(\Delta x)^2 f_{xx}(x, y) + 2\Delta x\Delta y f_{xy}(x, y) + (\Delta y)^2 f_{yy}(x, y)\right] \\
&+ \frac{1}{3!}\Big[(\Delta x)^3 f_{xxx}(x, y) + 3(\Delta x)^2\Delta y f_{xxy}(x, y) \\
&+ 3\Delta x(\Delta y)^2 f_{xyy}(x, y) + (\Delta y)^3 f_{yyy}(x, y)\Big] + \cdots
\end{aligned}$$

This model series was further generalized for a real function in n variables, $f(x_1, \ldots, x_n) = \sum_{j=0}^{\infty} \left\{ \frac{1}{j!} \left[\sum_{k=1}^{n} (x_k - a_k) \frac{\partial}{\partial x'_k} \right]^j f(x'_1, \ldots, x'_n) \right\}_{x'_1=a_1,\ldots,x'_n=a_n}$ Rewriting the equation rendered $f(x_1 + a_1, \ldots, x_n + a_n) = \sum_{j=0}^{\infty} \left\{ \frac{1}{j!} \left(\sum_{k=1}^{n} a_k \frac{\partial}{\partial x'_k} \right)^j f(x'_1, \ldots, x'_n) \right\}_{x'_1=x_1,\ldots,x'_n=x_n}$. We immediately determined $f(x_1, x_2) = \sum_{j=0}^{\infty} \left\{ \frac{1}{j!} [(x_1 - a_1) \frac{\partial}{\partial x'_1} + (x_2 - a_2) \frac{\partial}{\partial x'_2}]^j f(x'_1, x'_2) \right\}_{x'_1=a_1 x'_2=a_2} = f(a_1, a_2) + [(x_1 - a_1) \frac{\partial f}{\partial x_1} + (x_2 - a_2) \frac{\partial f}{\partial x_2}] +$ which occurred in the, county-level, estimator determinant, prognosticated, TB model output $f(x_1 + a_1, x_2 + a_2, x_3 + a_3) = \sum_{j=0}^{\infty} \left\{ \frac{1}{j!} \left(a_1 \frac{\partial}{\partial x'_1} + a_2 \frac{\partial}{\partial x'_2} + a_3 \frac{\partial}{\partial x'_3} \right)^j f(x'_1, x'_2, x'_3) \right\}_{x'_1=x_1 x'_2=x_2 x'_3=x_3}$, which converted in vector form to $f(\mathbf{r} + \mathbf{a}) = \sum_{j=0}^{\infty} \left[\frac{1}{j!} (\mathbf{a} \cdot \nabla_{\mathbf{r}'})^j f(\mathbf{r}') \right]_{\mathbf{r}'=\mathbf{r}}$ The zeroth- and first-order terms were $f(\mathbf{r})$ and $(\mathbf{a} \cdot \nabla_{\mathbf{r}'}) f(\mathbf{r}')\big|_{\mathbf{r}'=\mathbf{r}}$, respectively. The second-order term in the model was $\frac{1}{2}(\mathbf{a} \cdot \nabla_{\mathbf{r}'})(\mathbf{a} \cdot \nabla_{\mathbf{r}'}) f(\mathbf{r}')\big|_{\mathbf{r}'=\mathbf{r}} = \frac{1}{2}\mathbf{a} \cdot \nabla_{\mathbf{r}'} \left[\mathbf{a} \cdot (\nabla f(\mathbf{r}')) \right]_{\mathbf{r}'=\mathbf{r}} = \frac{1}{2}\mathbf{a} \cdot \left[\mathbf{a} \cdot \nabla_{\mathbf{r}'} (\nabla_{\mathbf{r}'} f(\mathbf{r}')) \right]\big|_{\mathbf{r}'=\mathbf{r}}$, so the first few terms of the expansion were expressable as $f(\mathbf{r} + \mathbf{a}) = f(\mathbf{r}) + (\mathbf{a} \cdot \nabla_{\mathbf{r}'}) f(\mathbf{r}')\big|_{\mathbf{r}'=\mathbf{r}} + \frac{1}{2}\mathbf{a} \cdot \left[\mathbf{a} \cdot \nabla_{\mathbf{r}'} (\nabla_{\mathbf{r}'} f(\mathbf{r}')) \right]\big|_{\mathbf{r}'=\mathbf{r}} f(\mathbf{r} + \mathbf{a}) = f(\mathbf{r}) + (\mathbf{a} \cdot \nabla_{\mathbf{r}'}) f(\mathbf{r}')\big|_{\mathbf{r}'=\mathbf{r}} + \frac{1}{2}\mathbf{a} \cdot \left[\mathbf{a} \cdot \nabla_{\mathbf{r}'} (\nabla_{\mathbf{r}'} f(\mathbf{r}')) \right]\big|_{\mathbf{r}'=\mathbf{r}}$.

As previously mentioned, in vector calculus, the Jacobian matrix is the matrix of all first-order partial derivatives of a vector-valued function. Specifically, suppose $F : \mathbb{R}^n \to \mathbb{R}^m$ is a function which takes as input real n-tuples and produces as output for an empirical geosampled dataset of stratifiable, TB-related m-tuples. Such a function here was given by m, real-valued, component functions, $F_1(x_1, \ldots, x_n), \ldots, F_m(x_1, \ldots, x_n)$. The partial derivatives of all the functions with respect to the non-asymptotical, hyper/hypoendemic, transmission-oriented, prognosticative county level variables $x_1, \ldots, x_n$ was organized in an m-by-n matrix $J(x) = \begin{bmatrix} \frac{\partial f_1(x)}{\partial x_1} & \frac{\partial f_1(x)}{\partial x_2} & \cdots & \frac{\partial f_1(x)}{\partial x_n} \\ \frac{\partial f_2(x)}{\partial x_1} & \frac{\partial f_2(x)}{\partial x_2} & \cdots & \frac{\partial f_2(x)}{\partial x_n} \\ \vdots & \vdots & \ddots & \vdots \\ \frac{\partial f_n(x)}{\partial x_1} & \frac{\partial f_n(x)}{\partial x_2} & \cdots & \frac{\partial f_n(x)}{\partial x_n} \end{bmatrix}$ whereby, the Jacobian matrix J of F, was $J = \begin{bmatrix} \frac{\partial F_1}{\partial x_1} & \cdots & \frac{\partial F_1}{\partial x_n} \\ \vdots & \ddots & \vdots \\ \frac{\partial F_m}{\partial x_1} & \cdots & \frac{\partial F_m}{\partial x_n} \end{bmatrix}$. The matrix entries were then functions of $x_1, \ldots, x_n$, which were expressable by $J_F(x_1, \ldots, x_n)$ and $\frac{\partial(F_1, \ldots, F_m)}{\partial(x_1, \ldots, x_n)}$. We employed the Harvard autograd library where grad and Jacobian took a function as their argument:

```
x = np.array([5,3], dtype=float)
def cost(x):
return x[0]**2 / x[1] - np.log(x[1])
gradient_covariate = grad(cost)
jacobian_covariate = jacobian(cost)
gradient_covariate(x)
jacobian_covaraiet(np.array([x,x,x]))
```

We used the Jacobian method available for matrices in sympy:

```
from sympy import sin, cos, Matrix
from sympy.abc import rho, phi
X = Matrix([rho*cos(phi), rho*sin(phi), rho**2])
Y = Matrix([rho, phi])
X.jacobian(Y)
```

In numerical analysis, the secant method is a root-finding algorithm that employs a succession of roots of secant lines to better approximate a root of a function f (Mordecai 1976). Each improvement in our prognosticative epidemiologic TB model was taken at the scaled-up, capture point, sentinel sites where the approximating line crossed the axis. We employed an approximating line based on georeferenced, scaled-up, hyper/hypoendemic, aggregation/non-aggregation-oriented, capture point, county-level, computable, discrete integrals. We employed two root estimations of root α, x_0 and x_1. Subsequently, we generated an epidemiologic, specified, TB-related, estimator determinant, linear function $q(x) = a_0 + a_1 x$ using $q(x_0) = f(x_0)$, $q(x_1) = f(x_1)$. This line was a secant line. The secant method was implementable employing the Wolfram Language when:

```
SecantMethodList[f_, {x_, x0_, x1_}, n_] :=
NestList[Last[#] - {0, (Function[x, f][Last[#]]*
Subtract @@ #)/Subtract @@ Function[x, f] /@ #}&, {x0, x1}, n
```

We described these data employing:

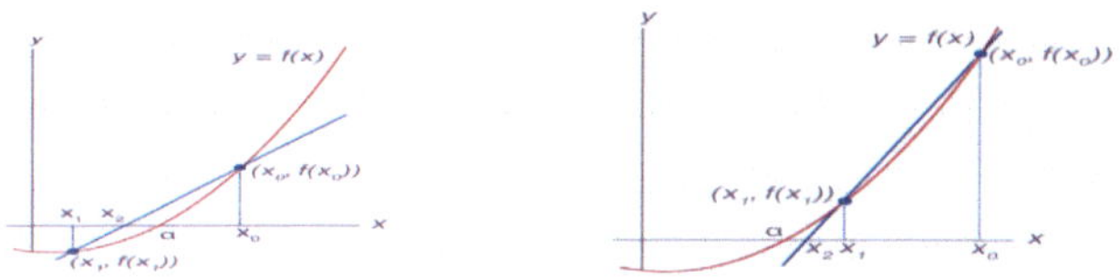

Since the initial, stratified, georeferenced, capture point, sentinel site, potentially, eigenizable, estimator determinant, county-level, hyper/hypoendemic, geosampled, county-level, scalable, discrete, integer values x_0 and x_1 were close enough to the root, the secant method iterated x_n and converged to a root of function f. The order of convergence was given by φ. The convergence was superlinear. This solution unfortunately was only valid under certain technical requirements, such as f being two times continuously differentiable and the root being simple (i.e., having multiplicity 1) in the capture point, sentinel site, epidemiological, TB-related, prognosticated model output.

An approximate initial value of $B_0 = I * x$ was subsequently employable to achieve rapid convergence in the stratified, epidemiologic vulnerability-oriented, prognosticative county model. We obtained explicit non-asymptotic rates of superlinear convergence for the standard quasi-Newton methods, which were based on the updating formulas from the Broyden class. for the epidemiologic, county-level, TB vulnerability model, we obtained the rates respectively where k was the iteration counter, n was the dimension of the problem, μ was the strong convexity parameter, and L was the Lipschitz constant of the gradient. The unknown x_k

was updated applying the Newton's step which we calculated using the Hessian matrix $B_k \Delta x_k = -\alpha_k B_k^{-1} \nabla f(x_k)$ with α chosen to satisfy the Wolfe conditions: $x_{k+1} = x_k + \Delta x_k$. Pyxu operators equipped with lipschitz and diff_lipschitz attributes stored the Lipschitz constants of the county-level TB, signature maps and their derivatives respectively. These constants were leveraged under the hood by Pyxu to auto-tune the step sizes in various optimization algorithms employing:

```
# Access Lipschitz constant of an operator 'op
L = op.lipschitz
```

We estimated the Lischiltz constants for the user-defined or complicated operators where Lipschitz constant were unknown, so we could estimate them using the estimatelipschitz() method.

The Frobenius and spectral norms were subsequently generated by:

```
from pyxu.abc import LinOp
import numpy as np
rand_op = LinOp.from_array(np.random.random((10000, 10000)))
[2]:
rand_op.lipschitz
[2]:
inf
[3]:
%%timeit
rand_op.lipschitz = rand_op.estimate_lipschitz(method="trace")
1.12   s ± 91.1 ms per loop (mean ± std. dev. of 7 runs, 1 loop
each)
[4]:
rand_op.lipschitz # Rough estimate
[4]:
5773.825067429628
[5]:
%%timeit
rand_op.lipschitz = rand_op.estimate_lipschitz(method="svd",
tol=1e-2)
1.38    s ± 220 ms per loop (mean ± std. dev. of 7 runs, 1 loop
each)
rand_op.lipschitz # Tighter estimate
4999.816657444895
rand_op = LinOp.from_array(np.random.random((10000, 10000)))
[2]:
rand_op.lipschitz
[2]:
inf
[3]:
%%timeit
rand_op.lipschitz = rand_op.estimate_lipschitz(method="trace")
1.12   s ± 91.1 ms per loop (mean ± std. dev. of 7 runs, 1 loop
each)
[4]:
rand_op.lipschitz # Rough estimate
[4]:
5773.825067429628
```

```
[5]:
%%timeit
rand_op.lipschitz = rand_op.estimate_lipschitz(method="svd",
tol=1e-2)
1.38   s ± 220 ms per loop (mean ± std. dev. of 7 runs, 1 loop
each)
rand_op.lipschitz # Tighter estimate
4999.816657444895
```

While the propagated constants are usually good enough for step size, mcall estimate_lipschitz() on arithmetic-produced operators force-compute a tighter Lipschitz constant [www.python.org].

```
op = op1 * op2 # arithmetic-induced operator
op.lipschitz # => 50 (example value from cheap Lipschitz
propagation.)
op.lipschitz = op.estimate_lipschitz() # => 3 (re-compute a
Lipschitz constant
```

The operationizable, prognosticated, epidemiologic, risk model was subsequently written as: $X_t = c + \sum_{i=1}^{p} \varphi_i X_{t-i} + \varepsilon_t$ where $\phi_1, \ldots, \phi_p$ were the semi-parameterizable, empirical, diagnostically stratifiable, clinical, environmental and socio-demographic, TB-related, forecasted, eigenizable, epidemiological estimator determinants, c was a constant and ε_t was the white noise. When coupled with regression and the normal probability, an autoregressive specification resulted in a covariation term characterizing latent, non-zero, eigen-autocorrelation by denoting the autoregressive parameter with ρ at a conditional, autoregressive, covariance specification in the stratified, estimator, determinant, prognosticated, vulnerability model output. Here this specification involved the matrix $(\mathbf{I} - \rho\mathbf{C})$ where $\mathbf{I}$ was an n-by-n identity matrix. In an autoregressive expression: however, the optimal response variable is on the left-side of the equation, while the spatial lagged version of the variable is on the right side (Griffth 2003). Therefore, one of the main objectives in this experiment was to bring the unlagged, endogenous, scaled-up, capture point, county-level, sentinel site, prognosticative, stratified, TB, clinical, sociodemographic and environmental explanatory variable, y, exclusively to the left-hand side of the regression equation to decorrelate the stratified, georeferenced, geosampled, estimator determinants. This was accomplished by expanding the weighted regression coefficient matrix term: $(I - \rho V)^{-1} = \sum_{k=0}^{\infty} \rho^k V^k$ as an infinite power series which in this experiment was conducted successfully under the assumption that the underlying spatial process in the asymptotical, estimator, determinant, empirical dataset was temporally non-stationary. The simultaneous autoregressive error model was subsequently rewritten as $y - \rho Vy = X\beta - \rho VX\beta + \varepsilon$. in pandas Series.autocorrn. This method computed the Pearson correlation between the diagnostically stratified, georeferenced, county-level, capture point, sentinel sites and its shifted self. The procedure rendered robust, epidemiologic forecasts of geolocations of the georeferenced, county-level, hyper/hypo-endemic, aggregation/non-aggregation-oriented, potentially eigenizable, estimator determinant, scaled-up, capture point, sentinel site, geolocations and confidence limits with future stratifiable, discrete integer, values of the independent variables included in the stratified output dataset. Our probabilistic, county-level, autoregressive, error model considered all departures from the trend in producing the forecasts [i.e., georeferenceable, capture point, sentinel site, TB-related, hyper/hypoendemic, aggregation/non-aggregation-oriented geolocations and their respective estimator determinants].

3 Eigen-Bayesian Semi-parametric GARCH Autocorrelation Approximation Matrix

We generated a fully eigen-Bayesian treatment of the probabilistic matrix factorization model in which model capacity was controlled automatically by integrating over the geosampled, empirical, county-level, TB-related, model parameters and hyperparameters. We introduced priors for the hyperparameters and maximized the log-posterior of the model over both parameters and hyperparameters which allowed model complexity to be controlled automatically based on the training data [i.e., capture point, hyper/hypoendemic, sentinel site, scaled-up, epidemiological, stratified, TB-related, county-level, aggregation/non-aggregation-oriented, eigenized, estimator determinants]. We employed a joint optimization technique to predictively geolocate, eigen-orthogonalized, county-level, georeferenceable, TB-related, hyper/hypoendemic, aggregation/non-aggregation, geo-spatiotemporally autocorrelated, estimator determinant, sentinel site, scalable county-level capture points.

We began with a "prior distribution" based on the relative likelihoods of the stratified, hyper/hypoendemic, clinical, sociodemographic and environmental county-level, time series, dependent, TB-related, estimator determinants. In practice, it is common to assume a uniform distribution over the appropriate range of values for the prior distribution [25]. Next, we calculated the likelihood of the observed non-asymptotic distribution as a function of the semi-parameterizable, potential, eigen-orthogonalizable, estimator, determinant, capture point, stratified, sentinel site, geosampled, discrete integer values by multiplying the likelihood function by the prior distribution which was subsequently normalized to obtain a unit probability over all possible independent, simulated, estimator determinants (i.e., posterior distribution). The mode of the distribution approximated probability intervals in the empirical, estimator determinant, county-level dataset. In geo-spatiotemporal eigen-Bayesian inference, a probability interval is a probabilistic region around a posterior moment and is similar in use to a frequentist CI [44].

Garc is a python library and command line tool for collecting JSON data from Gab.com. In July 2019, Gab switched its platform to a fork of Mastodon. This package was originally written prior to the switch, but has been updated to work with the new functionality of Mastodon. Garc which is built based on the twarc project published by the Documenting the Now project. Inspiration for structure, usage and outputs are from twarc, and garc is intended to be used for similar purposes. Garc is still very much a work in progress, and is being constantly updated to add deeper functionality and as new features and changes are implemented by Gab. Gab's API was historically relatively sparsely documented, so things may change without warning and break searches. This current version relies on these potentially unreliable API routes.

In our semi-parametric, Markovian, eigen-Bayesian, county-level, prognosticative, epidemiological, TB model, θ was a parameter that was unknown and had to be inferred from the geosampled, capture point epidemiologic data. The joint probability of the data was optimally quantifiable by employing: $p(\mathbf{X}|\theta) = p(x_1, \ldots x_n|\theta) = \prod_{i=1}^{n} p(x_i|\theta)$; whereby, $p(\mathbf{X}|\theta, \alpha) = p(\mathbf{X}|\theta)$ and $p(x_i \mid \theta,\ \alpha) = p(x_i \mid \theta)$ was conditionally independent of the hyperparameters. Eigen-Bayesian Markovian inference determined the posterior distribution of the parameter $p(\theta|\mathbf{X}, \alpha)$ by employing:

$$\begin{aligned} p(\theta\ |\mathbf{X}, \alpha) &= \frac{p(\theta, \mathbf{x}, \alpha)}{p(\mathbf{x}, \alpha)} \\ &= \frac{p(\theta, \mathbf{x}, \alpha)}{\int_{\theta} p(\theta, \mathbf{x}, \alpha) d\theta} \\ &= \frac{p(\mathbf{x}|\ \theta, \alpha) p(\theta \mid \alpha)}{\int_{\theta} p(\mathbf{X}|\ \theta, \alpha) p(\theta \mid \alpha) d\theta} \\ &= \frac{p(\mathbf{x}|\ \theta) p(\theta \mid \alpha)}{\int_{\theta} p(\mathbf{x}|\ \theta) p(\theta \mid \alpha) d\theta} \\ &= \frac{\left[\prod_{i=1}^{n} p(x_i|\ \theta)\right] p(\theta \mid \alpha)}{\int_{\theta} \left[\prod_{i=1}^{n} p(x_i|\ \theta)\right] p(\theta \mid \alpha) d\theta}. \end{aligned}$$

For the fixed regression parameters, a suitable choice was the diffuse prior, (i.e., $p(\gamma)$ const., but a weakly informative Gaussian prior was also possible). A second-order Gaussian random walk prior was employed to allow enough flexibility while penalizing abrupt changes in the function. The second-order random walk model is commonly used for smoothing data and for quantifying response functions as it is computationally efficient due to the Markov properties of the joint (intrinsic) Gaussian density (Gelman et al. 2013).

From a BayesPy plot generated, it was apparent that less of the upper tail was includable around the highest posterior density in the quantile-based probability interval. The high probability distribution (HPD) interval does not produce equal tails when inappropriate [26]. However, both the quantile-based probability interval and HPD interval did take the prior distribution into account in the epidemiologic, prognosticated, empirical, TB model output. [i.e., potential, georeferenced, hyper/hypoendemic, transmission-related, scaled-up, county-level, vulnerability-oriented, aggregation/non-aggregation, stratified, estimator, determinant, sentinel site capture points]. These intervals represented the eigen-Bayesian analogue of CIs in the model.

Subsequently, we estimated the interpolated, scaled-up, mixed, county-level, hyper/hypoendemic, capture point, sentinel site, vulnerability-oriented, prognosticative, stratified, model estimator determinants by employing a Laplace quadrature.

The recently developed quadrature by expansion (QBX) technique accurately evaluated the model layer potentials with singular, hypersingular kernel in the integral equation reformulations of multiple partial differential equations. A statement was representable by the differential equation $y' = ky(100 - y)$ and initial condition $y(1) = 1$ where y' = the rate of change of the number of georeferenceable, hyper/hypoendemic, aggregation/non-aggregation-oriented geolocations in a county with respect to time y = the number of potential infected people; t = time in days; and k = the constant of proportionality.

The GARCH model and its variants were subsequently employed to model and correct for geo-spatiotemporal heteroscedasticity. The georeferenced empirical dataset of scaled-up, capture point, sentinel site, prognosticative, eigenizable, estimator determinant, GARCH model with autoregressive errors was illustratable as;

$$
\begin{aligned}
y_t &= \mathbf{x}_t'\beta + v_t \\
v_t &= \varepsilon_t - \varphi_t v_{t-1} - \ldots - \varphi_m v_{t-m} \\
\varepsilon_t &= \sqrt{h_t} e_t \\
h_t &= \omega + \sum_{i=1}^{q} \alpha_i \varepsilon_{t-i}^2 + \sum_{j=1}^{p} \gamma_j h_{t-j} \\
e_t &\sim \mathrm{IN}(0, 1)
\end{aligned}
$$

This model combined the mth-order autoregressive error model with the GARCH (p, q) variance model. The paradigm was denotable as the AR (m)-GARCH (p, q) regression model. The tests for the presence of ARCH effects (namely, Q and LM tests, from [56] and from Wong and Li (1995) helped determine the order of the model appropriate for the scalable, county-level, hyper/hypoendemic, aggregation/non-aggregation-oriented, geosampled, stratified, capture point, sentinel site, georeferenced, empirical, epidemiological, geo-spatiotemporal data. For example, the Lagrange multiplier (LM) tests examined here were significant ($p < 0.0001$) which indicated that a very high-order ARCH model was needed to model the heteroscedasticity. The basic ARCH (q) model $(p = 0)$ is a short memory process in that only the most recent q squared residuals are used to estimate the changing variance [25].

Generalized autoregressive conditional heteroskedasticity [GARCH] model is a statistical model for time series data that describes the variance of the current error term or innovation as a function of the actual sizes of the previous time periods' error terms |(Box and Jenkins 1985); often the variance is related to the squares of the previous innovative. When the potentially eigenizable, time series dependent, scaled-up, county-level, stratified, TB-related, estimator, determinant, epidemiologic data was used in the regression analysis, the error term was not independent through time. Here the errors in the capture point, sentinel site model were serially correlated (autocorrelated). If the error term is autocorrelated the efficiency of OLS parameter estimates is adversely affected, and standard error estimates are biased (Griffith 2003). The eigenizable, county-level, stratified, TB-related, determinant models did not require the estimation of many lag coefficients. The autoregressive conditional heteroskedasticity model was designed to improve models by replacing assumptions of constant volatility with conditional volatility [26]. Engle and others working on GARCH models recognized that past data influences future data—that is the definition of autoregressiveness. In our prognosticative, epidemiologic, vulnerability-related, signature, county-level, scalable, TB model the conditional heteroscedasticity portion of the observable fact was that volatility in the paradigm was nonconstant. Substituting this transformation employing the geosampled, hyper/hypoendemic, aggregation/non-aggregation-oriented, stratified, clinical, sociodemiographic and environmental, potentially, geo-spatiotemporally eigen-geospatially, filterable estimator determinants rendered:

$$
\begin{aligned}
y &= (I - \rho V)^{-1}[X\beta - \rho V(X\beta) + \varepsilon], \\
y &= \sum_{k=0}^{\infty} \rho^k V^k (X\beta - \rho V X\beta + \varepsilon), \\
y &= \sum_{k=0}^{\infty} \rho^k V^k X\beta - \sum_{k=0}^{\infty} \rho^{k+1} V^{k+1}(X\beta) + \sum_{k=0}^{\infty} \rho^k V^k \varepsilon, \\
y &= X\beta + \underbrace{\sum_{k=1}^{\infty} \rho^k V^k X\beta - \sum_{k=1}^{\infty} \rho^k V^k (X\beta)}_{=0} + \sum_{k=0}^{\infty} \rho^k V^k \varepsilon, \\
y &= X\beta + \underbrace{\sum_{k=1}^{\infty} \rho^k V^k \varepsilon}_{\textit{misspecification term}} + \varepsilon.
\end{aligned}
$$

The misspecification term $\sum \rho^k V^k \varepsilon$ $(k = 1, \ldots, \infty)$ meanwhile remained uncorrelated with the georeferenced, diagnostic, stratified, hot/cold spot, county-level, geosampled, transmission-related, estimator determinants. We employed variable, X, as the standard OLS assumption of the disturbances, where ε was independent of the non-asymptotical, stratified, aggregation/non-aggregation-oriented, potentially, eigenizable, estimator determinant, hyper/hypoendemic, prognosticative variables generated from the estimation process. We noted that the spatial lag model was expressable as $(I - \rho V)y = X\beta + \varepsilon$. Substituting the transformation rendered:

$y = \sum_{k=0}^{\infty} \rho^k V^k (X\beta + \varepsilon)$ and $y = X\beta + \underbrace{\sum_{k=1}^{\infty} \rho^k V^k (X\beta + \varepsilon)}_{\textit{misspecification term}} + \varepsilon$. The misspecification term $\sum \rho^k V^k (X\beta + \varepsilon)$ $(k = 1, \ldots, \infty)$ included the exogeneous variables X. Consequently, the stratified, geosampled, hyper/hypoendemic, transmission-oriented, capture point, county-level, = TB-related, stratifiable, hot/cold spot, estimator determinant estimators obtained employing the new time-series, dependent, prognosticative variables were correlated with the misspecification term. Under this condition, standard OLS results for the basic regression model $y = X\beta + \varepsilon^*$ generated from the geosampled, potentially, eigenizable, estimator determinants provided biased estimates $\hat{\beta}$ of the underlying regression parameters β.

The correlation, or lack thereof, between the georeferenced, temporal heteroscedastic, non-Gaussian, hyper/hypoendemic, transmission-oriented, epidemiological, geosampled, stratified, clinical, environmental and sociodemographic, aggregation/non-aggregation-oriented, prognosticative variables and the misspecification terms in the scaled-up, stratified, autoregressive, county-level, risk model were employed to design spatial proxy variables so the properties of either model could be satisfied. Misspecification of the main exposure variable and other eigen-spatial, filterable, geo-spatiotemporal, estimator determinants may not be uncommon in prognosticative, time–series, dependent, hyper/hypoendemic, multivariate, TB-related, estimator determinant, county-level, vulnerability regression models.

Our first step for constructing a robust, georeferenceable, time–series, dependent, eigen-geospatially filterable, hyper/hypoendemic, epidemiologic, prognosticative, TB-related, county-level, risk model in Python –related, covariance matrix was to specify the uncertainty in the prognosticative, vulnerability model outpu In this experiment the multivariate normal distribution accepted an array representing a covariance matrix using the following code:

from scipy import stats as:

```
import numpy as np
d = [1, 2, 3]
A = np.diag(d) # a diagonal covariance matrix
x = [4, -2, 5] # a point of interest
dist = stats.multivariate_normal(mean=[0, 0, 0], cov=A)
dist.pdf(x)
4.9595685102808205e-08
```

The calculations were performed in a very generic way that did not take advantage of any special properties of the covariance matrix. Because our covariance matrix was diagonal, we used "Covariance.from_diagonal" to create an object representing the covariance matrix, and multivariate normal. We used this methodology to compute the probability density function more efficiently. Log of the pseudo-determinant of the covariance matrix was generated thereafter using log_pdet:

```
cov = stats.Covariance.from_diagonal(d)
dist = stats.multivariate_normal(mean=[0, 0, 0], cov=cov)
dist.pdf(x)
4.9595685102808205e-08
```

If an estimated iterative, interpolatable, time-series, dependent, aggregation/non-aggregation-oriented, semi-parameterizable, georeferenceable, estimator determinant, vulnerability-oriented model output is mis -pecified all forecasted values will be biased and inconsistent (Griffith 2003). In operationalizable, regression-based, vulnerability-oriented, epidemiologic, empirical, TB-related, prognosticative error-in variable models, the term misspecification covers a broad range of modeling uncertainties including discretizing, geosampled, continuous, prognosticative variables [41].

We considered two different projection matrices, $M_{(1)} \equiv I - 1\left(1^T 1\right)^{-1} 1^T$ and $M_{(X)} \equiv I - X\left(X^T X\right)^{-1} X^T$. for optimally, heuristically, optimizing, the georeferenced, geosampled, time series, dependent, capture point, potentially, geo-spatiotemporal, eigenizable, estimator, determinant, empirical data. The projection matrix $M_{(1)}$ is a special case of the more general projection matrix $M_{(X)}$ [42]. The general projection matrix $M_{(X)}$ in the model included constant unity vector 1 and multiple, additional, operationalizable, stratified explanatory regressors. An eigendecomposed dataset of eigenfunction eigenvectors $\{e_1, \ldots, e_n\}_{SAR}$ was subsequently extractable from the regressed quadratic.

$$\{e_1, \ldots, e_n\}_{SAR} \equiv evec\left[M_{(X)} \frac{1}{2}\left(V + V^T\right) M_{(X)}\right], \tag{1}$$

which was designed eigen-orthogonal to the exogeneous variable X. The projection matrix $M_{(X)}$ imposed this constraint. In contrast, the set of eigenvectors $\{e_1, \ldots, e_n\}_{Lag}$ was extractractable from

$$\{e_1, \ldots, e_n\}_{Lag} \equiv evec\left[M_{(1)} \frac{1}{2}\left(V + V^T\right) M_{(1)}\right]. \tag{2}$$

These two different sets of eigenvectors were employed to establish a basis for constructing an unbiased, regression-based, prognosticative distribution model. Both expressions were solely definable in terms of the potentially eigen-spatially filterable, georeferenced, capture point, sentinel site, county-level, estimator, determinant information in the empirical geosampled county-level, CDC dataset. This model feature in Python enabled us to employ the eigen-spatial filtering approach for deriving precise hyper/hypoendemic, aggregation/non-aggregation-oriented prognostications from the archived, time series, dependent, regressed sentinel site, capture point, county-level, risk-related, hyper/hypoendemic, aggregation/non-aggregation-oriented, stratified, estimator determinants y in eigenvector eigen-Bayesian eigen-geospace. The associated sets of eigenvalues $\{\lambda_1, \ldots, \lambda_n\}_{Lag}$ and $\{\lambda_1, \ldots, \lambda_n\}_{SAR}$, with $\lambda_i \geq$

λ_{i+1}, ranges were analyzed in lagplot() function provided by pandas for properly standardizing adjacent link matrices V which in this experiment was cartographically relatable to the irregular discrete tessellations rendered from the prognosticated, clinical, environmental and sociodemiographic, hyper/hypoendemic, aggregation/non-aggregation-oriented, georeferenced, stratified, transmission-related, geosampled, TB specified, eigenized, estimator determinants.

We generated non-zero, geo-spatiotemporal, latent, eigen-autocorrelation indices in this experiment employing the geosampled, estimator determinants derived from the georeferenced, county-level, scaled-up, stratified, TB-related epidemiologic, data capture points using Moran's indices (I). In simple terms, it was a way to quantify how closely the discrete values were clustered together in a 2-D space. Python measured how closely the clustered TB related covariate different features are on a map like household income, level of education, e tc scanpy.metrics.morans_i]. To calculate Moran's I we employed.

```
(N/W)*ΣΣwij(xi-x)(xj-x)/Σ(xi-x)2 where:
N: The number of county units indexed by i and j
W: The sum of all wij
x: The variable of interest (household income, years of
schooling, etc.)
x: The mean of x
wij: A matrix of spatial weights
```

We assumed that the simple terms could quantify how closely the sampled, eigenized, estimator determinant, geosampled, epidemiological, discrete, integer values were clustered together in a 2-D space. The upper and lower bounds for our eigen-decomposition, model spatial matrix was constructable employing Morans I which here was given by $\lambda_{\max}(\mathrm{n}/1^{\mathrm{T}}\mathrm{W}\ \ 1)$ and $\lambda_{\min}(\mathrm{n}/1^{\mathrm{T}}\mathrm{W}1)$ where $\lambda_{\max}$ and $\lambda_{\min}$ were the extreme eigenvalues of $\Omega = HWH$.

The components of each aggregation/non-aggregation-oriented, geo-spatiotemporal, dependent, TB-related, hyper/hypoendemic, eigendecomposed, capture point, sentinel site, non-zero, eigen-orthogonalized non-frequentistic eigenvectors e_i, where subsequently mapped in oscanpy.metrics.morans_into an underlying geo-spatiotemporal discrete tessellation which exhibited a distinctive topographic pattern ranging from positive spatial autocorrelation [PSA] (i.e., similar values of log-transformed, georeferenced, hyper/hypoendemic, aggregation/non-aggregation-oriented, stratified, count data clustering in eigenvector eigen-Bayesian eigen-geospace) for $\lambda_i > E(I)$ to negative spatial autocorrelation [NSA] (i.e., dissimilar log-values clustering in eigen-geospace) for $\lambda_i > E(I)$. Each geosampled, county-level hyper/hypoendemic, aggregation/non-aggregation-oriented, eigen-spatial filtered, eigenvector estimator determinant, was mappable where $E(I)$ was the expected value of Moran's I under the assumption of (a) spatial independence and (b) as outputs from related projection matrices $M_{(1)}$ or $M_{(X)}$, respectively.

Interestingly, we noted that the associated Moran's I eigen-autocorrelated, eigen-decomposed, stratified, discrete integer values of each eigen-spatial filter, eigenvector e_i rendered from the scaled-up, county-level, time–series, dependent, sentinel site, capture point, epidemiologic, TB-related, vulnerability-oriented, eigen-Bayesian Markovian, prognosticative. semi-parametric eigen-algorithmic model was equal to its associated eigenvalue $\lambda_i = \left[e_i^T\left(V + V^T\right)e_i\right]/\left(2e_i^T e_i\right)$, but only if V was scalable to satisfy $\left[1^T\left(V + V^T\right)1\right]/2 = n$. Moran's eigen-autocorrelation is an extension of Pearson's product moment correlation coefficient (Griffith 2003) which in this experiment was denotable as I to measure the amount of latent zero/non-zero autocovariance in the multivariate, geosampled, county-level, TB stratified empirical datasets of georeferenced, aggregation/non-aggregation-oriented, hyper/hypoendemic, TB-related, non-asymptotically eigenizable, eigen-spatial, filter, stratified, estimator, determinant weightages.

We employed the Pearson's correlation coefficient for summarizing the, autocovariance terms quantitated between the non-asymptotical, unbiased, regressively, scalable, geo-spatiotemporal, geosampled, TB-related, grid-stratifiable, clinical, environmental, stratified, eigen-sociodemographic, eigen-orthogonalized, estimator determinants. We defined the covariance of the geosampled, epidemiologic dataset of stratified, semi-parameterizable estimator, determinants divided by the product of their standard deviations employing $\rho_{X,Y} = \frac{\operatorname{cov}(X,Y)}{\sigma_X \sigma_Y} = \frac{E[(X-\mu_X)(Y-\mu_Y)]}{\sigma_X \sigma_Y}$. The formula defined the capture point, county-level, eigen-autocorrelated, prognosticative, scaled-up, eigen-spatial filtered, estimator, determinant, correlation coefficients.

Thereafter, the eigendecomposition, eigen-spatial filtering approach added a minimally sufficient set of eigen-orthogonalized, eigenvectors as proxy variables to the set of evidential prognosticators in the county-level, vulnerability-oriented, county-level, capture point, sentinel site model by inducing mutual independence in the scaled-up, geo-spatiotemporal, geosampled, georeferenced, estimator determinants in eigen-Bayesian, on-frequentistic, conditional, autoregressive, eigenvector eigen-geospace. The semi-parametric, Markovian, hot/cold spot, residual, county-level, TB-related, scalable data in scanpy._ revealed independent, stratifiable, prognosticative, sentinel site, county-level, variable, clustering tendencies. The spatial pattern in the eigenvectors was synthetic. In the county-level, vulnerability, forecast, epidemiological model, positive, global eigen-autocorrelation in the local patterns of the parameters exhibited only positive local autocorrelation and vice versa for negative global eigen-autocorrelation. The stratified hot/cold spot explanators e_i and e_j within each set of eigenvectors were mutually eigen-orthogonal employing symmetry transformation $\frac{1}{2}\left(V + V^T\right)$ which in this experiment was expressible employing a quadratic form as revealed in Eqs. (1) and (2).

As mentioned previously, the eigen-spatial filter eigenvectors of specification (1) were eigen-orthogonal to the georeferenced, hyper/hypoendemic aggregation/non-aggregation-oriented, geosampled, county-level, epidemiologic, hot/cold spot, X of the prognosticated, county-level, regression-related, TB, estimator determinant, vulnerability model. This paradigm was constructed employing the transmission-oriented, operationalizable, eigenized, stratified, scalable, clinical, sociodemographic and environmental, county-level, georeferenced, estimator determinants.

Conversely, the eigenvectors of specification (2) were eigen-orthogonal only to the constant unity vector 1 in X. This eigen-orthogonality had implications for predictive signature modeling misspecification terms in the epidemiological model as they allowed linking each collection of the eigenvectors to its specific regressable, county-level, hyper/hypoendemic aggregation/non-aggregation-oriented, georeferenceable, epidemiologic, TB predictive vulnerability model residuals by letting E_{SAR} be a matrix whose vectors were subsets of $\{e_1, \ldots, e_n\}_{SAR}$.

We also considered within-group estimation of higher-order, autoregressive, probabilistic, panel models employing the georeferenced, time series, dependent, vulnerability-oriented, TB-related stratified regressors and fixed effects where the lag order was possibly mis-specified. Even when disregarding the misspecification bias, the fixed-effect bias formula regressed differently from the correctly specified case though its asymptotic order remained the same under stationarity. A linear combination of this subset was approximated by employing the misspecification term of the clustering/non-clustering, hyper/hypoendemic, epidemiological, capture point, sentinel site, prognosticated, model output which was expressable as.

$$\left(E_{SAR}\gamma \approx \sum_{k=1}^{\infty} \rho^k V^k \varepsilon \right). \tag{3}$$

The linear combination $E_{SAR}\gamma$ remained eigen-orthogonal to the geosampled prognosticated exogeneous variables X and, consequently, the estimated geosampled, stratified, clinical, sociodemographic, and environmental, TB county-level, explanatory predictor variables $\hat{\beta}$ were unbiased. Further, as a property of the OLS estimator, the approximated term $E_{SAR}\gamma$ was also eigen-orthogonal to the model residuals $\hat{\varepsilon}$. The model $y = X\hat{\beta} + E_{SAR}\hat{\gamma} + \hat{\varepsilon}$ eigen-decomposed the, stratified prognosticative variable y into a systematic trend component, a stochastic signal component and white-noise residuals. The term $E_{SAR}\hat{\gamma}$ removed variance inflation in the MSE term attributable to latent, geo-spatiotemporal, non-zero autocorrelation embedded in the geosampled, hyper/hypoendemic, aggregation/non-aggregation-oriented, prognosticated, capture point, sentinel site, scalable, eigen-spatial filtered, estimator determinants.

Alternatively, for the spatial lag model (3), a georeferenced, transmission-related, autoregressive, predictive, vulnerability-oriented, geo-spatiotemporal, county-level TB model was constructable employing E_{Lag} which was a matrix of those eigen-spatial filter, eigenvectors that were a subset of $\{e_1, \ldots, e_n\}_{Lag}$. The approximation of the misspecification term became $E_{Lag}\gamma \approx \sum_{k=0}^{\infty} \rho^k V^k (X\beta + \varepsilon)$ Since $E_{Lag}\gamma$ was correlated with the eigenfunction variables X, its incorporation into the scaled-up, capture point, sentinel site, county-level, stratified, prognosticative risk model corrected the bias of estimated plain OLS parameters $\hat{\beta}$ in the latent, spatial lag, hot/cold spot, estimator determinant, summary diagnostics. The model

$y = X\hat{\beta} + E_{Lag}\hat{\gamma} + \hat{\varepsilon}$ was constructable from the transmission–related, estimator determinants which in this experiment was derivable from an eigendecomposition of the spatial lag model. However, for the georeferenced, scaled-up, capture point, hyper/hypoendemic, aggregation/non-aggregation-oriented, county-level, prognosticative, eigenized, estimator determinants, we noted that the trend and the time-series, signals were no longer uncorrelated and the MSE was deflated.

The set of eigenvectors $\{e_1, \ldots, e_n\}_{Lag}$ of the spatial lag model (3) was calculable in Python independent of the hyper/hypoendemic, county-level, eigendecomposed prognosticative county-level variables X. This calculation was dependent on the underlying spatial link matrix V. We found that this eigen-spatial filtering eigen autocorrelation approach was more adaptable to an exploratory specification search of the relevant scalable, TB-related, capture point, georeferenced, sentinel site, hot/cold spots, and spatial predictions rather than with the regressed temporal shifting, prognosticatable, stratified, clinical, sociodemographic and environmental, aggregation/non-aggregtation-oriented, geosampled, explanatory, eigenized, estimator determinant, discrete, integer values in the risk model output. In contrast, for the simultaneous autoregressive model (2), the eigenvectors $\{e_1, \ldots, e_n\}_{SAR}$ depended through the projection of $M_{(X)}$ on the approximated capture point, sentinel site, geosampled variables X. Thus, any change in the underlying model structure required a recalculation of the, eigen-spatial filters for generating robust tessellations. Eigen-spatial filtering of either the spatial lag model or the simultaneous autoregressive model with a common factor constraint, thereafter, only required identification of one set of selected eigenvectors, namely, E_{SAR} or E_{Lag}, respectively. The relevant set of eigenvectors was applied simultaneously to the multivariate, capture point, sentinel site, scaled-up, epidemiologic, aggregation/non-aggregation-oriented, georeferenced, hyper/hypoendemic, prognosticative, hot/cold spot, stratified, estimator determinants.

For the generic autoregressive model (1), however, eigen-spatial filtering was applied individually to each geosampled, hyper/hypoendemic, aggregation/non-naggregation-oriented, eigenized, estimator determinant. The generic specification of autoregressive spatial models associated a specific spatial lag factor with the y variable and other lag factors for each additional, operationalizable, transmission-oriented, hyper/hypoendemic, stratified, epidemiological, TB-related, aggregation/non-aggregation-oriented, estimator determinant. The eigenvectors $\{e_1, \ldots, e_n\}_{Lag}$ were employed to filter latent non-zero/zero, geospatiotemporal, eigen-autocorrelation coefficients embedded in the generic, scaled-up, county-level, georeferenced, prognosticative, epidemiologic, capture point, risk model for each geosampled, hyper/hypoendemic, georeferenced, aggregation/non-aggregation-oriented, sentinel site, semi-parameterizable, eigenizable, clinical, sociodemographic and environmental stratified estimator determinant.

The next step was to identify suitable and p subsets of eigen-orthogonalizable, hyper/hypoendemic, eigen-autorelatable eigen-spatial filters E_{SAR} and E_{Lag} in the time-series, dependent, georeferenced, aggregation/non-aggregation-oriented, hyper/hypoendemic, stratified, vulnerability-oriented, model specifications (1) or (2). We assumed that a particular subset of eigendecomposed clinical, environmental and sociodemographic, TB-related, eigen-spatial filter eigenvectors were suitable for optimally, heuristically, optimizing, hot/cold spot, diagnostic, stratified, scaled-up, capture point, county-level, sentinel sites if the residuals $\hat{\varepsilon}$ of the resulting prognosticative, eigen-autocorrelation model became independent with respect to the underlying geosampled spatial structure V. Here, parsimony in the model estimation was definable as the smallest possible subset of eigen-spatial filtered eigenvectors which led to independence in the epidemiologic, county-level, TB model, scalable, county-level capture point, sentinel site, stratified, estimator determinant derivatives. We noted that the patterns of different non-zero, eigenvectors expressed independent and filtered residual autocorrelation as interpreted from the hyper/hypoendemic, aggregation/non-aggregation-oriented regressed model derivatives. This was formalized by an eigenized, estimator determinant associated to the scalable, sentinel site, capture point, georeferenced, epidemiologic, empirical, TB, stratified, transmission-related, scaled-up, county-level, epidemiologic, hyper/hypoendemic foci. Positive and negative, latent, geo-spatiotemporal, autocorrelation pseudo-R^2 values were subsequently reported by employing GLMM estimation results from Python (Fig. 1).

Suppose A $\in$ Rn $\times$ **n** is an invertible square matrix and u,v $\in$ Rn are column vectors. Then $A + uvT$ is invertible if $1 + \text{vTA} - 1\text{u} \neq 0$ (Dieudonné 1960). In this case, $(\text{A} + \text{uvT}) - 1 = \text{A} - 1 - \text{A} - 1\text{uvTA} - 11 + \text{vTA} - 1\text{u}$. Here, uvT was the outer product of two vectors u and v in the epidemiologic, capture point, sentinel site, empirical, scalable, county-level, prognosticative model. The general form shown here is the one published by Morrison 1950 (Morrison 1950). To prove that the backward direction $(1 + \text{vTA} - 1\text{u} \neq 0 \Rightarrow \text{A} + \text{uvT}$ was invertible with inverse given in the epidemiological, TB model residual forecasts, we verified the properties of the inverse. A matrix Y (in this case the right-hand side of the Sherman–Morrison formula) was the inverse of a matrix X (in $A + uvT$), if and only if, $\text{XY} = \text{YX} = \text{I}$ occurred in the county-level, TB-related, stratified, estimator, determinant, prognosticated, epidemiologic, model output.

We first verified if the right hand side (Y) of the TB, risk model satisfied $XY = I.XY = (A+uvT)(A-1-A-1uvTA-11+vTA-1u) = AA-1+uvTA-1-AA-1uvTA-1+u\,vTA-1uvTA-11+vTA-1u = I+uvTA-1-uvTA-1+uvTA-1uvTA-11+vTA-1u = I+uvTA-1-u(1+vTA-1u)vTA-11+vTA-1u = I+uvTA-1-uvTA-1$. We showed that $YX = I$ in a similar way as above where $YX = (A-1-A-1uvTA-11+vTA-1u)(A+uvT) = I$. We noted that the square matrices X and Y in the prognosticated model renderings $XY = I$ was equivalent to $YX = I$.) ($\Rightarrow$) Reciprocally, if $1 + vTA - 1u = 0$ occurred in the epidemiologic, forecasted, model derivatives then via the matrix determinant lemma, $\det(\text{A}+uv\text{T}) = (1+vTA-1u)\det(\text{A}) = 0$, so $(A+uvT)$.We assumed this equation was not invertible. Subsequently to solve $Bx = b$ when B was an $n \times n$ matrix we employed Gaussian elimination with partial pivoting requires O(n 3) operations. If B has a special form, [e.g., triangular or tridiagonal], the computational complexity can be reduced to O(n 2) or O(n) (Higgins 1940).

We sought to optimally, heuristically, optimize, targeting and prioritizing, georefernceable, aggregation/non-aggregation-oriented, hyper/hypoendemic, stratified, hot/cold spot, eigen-orthogonalizable, estimator determinants, employing a

```
def RunAnalyseMLMlogit(dataset, outcomevars, meeneemvars, randintercept, randslope):

    from rpy2.robjects import pandas2ri
    from rpy2.robjects.packages import importr
    base = importr('base')
    stats = importr('stats')
    lme4 = importr('lme4')

    #data
    with SavReaderNp(dataset) as reader_np:
      array = reader_np.to_structured_array()

    df = pd.DataFrame(array)

    variabelen = ' '.join(outcomevars) + ' ~ ' + '+'.join(meeneemvars)
    randintercept2 = ['(1|'+i+')' for i in randintercept]
    intercept = '+'.join(randintercept2)
    randslope2 = ['(1+'+meeneemvars[0]+'|'+i+')' for i in randslope]
    slope = ' '.join(randslope2)

    pandas2ri.activate()
    r_df = pandas2ri.py2ri(df)

    #model
    #random intercepts + random slopes
    if len(randslope) > 0:
      formula = variabelen + '+' + intercept + '+' + slope

    #only random intercepts
    else:
      formula = variabelen + '+' + intercept

    model = lme4.glmer(formula, data=r_df, family= 'binomial')
    resultaat = base.summary(model).rx2('coefficients')
    uitkomst = base.summary(mod
```

Fig. 1 The GLMM code of the county-level prognostictiave estimator determinant TB model in r2py

symmetric solution $(B^T = B)$, whereby the variants were motivated by finding an update B_{k+1} that was as close as possible to B_k. That is $B_{k+1} = \arg\min_B \| B - B_k \|_V$, where V was a positive definite matrix in the model. In our quasi-Newton methods, instead of computing the actual Hessian, we just approximated it with a positive-definite matrix. A positive definite matrix is a symmetric matrix with all positive eigenvalues [25]. Note that since we employed a symmetric matrix in the model all the potentially geo-spatiotemporal, eigendecomposed, estimator, determinant eigenvalues were real, i.e., positive, or negative. We assumed that this matrix can be used to reveal. geosampled, scalable, georeferenceable, TB-related, capture point, sentinel site, hot/cold spot, stratified, estimator, determinant properties which included the finite termination property, invariance, global convergence and local superlinear convergence properties. In this experiment this matrix solution was definable by Broyden's method.

In this experiment a positive defined matrix solution was definable by Broyden's method [see Fig. 1]. Broyden's method is a generalization of the secant method to nonlinear systems (Broyden 1965) (Fig. 2).

The idea behind employment of Broyden's method was to compute the first iteration, and to conduct a rank-one update at the other iterations. We also considered a stationary, non-negative, autoregressive, risk model which was given by $x_t = b_1 x_{t-1}$,

```
function [xv,it]=broyden(x,f,n,tol)
% Broyden's method for solving a system of n non-linear equations
% in n TB county variables.
%
% Example call: [xv,it]=broyden(x,f,n,tol)
% Requires an initial approximation column vector x. tol is required
% accuracy. User must define function f, for example see page 115.
% xv is the solution vector, parameter it is number of iterations
fr=zeros(n,1); it=0; xv=x;
%Set initial Br
Br=eye(n);
fr=feval(f, xv);
while norm(fr)>tol
 it=it+1;
 pr=-Br*fr;
 tau=1;
 xv1=xv+tau*pr; xv=xv1;
 oldfr=fr; fr=feval(f,xv);
 %Update approximation to Jacobian using Broydens formula
 y=fr-oldfr; oldBr=Br;
 oyp=oldBr*y-pr; pB=pr'*oldBr;
 for i=1:n
  for j=1:n
   M(i,j)=oyp(i)*pB(j);
  end;
 end;
 Br=oldBr-M./(pr'*oldBr*y);
end;
x0 = [-0.5 1.4];
f = @(x,y) [(x+3)*(y^3-7)+18; sin(y*exp(x)-1)];
[xv, it] = broyden(x0, f, 15, 1e-9);
Error in hw10>@(x,y)[(x+3)*(y^3-7)+18;sin(y*exp(x)-1)]
Error in hw10>broyden (line 55)
fr=feval(f, xv);
Error in hw10 (line 9)
[xv, it] = broyden(x0, f, 15, 1e-9);
```

Fig. 2 The code for the Broyden's method foremploying a quasi-Newton root-finding algorithm in the k-specified, TB variables

$+ \ldots + b_p x_{t-p} + e_t$, where the e_t were the i.d.d., geosampled, hyper/hypoendemic, operationalizable, potentially iterative interpolatable, time-series dependent, aggregation/non-aggregation-oriented, stratified, TB, county-level, estimator determinants and $b_1, \ldots, b_p$ were the non-negative parameters. In this model, all the roots of the equation $1 - b_{1u} - \ldots - b_p u^p = 0$ were outside the unit circle. The stationary solution of the autoregressive model is called a stationary non-negative autoregressive process (Rudin 1976). We let $x_1, x_2, \ldots, x_n$ be an example of a stationary, non-negative, autoregressive process in the scalable, estimator, determinant, capture point, sentinel site, vulnerability-oriented, county-level analyses. Under very generalizable conditions, strongly consistent, non-asymptotical, unbiased, semi-parameterizable, non-Gaussian, estimator determinants delineating stratified, georeferenceable, hyper/hypoendemic, aggregation/non-aggregtation-oriented, transmission-related, operationalizable, time series sequence $b_1, b_2, \ldots, b_p$ studied for application to predictive, TB-related, epidemiologic, signature risk models for optimally, heuristically, optimizing, targeting and prioritizing, scalable, georeferenceable, hot/cold spot, capture point, county-level, sentinel sites in the literature [e.g., Dodd et al. 2023]. A new procedure here was proposed to estimate not only $b_1, b_2, \ldots, b_p$ but also b_0 which we assumed would be deductible by the essential lower bound of a hyper/hypoendemic, capture point transmission-related, georeferenceable, aggregation/non-aggregation-oriented, prognosticative explanatory variable e_t.

We assumed that the refined, geosampled, county-level, hot/cold spot, geolocation, stratified, estimator determinants obtained employing the new algorithmic procedure in Python were consistent derivatives of $b_0, b_1, \ldots, b_p$.This was the weakest condition generalizable guaranteeing that the non-negative, stationary, autoregressively, operationalizable, time series, dependent, prognosticative, TB model had a stationary non-degenerative solution. A non-negative matrix was built. If matrix A is nonnegative, $A \geq 0$ all its entries are nonnegative (Godement 1998a).

A well-known property of an M matrix M is that the inverse is element-wise non-negative, which in this experiment we wrote as $M^{-1} > 0$. in Python for quantifying the non-Gaussian biasing uncertainties in the geosampled, georeferenced, empirical, epidemiologic, stratified dataset of georeferenced, hyper/hypoendemic, agrgegation/non-aggregation-oriented, estimator determinants. We considered element-wise perturbations of non-symmetric tridiagonal M-matrices in Python and obtained bounds on the perturbations so that the unquantitated, non-negative, inverse estimates attained during the forecast signature modeling procedure persisted in the endemic, transmission-related, capture point, sentinel site, county-level, prognosticated, epidemiologic, county-level TB data. Sufficient bounds were written in terms of decay estimates which characterized the deterioration of the elements of the inverse unperturbed matrix. Results for the generalized symmetric matrices and symmetric Toeplitz matrices were subsequently definable. In linear algebra, a Toeplitz matrix or diagonal constant matrix is a matrix where each descending diagonal from left to right is constant [14]. Importantly, a matrix equation of the form is called a Toeplitz system, if A is a Toeplitz matrix (Godement 1998a) Further, if A is a Toeplitz matrix then the system has only 2n − 1 df rather than n^2 (Hazewinkle 2000).

The first type of prognosticated, scalable, capture point, sentinel site, stratified, clinical, sociodemographic or environmental, estimator determinant, discrete integer value was obtainable only from the structural part of the epidemiologic, capture point, sentinel site, prognosticative, epidemiological, county-level, scalable TB model $\mathbf{x}_t'\mathbf{b}$. These were useful in predicting values of new response scalable, time series, county-level, capture point, sentinel site, hyper/hypoendemic, aggregation/non-aggregation-oriented, estimator determinants. The predicted model values, residuals, and upper and lower confidence limits for the These confidence limits are employable for estimation of the mean of the dependent variable [i.e., county-level prevalence], $\mathbf{x}_t'\mathbf{b}$ in an epidemiologic, prognosticative, hyper/hypoendemic, georeferenceable, stratifiable, estimator determinant, scalable, county or district-level TB model where $\mathbf{x}_t$ was the column vector of the independent variables at a capture point, hot/cold spot, sentinel site, geosampled, observational, evidential prognosticator. Our assumption was that estimator determinant, discrete integer values could be computable as $\lambda' = X'p$ and the upper and lower confidence limits were $\hat{u}_t = \hat{y}_t + t_{\alpha/2}\text{v}$ and $\hat{l}_t = \hat{y}_t - t_{\alpha/2}v$ where V_2 is an estimate of the variance of $\hat{y}_t$ and $t_{\alpha/2}$ was the upper $\alpha/2$ percentage point of the t distribution $\text{Prob}(T > t_{\alpha/2}) = \alpha/2$ Furthermore, we assumed a stratifiable, georeferenceable hyper/hypoendemic, aggregation/non-aggregation-oriented, estimator determinant derivable from a t distribution with q df with a value of α could be set with a specific option in Python. The df parameter, q, would be subsequently taken to be the number of geosampled, scalable, hot/cold spot, epidemiologic, capture point, sentinel site observations minus the number of free parameters in the final prognosticative, hyper/hypoendemic, georeferenceable, stratifiable, estimator determinant, satisfiable, county-level, prognostication model.

We extended the Neyman-Pearson lemma to settings involving composite hypotheses with monotone likelihood ratios in the georeferenced, scaled-up, sentinel site, vulnerability-oriented, capture point, epidemiological, TB, stratified model for, optimally, heuristically, optimizing and targeting, hyper/hypoendemic, aggregation/non-aggregation-oriented, county-level, hot/cold spots and their estimator determinants. The Neyman–Pearson lemma is part of the Neyman–Pearson theory of statistical testing, which introduced concepts like errors of the second kind, power function, and inductive behavior. The previous Fisherman theory of significance testing postulated only one hypothesis. By introducing a competing hypothesis, the Neyman–Pearsonian flavor of statistical testing allow investigating the two types of errors. The trivial cases where one always rejects or accepts the null hypothesis are of little interest, but it does prove that one must not relinquish control over one type of error while calibrating the other. Neyman and Pearson accordingly proceeded to restrict their attention to the class of all level tests while subsequently minimizing type II error, traditionally denoted by their seminal paper of 1933, including the Neyman–Pearson lemma, which comes at the end of this endeavor, not only showing the existence of tests with the most power that retain a prespecified level of type I error but also providing a way to construct such tests. The Karlin-Rubin theorem extends the Neyman–Pearson lemma to settings involving composite hypotheses with monotone likelihood ratios (Hosmer and Lemeshew 2002). Neyman and Pearson accordingly can proceed to restrict attention to the class of all alpha level tests while subsequently minimizing type II error which in our epidemiological prognosticative, county-level TB model was denotable by beta.

In scaled-up, county-level, capture point, sentinel site, epidemiologic, TB-related, time series, dependent, estimator determinant analysis, it may be of interest to characterize the heterogeneousness of the dependent variable around the regression function which we assumed may be describable by a eigen-Bayesian, semi-parametric, Markovian, eigen-algorithmic, TB-relared, estimator determine, probability distribution. In so doing any erroneous, multicollinear and/or skew heteroscedastic, hyper/hypo-endemic, aggregation/non-aggregtaion-oriented, georeferenceable, hot/cold spot, aggregation/non-aggregation-oriented propensities may be also quantifiable in a second order, eigenfunction, eigendecomposition, eigen-algorithm which may involve the asymptotic theory.

In asymptotic theory, the standard approach is $n \to \infty$ [38]. For some statistical, prognosticative, vulnerability-oriented, county-level, eigen-Bayesian, semi-parametric, Markovian, eigen-algorithmic, county-or district-level, scalable, epidemiologic, prognosticative TB models, slightly different approaches of asymptotics may be useful for quantitating uncertainty-oriented, georeferenceable, geosampled, geolocation, erroneous estimator determinants from non-frequentistic, georeferenced, scalable, aggregation/non-aggregation-oriented, time series, dependent, clinical, sociodemographic and or environmental, stratifiable, hyper/hypoendemic, capture point sentinel sites. For example, with geosampled, county-level, stratifiable, sentinel, site, TB-related, epidemiologic data, it is commonly assumable that one dimension in the epidemiologic, data capture point, remains fixed, whereas the other dimension grows: T = constant and $N \to \infty$, or vice versa [42]. The asymptotic theory proceeds by assuming that it is possible to keep collecting additional data, for example seasonal, geosampled, georeferenced, scaled-up, county-level, stratifiable, geo-spatiotemporal, TB-related, clinical, sociodemographic and or environmental, hyper/hypoendemic, aggregation/non-aggregation-oriented, estimator determinants; hence, the sample size can grow infinitely, i.e., $n \to \infty$. Under this assumption, many vulnerability-oriented, transmission-oriented TB-related, statistically, regressable, county-level, scalable, georeferenceable, capture point, sentinel sites may be obtainable that are unavailable for samples of non-finite size. In many cases, highly accurate results for non-finite samples may be obtainable via numerical methods though asymptotic analysis which may be useful for prognosticatively, geo-spatiotemporally, non-frequentically vulnerability modeling, a georeferenced, scaled-up, capture point, eigen-Bayesian, semi-parametric, Markovian, algorithmic, sentinel site, county-level, TB, stratifiable, hot/cold spot and its respective, county-level, estimator determinants. Besides the standard approach to asymptotics in the model other alternative approaches may exist for optimally, heuristically, optimizing, geolocating, hyper/hypoendemic, aggregation/non-aggregation-oriented, hot/cold spot sites and their estimator determinants.

Within the local non-asymptotic normality framework, it may be assumable that the value of a "true geosampled, stratifiable, clinical, sociodemiographic or environmental parameter" in the model varied slightly with n, such that the n-th model corresponds to $\theta_n = \theta + h/\sqrt{\boxed{n}}$. This approach allowed us ist to study the transmission-related, hyper/hypoendemic, geosampled, estimator determinants regressively non-biasedly.

When stratified, estimator determinant, scalable, county or district-level, TB-related, statistical, estimator determinant tests are studied for their power to distinguish against the alternatives that are close to the null hypothesis [which may be deductible within the so-called "local alternatives" framework], then the null hypothesis should be $H_0: \theta = \theta_0$ and the alternative hence should be $H_1 : \theta^{\circ} = \theta_0^{\circ} + q/\sqrt{\boxed{n}}$ [42]. This approach may be usable for the unit root tests in these diagnostic summaries. In statistics, a unit root test tests whether a time series variable is non-stationary and possesses a unit root [38]. The null hypothesis in a scaled-up, epidemiologic, capture point, sentinel site, county-level, stratifiable, TB-related, prognosticative estimator determinant, semi-parametric, eigen-Bayesian, Markovian, eigen-algorithmic model may be definable as the presence of a unit root whenst the alternative hypothesis is either stationarity, trend stationarity or explosive root depending on the test employed.

There are models where the dimension of the parameter space Θ_n slowly expands with n, reflecting the fact that the more geosampled, hyper/hypoendemic, aggregation/non-aggregation-oriented, stratifiable, TB-related, hot/cold spot, observational, evidential county-level prognosticators there are, the more structural effects that can be feasibly incorporated into an eigen-Bayesian, semi-parametric, Markovian, eigen-algorithmic mon-frequentistic model. For example, in a kernel density estimation regression model, an additional parameter is assumable—the bandwidth h. Hence, in an epidemiologic, scaled-up, capture point, sentinel site, county-level, eigen-Bayesian, semi-parametric, Markovian, eigen-algorithmic, TB-related, prognosticative, estimator determinant, vulnerability-oriented model, it may be typically assumable that $h \rightarrow 0$ as $n \rightarrow \infty$. The rate of convergence must be chosen carefully, e.g., $h \propto n^{-1/5}$.

In this experiment, we also took an alternative approach to the eigen-Bayesian inference by employing a multivariate Student's t-distribution that approximated the posterior density of the geo-spatiotemporal estimator determinant inference. We compared the performance of the MCMC and other sampling methods. We assumed that the overall performance was measurable in terms of statistical errors obtained for the same size of the Monte Carlo data. The eigen-Bayesian inference of the prognosticative, epidemiologic, scaled-up, capture point, sentinel site, georeferenceable, hyper/hypoendemic, aggregation/non-aggregation-oriented model, estimator determinants was performed by the MCMC method. We assumed that by so doing we would be able to quantitatively iteratively interpolate the geosampled, georeferenced, hyper/hypo-endemic, aggregation/non-aggregation-oriented, statistically significant, hot/cold spot, scaled-up, epidemiological, data, capture point sentinel sites.

Subsequently we generated a GARCH model for post treatment of the semi-parametric, eigen-Bayesian, Markovian, eigenvector eigen-Bayesian eigen-geospace, uncertainty-oriented temporal non-homoscedascity in the hyper/ hypoendemic, aggregation/non-aggregation-oriented, capture point, sentinel site, scalable, stratifiable, TB model, geosampled estimator determinants. Our model incorporated time series-specific, hidden, Markovian, chain processes that drove the epidemiologic, county-level, TB-related, GARCH, transformed parameters. The semi-parametric, eigen-Bayesian, Markovian, eigen-algorithmic inference of the GARCH model included checking the Weighted Ljung-Box test on standardized residuals and squared residuals to verify if the scaled-up, county-level, capture point, sentinel site model was adequate in describing the linear dependence in the empirical, epidemiologic, geosampled, stratified, hyper/hypo-endemic, aggregation/non-aggregation-oriented, estimator determinants.

The epidemiologic, county-level, scaled-up, capture point, sentinel site, eigen-Bayesian, eigen-algorithmic Markovian, GARCH, hyper/hypoendemic, aggregation/aggregation-oriented, transmission-related, estimator determinant, vulnerability-oriented, analytical process was written in python $\varepsilon_t^2 = \omega + \sum_{i=1}^{n} (\alpha_i + \gamma_i)\varepsilon_{t-i}^2 - \sum_{j=1}^{p} \gamma_j \eta_{t-j} + \eta_t$ where $\eta_t = \varepsilon_t^2 - h_t$ and $n = \max(p, q)$. We first loaded the epidemiological geosampled, TB, time series data The standard errors were not shown for transformed county-level variables. We passed through a transformed = False argument to summary to obtain this information for untransformed variables.

We plotted the GARCH latent stratified, temporally heteroskedastic, hyper/hypoendemic, aggregation/non-aggregation-oriented, TB county-level variables with **plot_z()**:model.plot_z(figsize = (115,5)) We plotted the fit with **plot_fit()**:The plot predictions of future conditional volatility with **plot_predict()**:model.plot_predict(h = 100 Wen view how well we predicted using in-sample rolling prediction with **plot_predict_is()**:model.plot_predict_is(h = 50,figsize = (15,5)).

We assumed that this representation would reveal that the squared residual ε_t^2 in the epidemiological, estimator determinant, diagnostically stratifiable, TB, county-level, prognosticative model followed an ARMA (n, p) process. Then for any $d > 0$, the conditional expectations in the prognosticative, hyper/hypoendemic, aggregation/non-aggregation-oriented, capture point, sentinel site, county-level, scalable, epidemiologic, prognosticative model would be as follows: $\mathbf{E}\left(\varepsilon_{t+d}^2 \mid \Psi_t\right) = \omega + \sum_{i=1}^{n} (\alpha_i + \gamma_i)\mathbf{E}\left(\varepsilon_{t+d-i}^2 \mid \Psi_t\right) - \sum_{j=1}^{p} \gamma_j \mathbf{E}\left(\eta_{t+d-j} \mid \Psi_t\right)$ or so we assumed. Further we assumed that the d-step-ahead prediction error, $\xi_{t+d} = y_{t+d} - y_{t+d|t}$, would have the conditional variance $\mathbf{V}(\xi_{t+d} \mid \Psi_t) = \sum_{j=0}^{d-1} g_j^2 \sigma_{t+d-j|t}^2$ where $\sigma_{t+d-j|t}^2 = \mathbf{E}\left(\varepsilon_{t+d-j}^2 \mid \Psi_t\right)$.Coefficients in the conditional d-step prediction error variance are calculable recursively employing the formula $g_j = -\varphi_1 g_{j-1} - \ldots - \varphi_m g_{j-m}$ where $g_0 = 1$ and $g_j = 0$ if $j < 0$; $\varphi_1, \ldots, \varphi_m$ were the stratified, scaled-up, capture point, sentinel site, autoregressive, stratified, geosampled, TB-related parameters [42].

4 Results

Initially, we constructed a Poisson regression model employing the empirical epidemiologic dataset of scaled-up, county-level, capture point, sentinel site, hyper/hypoendemic, transmission-related, CDC archived, stratified, TB-related, estimator, determinant geosampled, count, valued, discrete integers. in import numpy as np.

Our model was generalized by introducing an unobserved heterogeneity term for each county-level observation. The weights were assumed to differ randomly in a manner that was not fully accountable for by the other geosampled estimator determinants. In this experiment this process was formulated as $E(y_i|x_i, \tau_i) = \mu_i\tau_i = e^{x_i'\beta+\varepsilon_i}$ where the unobserved heterogeneity term $\tau_i = e^{\varepsilon_i}$ was independent of the vector of the stratified, clinical, sociodemographic and environmental regressor x_i. Subsequently the distribution of y_i was conditional on x_i and had a Poisson specification with conditional mean and conditional variance $\mu_i\tau_i : f(y_i|x_i, \tau_i) = \frac{\exp(-\mu_i\tau_i)(\mu_i\tau_i)^{y_i}}{y_i!}$.

We subsequently let $g(\tau_i)$ be the PDF of τ_i in the county-level, regression TB model. In so doing, the distribution $f(y_i|x_i)$ was no longer conditional on τ_i Instead it was obtained by integrating $f(y_i|x_i\tau_i)$ with respect to $g(\tau_i) : f(y_i|x_i) = \int_0^\infty f(y_i|x_i\tau_i)g(\tau_i)d\tau_i$ We found that an analytical solution to this integral existed in our Possionian TB, county-level, epidemiological model output when τ_i was assumed to follow a Gamma distribution. The model also revealed that y_i, was the vector of the stratified, TB-related, estimator, determinant, count variable, discrete, integer values while x_i, was independently Poisson distributed with $P(Y_i = y_i|x_i) = \frac{e^{-\mu_i}\mu_i^{y_i}}{y_i!}, y_i = 0, 1, 2, \ldots$ and the mean parameter—that is, the mean number of scaled-up, capture point, sentinel site, county-level, estimator determinant, sampling events in the CDC archived, empirical, epidemiologic dataset. This was given by $\mu_i = \exp\left(x_i'\beta\right)$ where β was a $(k+1) \times 1$ parameter vector. The intercept in the model was β_0 and the coefficients for the k regressors were denotable as $\beta_1, \ldots, \beta_k$. Taking the exponential of $x_i'\beta$ ensured that the mean parameter μ_i was non-negative. Thereafter, the conditional mean was provided by $E(y_i|x_i) = \mu_i = \exp\left(x_i'\beta\right)$.

The geosampled, capture point, sentinel site, stratified, TB-related, estimator determinants were analyzed employing $\ln\left[E(y_i|x_i)\right] = \ln(\mu_i) = x_i'\beta$. Note, that the conditional variance of the count random variable was equal to the conditional mean (i.e., equidispersion) in the forecasted, signature, model output [i.e., $V(y_i|x_i) = E(y_i|x_i) = \mu_i$]. In a log-linear model the logarithm of the conditional mean is linear [38]. The marginal effect of any stratified, geosampled, county-level, hyper/hypo-endemic, transmission-related, explanatory regressor in the epidemiologic TB model was provided by $\frac{\delta E(y_i|x_i)}{\delta x_{ji}} = \exp\left(x_i'\beta\right)\beta_j = E(y_i|x_i)\beta_j$ A one-unit change in the j th regressor in the model led to a proportional change in the conditional mean $E(y_i|x_i)$ of β_j.

In this experiment, the standard estimator for the Poissonian, epidemiologic, TB model was the ML estimator. Since the scaled-up, geosampled observations were independent, the log-likelihood function in the model was $= \sum_{i=1}^{N}\left(-\mu_i + y_i \ln \mu_i - \ln y_i'\right) = \sum_{i=1}^{N}\left(-e^{x_i'\beta} + y_i x_i'\beta - \ln y_i!\right)$. Given the empirical, epidemiologic, geosampled dataset of the county-level, capture point, scaled-up, hyper/hypoendemic, sentinel site, estimator determinants (i.e., θ) and an input vector x, the mean of the predicted Poissonian distribution was provided by $E(Y|x) = e^{\theta x}$. Thereafter, the Poisson distribution's PMF was provided by $p(y|x;\theta) = \frac{e^{y(\theta' x)} e^{-e^{\theta' x}}}{y!}$ The PMF in an empirical, epidemiologic, prognosticative, stratifiable, TB-related, estimator determinant, vulnerability-oriented model can be the primary means for defining a discrete probability distribution and, as such, it may be assumable that functions could exist for either scalar or multivariate, stratified, random variables, given that the distribution is discrete [42].

The prognosticated, county-level, scaled-up, capture point, sentinel site, explanatory, stratified, TB, estimator determinant, empirical, data consisted of m vectors $x_i i \in \mathbb{R}^{n+1}, i = 1, \ldots, m,$, along with a set of m values $y_1, \ldots, y_2 \in \mathbb{R}$ and the geosampled, hyper/hypoendemic, aggregation/non-aggregation-oriented, estimator determinants θ. The probability of attaining this set of the capture point, sentinel site, county-level, observational, evidential prognosticators was provided by the equation $p(y_1, \ldots, y_m | x_1, \ldots, x_m; \theta) = \prod_{i=1}^{m} \frac{e^{y_i(\theta x_i)} e^{-e^{e} x_i}}{y_i!}$. Consequently, we found the set of θ that made this probability as large as possible in the epidemiological regression model forecast, estimator, determinant dataset. To do this, the equation was first rewritten as a likelihood function in terms of θ: $p(y_1, \ldots, y_m | x_1, \ldots, x_m; \theta) = \prod_{i=1}^{m} \frac{e^{y_i(\theta' x_i)} e^{-e^{\theta' x_i}}}{y_i!}$. Note the expression on the right-hand side in the model had not actually changed. Next, we employed a log-likelihood [i.e., $\ell(\theta|X, Y) = \log L(\theta|X, Y) = \sum_{i=1}^{m} \left(y_i\left(\theta' x_i\right) - e^{\theta' x_i} - \log(y_i!)\right)$]].

Finding the maximum of a function in our scaled-up, prognosticative, capture point, sentinel site, county-level, epidemiologic, TB-related, regression model involved taking the derivative of a function and solving for the determinant being maximized. Because the logarithm is a monotonically increasing function in any prognosticative, estimator, determinant, vulnerability-oriented, regression-related model [39], the logarithm of a function in the TB model achieved its maximum value at the same scaled-up, county-level, capture point, geosampled, sentinel sites as the function itself, and, hence, the log-likelihood was employable in place of the likelihood in the ML estimation. Here we noted that this was easier to conduct when the model function being maximized was a log-likelihood rather than the original likelihood function. Notice that the parameters θ only appeared in the first two terms of each term in the summation. Therefore, given that we were only interested in discovering the optimal value for θ in the epidemiologic, county-level, stratified, capture point, sentinel site, prognosticative, regression model, we dropped the y_i! and simply wrote $\ell(\theta|X, Y) = \sum_{i=1}^{m} \left(y_i\left(\theta' x_i\right) - e^{\theta' x_i}\right)$. Thereafter, to find a maximum, we solved an equation $\frac{\partial \ell(\theta|X, Y)}{\delta \theta} = 0$ which had no closed-form solution. However, the negative log-likelihood (LL) [i.e., $-\ell(\theta|X, Y)$] was a convex function, and so standard convex optimization was applied to find the optimal value of θ.

We found that given the Poisson process in the scaled-up, sentinel site, capture point, prognosticative, TB-related, regression model, the limit of a binomial distribution was $p_p(n|N) = \frac{N!}{n!(N-n)!}p^n(1-p)^{N-n}$ Viewing the distribution as a function of the expected number of successes [i.e., $\nu \equiv Np$] in the epidemiological county-level model output, instead of the sample size n for fixed P, rendered the equation $p_{\nu/N}(n|N) = \frac{N!}{n!(N-n)!}\left(\frac{\nu}{N}\right)^n\left(1-\frac{\nu}{N}\right)^{N-n}$. Our model revealed that as the capture point, scaled-up, sentinel site, geosampled, TB-related size n become larger, the distribution approached P when the equations $\lim_{N\to\infty} p_p(n|N)$ $\lim\limits_{N\to\infty} \frac{N(N-1)\ldots(N-n+1)}{n!}\frac{\nu^n}{N^n}\left(1-\frac{\nu}{N}\right)^N\left(1-\frac{\nu}{N}\right)^{-n}$, $\lim\limits_{N\to\infty} \frac{N(N-1)\ldots(N-n+1)}{N^n} \quad \frac{\nu^n}{n!}\left(1-\frac{\nu}{N}\right)^N\left(1-\frac{\nu}{N}\right)^{-n}$. and $1\cdot\frac{\nu^n}{n!}\cdot e^{-\nu}\cdot 1$ and $\frac{\nu^n e^{-\nu}}{n!} 1\cdot\frac{\nu^n}{n!}\cdot e^{-\nu}\cdot 1$ and $\frac{\nu^n e^{-\nu}}{n!}$ aligned.

Note, the sample size n had completely dropped out of the probability function. The Possionian model output had the same functional form for the stratified, TB-related, estimator, determinant, geosampled, discrete integer values ν. Thereafter, as expected, the Poisson regression distribution was closer to being normalized [i.e., the sum of probabilities was more equal to 1], since $\sum_{n=0}^{\infty} P_\nu(n) = e^{-\nu}\sum_{n=0}^{\infty}\frac{\nu^n}{n!} = e^{-\nu}e^{\nu} = 1$ occurred in the forecasted model output. The ratio of probabilities was subsequently provided by the equation $\frac{P_\nu(n=i+1)}{P(n=i)} = \frac{\frac{\nu^{i+1}e^{-\nu}}{(i+1)!}}{\frac{e^{-\nu}\nu^i}{i!}} = \frac{\nu}{i+1}$.

Our capture point, sentinel site, scaled-up, county-level, regression model revealed that the stratified, Poissonian, estimator, determinant distribution reached a maximum when $\frac{dP_\nu(n)}{dn} = \frac{e^{-\nu}n(\gamma-H_n+\ln\nu)}{n!} = 0$ where g was the Euler-Mascheroni constant and H_n was a harmonic number, leading to the equation $\gamma - H_n + \ln\nu = 0$ which could not be solved exactly for n. The Euler-Mascheroni constant has many applications in real analysis and number theory and makes appearances in information theory [38].

We considered the Euler product $\zeta(s) = \prod_{k=1}^{\infty}\frac{1}{1-\frac{1}{p_k^s}}$ where $\zeta(s)$ was the Riemann zeta function and p_k was the k the prime. $\zeta(1) = \infty$. We noted that by taking the finite product up to $k = n$ in the county-level, epidemiological, scaled-up, capture point, sentinel site, stratified, estimator determinant, regression model and pre-multiplying by a factor $1/\ln p_n$, we were able to generate $n \to \infty$ which allowed rendering $\lim_{n\to\infty}\frac{1}{\ln p_n}\prod_{k=1}^{n}\frac{1}{1-\frac{1}{p_{k=y^y}}} = e^r$. The product was equivalent to 1.849071. In the prognosticative, empirical, TB model the Euler-Mascheroni constant represented the limit of the sequence $\gamma = \lim_{n\to\infty}\left(\sum_{k=1}^{n}\frac{1}{k} - \ln n\right) = \lim_{n\to\infty}(H_n - \ln n) =$ where H_n was the harmonic number and had the form $H_n = \sum_{k=1}^{n}\frac{1}{k}$ in the forecasts [i.e., stratified clinical, sociodemographic and environmental, scaled-up, transmission-related, capture point, sentinel site, hyper/hypo-endemic, transmission-related, estimator determinants]. The output also revealed that the Euler product attached to the Riemann zeta function $\zeta(s)$ represented the sum of the geometric series rendered from the transmission-related, hyper/hypo-endemic, scaled-up, capture point, sentinel site, estimator determinants as $\prod_p\left(1-p^{-s}\right)^{-1} = \prod_p\left(\sum_{n=0}^{\infty}p^{-ns}\right) = \sum_{n=1}^{\infty}\frac{1}{n^s} = \zeta(s)$. A closely related result was also obtainable by noting that $1 + \frac{1}{p_k} = \frac{1-\frac{1}{p_k^2}}{1-\frac{1}{p_k}}$. We considered the variation of the forecasted epidemiologic, model output when the

$+$ sign changed to a - in $\ln p_n$ which moved the denominator to the numerator thereby rendering $\lim_{n\to\infty} \ln p_n \prod_{k=1}^{n} \frac{1}{1+\frac{1}{p_k}} = \lim_{n\to\infty} \ln p_n \prod_{k=1}^{n} \frac{\frac{1}{1-\frac{1}{p_k^2}}}{\frac{1}{1-\frac{1}{p_k}}} = \frac{\prod_{k=1}^{\infty} \frac{1}{1-\frac{1}{p_k^2}}}{\lim_{n\to\infty} \frac{1}{\ln p_n} \prod_{k=1}^{t} \frac{1}{1-\frac{1}{p_k}}} = \frac{\zeta(2)}{e^{\gamma}} = \frac{\pi^2}{6e^{\gamma}} = 0.923563\ldots$

The moment-generating function of the Poisson distribution was given by M $= M = e^{-\nu} e^{\nu e^t} = e^{\nu(e^t-1)}_{,\mathrm{M}=} = \nu = \nu e^t e^{\nu(e^t-1)}$ and M $= M = \left(\nu e^t\right)^2 e^{\nu(e^t-1)} + \nu e^t e^{\nu(e^t-1)}$, when $R = \nu\left(e^t - 1\right), R' = \nu e^t$, so $R = R'(0) = \nu$. The raw moments were computable directly by summation which yielded an unexpected connection with the exponential polynomial $\varphi_n(x)$ and Stirling numbers of the second kind [i.e. $\phi_n(x) = \sum_{k=0}^{\infty} \frac{e^{-x} x^k}{k!} k^n = \sum_{k=1}^{n} x^k S(n, k)$]. This product was expressible employing the Dobiński's formula. In combinatorial mathematics, Dobinski's formula states that the number of partitions of a set of n members is $\frac{1}{e} \sum_{k=0}^{\infty} \frac{k^n}{k!}$ (Hosmer and Lemeshew 2000). This number has come to be called the nth Bell number$\mathrm{B_n}$, and the proof is renderable as an adaptation to probabilistic language as given by Rota (1964).

In our stratified prognosticative, regression-related, scaled-up, capture point, sentinel site, vulnerability-oriented, TB-specified, diagnostic, stratified estimator determinant, model, the formula $\phi_n(x) = \sum_{k=0}^{\infty} \frac{e^{-x} x^k}{k!} k^n = \sum_{k=1}^{n} x^k S(n, k)$ was viewed as a particular case, for $x = 0$, employing the relation $\frac{1}{e} \sum_{k=x}^{\infty} \frac{k^n}{(k-x)!} = \sum_{k=0}^{n} \binom{n}{k} B_k x^{n-k}$. The expression given by the model's Dobinski's formula was revealed as the n th moment of the Poissonian distribution with expected value 1. Dobinski's formula here was the number of partitions of a set of the geosampled estimator determinants (i.e.,n) which equaled the nth moment of the final distribution.

We employed the Pochhammer symbol $(\mathrm{x})_\mathrm{n}$ in the capture point, sentinel site, TB model to denote the falling factorial $(x)_n = x(x-1)(x-2)\ldots(x-n+1)$ (Abramowitz and Stegun 1972; Spanier 1987). If x and n are nonnegative integers, $0 \le n \le x$, then $(\mathrm{x})_\mathrm{n}$ is the number of one-to-one functions that map a size-n set into a size-x set (Koepf 1998). At this junction we let f be any function from a size-n set A into a size-x set B. Thus, in the epidemiologic, estimator determinant, TB model output u ∈ B.We then let $f^{-1}(\mathrm{u}) = \{\mathrm{v} \in \mathrm{A} : \cdot f(\mathrm{v}) = \mathrm{u}\}$. Then $\left\{f^{-1}(\mathrm{u}) : \mathrm{u} \in \mathrm{B}\right\}$ was a partition of A. This equivalence relation was the "kernel" of the function f in the TB model.

Any function from A into B factors in to one function that maps a member of A to that part of the kernel to which it belongs, and another function, which is necessarily one-to-one, that maps the kernel into B (Altman 1992). The first of these two factors were completely determinable by the partition π, which was the kernel in the stratified, estimator determinant, capture point, sentinel site, prognosticated, model output. The number of one-to-one functions from π into B in the model output was then $(\mathrm{x})_{|\pi|}$, when |π| was the number of parts in the partition π. Therefore, the total number of functions from a size-n set A into a size-x set B was $\sum_{\pi} (x)_{|\pi|}$ in the scaled-up, stratifiable, county-level, capture point, sentinel site, estimator determinant, epidemiological, TB, model forecasts when the index π ran through the set of all partitions of A. On the other hand, the number of functions from A into B was clearly x^n. Hence, we had $x^n = \sum_{\pi} (x)_{|x|}$. Since X was a stratified, Poissonian, distributed, random variable with expected value 1, the nth moment of this probability distribution was $E(X^n) = \sum_{\pi} E\big((X)_{|\pi|}\big)$ but all of the factorial moments $E((X)_k)$ of this probability distribution was equal to 1 in the prognosticated, TB-related, hypo/hyperendemic, transmission-related, model output. We noted we had, $E(X^n) = \sum_{\pi} 1$, which was the number of partitions of the set A in the output. Hence $v(1+v)$, $v\big(1+3v+v^2\big)$, and $v\big(1+7v+6v^2+v^3\big)$ were quantifiable in the county-level epidemiologic, scalable forecasts.

Thereafter, the central moments in the capture point, sentinel site, county-level, estimator, determinant. prognosticated, model output was computable as $v(1+3v)$ so the mean, variance, skewness, and kurtosis were $\frac{\mu_3}{\sigma^3} = \frac{v}{v^{3/2}} = v^{-1/2}$, $\frac{\mu_4}{\sigma^4} - 3 = \frac{v(1+3v)}{v^2} - 3$ and $\frac{v+3v^2-3v^2}{v^2} = v^{-1}$, respectively. The characteristic function for the Poisson distribution was revealed as $\phi(t) = e^{v(e^{t}-1)}$ and the CDF was $K(h) = v\big(e^h - 1\big) = v\big(h + \frac{1}{2!}h^2 + \frac{1}{3!}h^3 + \ldots\big)$ so $\kappa_r = v$ The mean deviation of the Poisson distribution mode was renderable by $MD = \frac{2e^{-v}v^{[v]+1}}{[v]!}$. We noted that the CDFs of the Poisson and chi-squared distributions were related in the capture point, sentinel site, stratified, TB, potentially, eigenizable, estimator, determinant, prognosticated, model output as $F_{\text{Poisson}}(k;\lambda) = 1 - F_{x^2}(2\lambda; 2(k+1))$ integer k and $P_r(X = k) = F_{x^2}(2\lambda; 2k) - F_{x^2}(2\lambda; 2(k+1))$.

The TB Poissonian distribution was expressable in terms of $\lambda \equiv \frac{v}{x}$ whereby, the rate of changes was equal to the equation $P_v(n) = \frac{(\lambda x)^n e^{-ix}}{n!}$. The moment-generating function of the Poisson distribution generated from the regressed, scaled-up, capture point, geosampled, TB, estimator determinants was also definable by $M(t) = e^{(v_1+v_2)(e^{t}-1)}$ Given a capture point, sentinel site, stratified, potentially eigenizable, estimator determinant, hyper/hypoendemic, transmission-related, geosampled, explanatory/non-explanatory, random variable x and a probability distribution function $P(x)$, if there exists $h > 0$ such that $M(t) \equiv \langle e^{tx} \rangle$ for $|t| < h$, where $\langle y \rangle$ denotes the expectation value of y, then $M(t)$ [i.e., the moment-generating function] [44].

Commonly, for a continuous distribution in an epidemiologic, regression-based prognosticative, signature, TB-related, capture point, sentinel site, scalable model $\int_{-\infty}^{\infty} e^{tx}P(x)dx \int_{-\infty}^{\infty} \big(1 + tx + \frac{1}{2!}t^2x^2 + \ldots\big)P(x)dx$ the equation. $1 + tm'_1 + \frac{1}{2!} + t^2 m'_2 + \ldots is$ is employable where m'_r is the raw moment [42]. In probability theory and statistics, the continuous uniform distributions are a family of symmetric probability distributions [38]. Such a distribution can describe a TB-related experimental model forecast where there is an arbitrary outcome that lies between certain bounds in a prognosticative, hyper/hypoendemic, capture point, sentinel site, county-level, scalable, georeferenceable model. Bounds in our county-level TB model were definable by the stratified parameters, which here was representable by the minimum and maximum values. For quantifying independent X and Y, the moment-generating function in our vulnerability-oriented, epidemiologic, TB-related, scaled-up, hyper/hypoendemic, capture point, sentinel site transmission-related, county model, we had to satisfy the equation $M_{x+y}(t) = \langle e^{t(x+y)} \rangle, \langle e^{tx} e^{ty} \rangle, \langle e^{tx} \rangle \langle e^{ty} \rangle$ and $M_x(t) M_y(t)$. In so doing, we were able to determine if the independent variables $x_1, x_2, \ldots, x_N$ had Poissonian distributions employing parameters $\mu_1, \mu_2, \ldots, \mu_N$ and $X = \sum_{j=1}^{N} x_j$. In our model output this was evident since the CDF was $K \equiv \sum_j K_j(h) = \left(e^h - 1\right) \sum_j \mu_j = \mu\left(\left(e^h - 1\right)\right.$.

We subsequently tested the scaled-up, capture point, sentinel site, county-level, epidemiological, TB model for overdispersion with a likelihood ratio test. This test quantitated the equality of the mean and the variance imposed by the Poissonian distribution against the alternative that the variance exceeded the mean. For the negative binomial distribution, the variance = mean+k mean^2(k >= 0, the negative binomial distribution reduces to Poisson when $k = 0$) (Haight 1967). The null hypothesis was H_0: $k = 0$ and the alternative hypothesis was H_a: $k > 0$. To carry out the test, we employed the following steps and then ran the model employing a negative binomial distribution and a record log-likelihood (LL) value. We then recorded LL for the Poisson model. We employed the likelihood ratio (LR) test, that is, we computed LR statistic, −2(LL (Poisson)—LL (negative binomial). The asymptotic distribution of the LR statistic had a probability mass of one half at zero and one half—chi-sq distribution with 1 d.f. To test the null hypothesis further at the significance level, we then employed the critical value of chi-sq distribution corresponding to significance level 2, that is we rejected H_0 if LR statistic $>^2$ (1 − 2, 1df).

We assumed that our potentially eigenizable, scaled-up, capture point, county-level, prognosticated, geosampled, stratified, sentinel site, epidemiologic, TB model, estimator determinants were quantifiable based on the log of the mean which in the model was a linear function of independent variables, log() = intercept + b1 ∗ X1 + b2 ∗ X2 + ... + b3 ∗ Xm. This log-transformation implied that this was the exponential function of independent variables, = exp(intercept + b1 ∗ X1 + b2 ∗ X2 + ... + b3 ∗ Xm). Instead of assuming, as before that the distribution of the geosampled, stratified, estimator, determinant, coefficient estimates [i.e., Y], was Poissonian, we assumed a negative binomial distribution. That meant, relaxing the generalized, Poissonian, regression, specification assumption about the equality of the mean and variance. We found that the variance of negative binomial was equal to+k2, where k >= 0 was a dispersion parameter. The ML method was subsequently employable to estimate k as well as the geosampled, stratified, estimator determinants derived from the regression-related, prognosticative, vulnerability-oriented, TB model for log (). The PMF of the negative binomial distribution with a non-homogenous, gamma, distributed mean in the capture point, sentinel site, scaled-up, county-level, epidemiological, prognosticated, TB-related, model output was subsequently expressable employing the eigenizable, estimator determinants as $f(k) \equiv \Pr(X = k) = \binom{k + r - 1}{k} (1 - p)^r p^k$ for optimally regressively quantitating the transmission-related, hyper/hypoendemic, estimator determinants $k = 0, 1, 2, \ldots$. In this equation, the quantity in parentheses was the binomial coefficient, which was equal to $\binom{k+r-1}{k} = \frac{(k+r-1)!}{k!(r-1)!} = \frac{(k+r-1)(k+r-2)\cdots(r)}{k!}$ This quantity was also alternatively written as $\frac{(k+r-1)\ldots(r)}{k!} = (-1)^1 \frac{(-r)(-r-1)(-r-2)\ldots(-r-k+1)}{k!} = (-1)^1 \binom{-r}{k}$ for describing the "negative binomialness' in our epidemiologic, prognosticative model.

An assumption that must be fulfilled on Poisson distribution is the mean value of data equals to the variance value (or so- called equidispersion) (Haight 1967). If the variance value is greater than the mean value, it is called overdispersion. Overdispersion occurred in our prognosticative model which may have been due to such factors as the presence greater variance of a county response variable caused by other variables [e.g., stratified, hyper/hypoendemic, clinical, sociodemographic and environmental, estimator, determinant unobserved heterogeneity]; the influence which leads to dependence of the probability of an even [hyperendemic TB transmission]t on previous events. The over-Poissonian variation in the regressed, scaled-up, capture point, sentinel sites may have also been due to the presence of outliers, or the existence of excess zeros on the response variable. In our model the overdispersion was probably due to the latter. If the equidispersion is not met due to excessive zeroes in the dependent variable, the Poisson regression is no longer appropriate to model the data (Haight 1967). Hence, we assumed the resulted model would yield biased parameter estimation and underestimated standard error, leading to invalid conclusions.

To handle overdispersion in the epidemiologic, prognosticative, county, TB model, initially the generalized Poisson regression model was also constructed employing. MLE, as well as a iterative solution using the Newton–Raphson method. The iterative estimation obtained was $\alpha\hat{}(\mathrm{t}+1) = \alpha(\mathrm{t}) - \mathrm{H} - 1(\mathrm{t})\mathrm{G}(\mathrm{t})$ and $\beta\hat{}(\mathrm{t}+1) = \beta(\mathrm{t}) - \mathrm{H} - 1(\mathrm{t})\mathrm{G}(\mathrm{t})$ where t represented the number of iterations required and $\boldsymbol{\alpha}$ was the dispersion parameter. The analysis resulted in the generalized Poisson regression model which here was expressible as $\mathrm{Y}\hat{}\mathrm{i} = \exp(\beta\hat{}0 + \beta\hat{}1\mathrm{X}1\mathrm{i} + \beta\hat{}2\mathrm{X}2\mathrm{i} + \ldots + \beta\hat{}\mathrm{pXpi}$.

The results from the count variable regression analyses revealed that the residual, explanatory, county-level, scaled-up, capture point, sentinel site, estimator determinants were highly significant, but virtually furnished no predictive power. Inclusion of indicator variables denoting the time sequence and the county geolocation spatial structure was then articulated with Thiessen polygons in Python which also failed to reveal meaningful covariates [i.e., georeferencable, scaled-up, capture point, sentinel site, hyper/hypoendemic, county-level, hot/cold spot and their aggregation/ non-aggregation-oriented, TB, estimator determinants.

Next, Pearson's chi-squared test was employed to assess three types of comparison in the archived, CDC, diagnostically stratified, county-level, model, estimator determinant, geosampled, empirical dataset: goodness of fit, homogeneity, and independence. A test of goodness of fit established whether an observed frequency distribution differed from a theoretical distribution in the prognosticated, epidemiological, empirical TB, model output. A test of homogeneity was able to compare the distribution of the scaled-up, capture point, sentinel site, county-level, TB prevalence counts for geolocating stratifiable, hyper/hypoendemic, aggregated/non-aggregated, capture point, sentinel site, hot/cold spots using the geosampled variables.

The GLIMMIX procedure displayed hyper/hypoendemic, transmission-oriented, fit statistics related to the marginal distribution as well as the conditional distribution $p(\mathbf{y}|\hat{\mathbf{g}}, \widehat{\mathbf{b}}, \hat{\phi})$. For ODS purposes, the name of the "Conditional Fit Statistics" table in python is "CondFitStatistics." Because the marginal likelihood was approximatable numerically for these methods, statistics based on the marginal, stratified, TB, estimator, determinant, model distribution were not available. Instead, the generalized Pearson chi-square statistic in the "Fit Statistics" table. A variance function was determined in Python which assumed that the conditional distribution of the data was unknown. Laplace or quadrature estimation was not possible. When we specified a variance function, the procedure assumed that the conditional distribution was normal. For example, we considered the following statements to fit a mixed model for delineating the county-level, capture point, sentinel site, geosampled, estimator determinant, count data. We briefly demonstrate how Python gpboost packages was used for inference and prediction with GLMMs.

The gpboost Python packages are available on PyPI which was installed as follows:

Python: pip install gpboost -U.

Estimation of GLMMs is a non-trivial task due to the fact that the likelihood (the quantity that should be maximized) cannot be written down in closed form. The current implementation of GPBoost (version 0.6.3) was based on the Laplace approximation. Model estimation in Python which in this experiment was generated as follow.

```
gp_model = gpb.GPModel(group_data=group_data,
likelihood="binary")
gp_model.fit(y=y, X=X)
gp_model.summary()
gp_model <- fitGPModel(group_data=group_data,
likelihood="binary",
y=y, X=X)
summary(gp_model)
```

where the TB county group data was a matrix or vector with categorical grouping variable(s) specifying the random effect's structure. The likelihood denoted the distribution of the response variable (e.g., likelihood = "binary" denoted a Bernoulli distribution with a probit link function) After estimation, the summary() function showed the estimated variance and covariance parameters of the random effects and the fixed effects coefficients b.

We then obtained p-values for fixed effects coefficients in GLMMs. Since the likelihood could not be calculated exactly in the first place, we had to rely on multiple non-asymptotic arguments. However, since the lme4 packages allowed for calculating approximate standard deviations and p-values, we were able to reveal how this can be done using gpboost relying on the same approach as lme4. In short, we had to enable "std_dev": True / std_dev = TRUE, to calculate approximate standard deviations when fitting the model, and then use approximate Wald tests as shown below.

```
gp_model = gpb.GPModel(group_data=group, likelihood="binary")
gp_model.fit(y=y, X=X, params={"std_dev": True})
coefs = gp_model.get_coef()
z_values = coefs[0] / coefs[1]
p_values = 2 * stats.norm.cdf(-np.abs(z_values))
print(p_values) # show p-values
```

Predictions were obtained by calling the predict() function as shown below.

```
pred = gp_model.predict(X_pred=X_test, group_data_pred=group_
test,
predict_var=True, predict_response=False)
print(pred['mu']) # predicted latent mean
print(pred['var']) # predicted latent varianc
```

The group_data_pred is a matrix or vector with categorical grouping variable(s) for which predictions are made [www.python.org] Here, X_pred was a matrix with fixed effects county-level, epidemiological TB covariate data for which predictions were made from.predict_var (boolean) indicated whether predictive variances were calculated in addition to the mean. The predict response (boolean) indicated whether the response y or the latent $Xb + Zu$ should be predicted. i.e., the random effects part is also predicted. If group_data_pred contains new, unobserved categories, the corresponding random effects predictions will be 0.

In this situation, the GLIMMIX procedure fit a mixed model with random intercepts, log link function, and variance function $\phi\mu$, assuming that the count variable was normally distributable, given the capture point, county-level, hot/cold spot, sentinel site, transmission-oriented, hyper/hypoendemic, specified, scaled-up, TB, county-level, regression, estimator, determinant, potentially eigenizable, random effects. The GLIMMIX procedure fit the statistical models to the geosampled, empirical, epidemiologic, TB data with correlations and nonconstant variability where the response was not necessarily normally distributed.

Our prognosticative, county-level, epidemiological TB model Y represented the $(n \times 1)$ vector of observed, geosampled, georeferenced, hyper/hypoendemic, aggregation/non-aggregation-oriented, scaled-up, capture point, sentinel site, geosampled, epidemiologic data and γ was a $(r \times 1)$ vector of random effects. The model fit by the GLIMMIX procedure assumed that $\mathrm{E}[\mathbf{Y}|\mathbf{g}] = g^{-1}(\mathbf{Xb} + \mathbf{Zg})$ where $g(\cdot)$ was a differentiable monotonic link function and $g^{-1}(\cdot)$ was its inverse. The matrix $\mathbf{X}$ was an $(n \times p)$ matrix of rank k, and z was an $n \times r$ design matrix for the random effects. The random effects were assumed to be normally distributed with mean 0 and variance matrix G.

The GLMM contained a linear mixed model inside the inverse link function. Here, this model component was the linear predictor, $\eta = \mathbf{X}\beta + \mathbf{Z}\gamma$. The variance of the observational, evidential, TB-related, estimator determinants conditional on the random effects was $\mathrm{Var}[\mathbf{Y}|\mathbf{g}] = \mathbf{A}^{1/2}\mathbf{R}\mathbf{A}^{1/2}$ The matrix $\mathbf{A}$ was a diagonal matrix and contained the variance functions of the model. The variance function expressed the variance of a response as a function of the mean. The matrix **R** was a variance matrix. If the conditional distribution of the capture point, scaled-up, county-level, sentinel site, geosampled, potentially, eigen-orthogonalizable, eigen-spatial filterable, empirical, TB data contained an additional scale parameter, it was either part of the variance functions or part of the **R** matrix. For example, the Gamma distribution in our model with mean μ had the variance function $a(\mu) = \mu^2$ and $\mathrm{Var}[Y|\mathbf{g}] = \mu^2\phi$. Unfortunately, the vulnerability-oriented, stratified, county-level GLMM, forecasted, estimator determinants assumed normal (Gaussian) random effects like the linear mixed models.

The relationship between county-level, TB prevalence and each geosampled, scaled-up, exploratory, individual, stratified, capture point, sentinel site, vulnerability-oriented regressors in the CDC epidemiologic, empirical dataset was subsequently investigated by employing a single, frequency, variable analysis in Python. The first line of the code began the command. The second line specified the fixed portion of each hypo/hyperendemic, transmission-oriented, county-level, geosampled, empirical, epidemiologic, risk model [i.e., the model without the random intercept, value (i.e., xb)] potentially, eigen-orthogonalizable, eigen-spatial filterable, estimator determinants in semiparametric, eigen-algorithmic, eigen-Bayesian eigenvector eigen-geospace. The third line of code created a value (i.e., rand) that was equal to the fixed part of the model (xb) plus a random intercept term u. The model statement specified the stratified, clinical, environmental and socio-demographic, eigenizable, estimator determinants which were distributed (~) normally with a mean of xb and variance s^2. The random statement defined the random effect u while quantitating the normally, distributed data with mean zero and a variance term. In so doing, s2u was optimally solvable. The level 2 units were identifiable by subject = id in Python.

Importantly, the last two lines of the command in Python were predictive statements. While traditionally only a single set of discrete integer values are created when constructing georeferenceable, hyper/hypoendemic, time series, dependent, epidemiological, signature, hot/cold spot, frequency models, we generated two from our scaled-up, county-level, vulnerability-oriented, model, estimator determinants. The procedure fit the county-level, nonlinear, mixed models—that is, models in which both fixed and random effects entered nonlinearly. These models have been employed in wide variety of applications in the literature, one of the most common being pharmacokinetic, overdispersed binomial data [e.g., Ludden 1991]. Python enabled us to specify a conditional distribution for the empirical, scaled-up, hyper/hypoendemic, aggregation/non-aggregation-oriented, county-level, capture point, sentinel site, epidemiologic, geosampled data (given the potentially, potentially, eigen-orthogonalizable, eigen-spatial filterable, estimator, determinant, stratifiable, random effects) having either a standard form (normal, binomial, Poisson) or a general distribution that we coded using SAS programming statements.

Python fit the nonlinear, mixed models by maximizing an approximation to the likelihood integrated over the random effects. GPBoost supported random coefficients such as random slopes or spatially varying coefficients in the TB model output. GPBoost is an approach and a software library aimed at combining tree-boosting with mixed-effects models and Gaussian Process (GP); hence the name

'GP + Tree-Boosting' [github.com]. Different integral approximations were available; the principal ones we employed in this experiment included adaptive Gaussian quadrature and a first-order Taylor series approximation. These alternative optimization techniques were carried out to determine the maximization; the default in the epidemiologic, prognosticative, county-level model was determined employing a dual quasi-Newton algorithm.

Successful convergence of the optimization problem resulted for determining the optimally regressable stratifiable, hyper/hypoendemic, potentially eigenizable, TB-related, estimator determinants along with their approximate standard errors which in this experiment was based on the second derivative matrix of the likelihood function. Python enabled employing the approximated model to construct predictions of arbitrary functions by employing empirical Bayes estimates of the random effects. We estimated arbitrary functions of the non-random, capture point, stratified, county-level, sentinel site parameters. Python computed approximate standard errors by employing the delta method. The predict statement rendered the diagnostic, stratified, hyper/hypoendemic, clinical, socio-demographic and environmental, time-series, dependent, discrete, integer values for the fixed portion of the model. The model identified xb, and output a dataset called 'Output-fixed'. The second statement revealed the scaled-up, county-level, hyper/hypoendemic, transmission-oriented, potentially eigenizable predicted values that included the estimate of the random intercept in addition to the fixed portion random estimates. Davidian and Giltinan [28] provide good overviews as well as general theoretical developments and examples of nonlinear mixed models.

Here we describe and compare several different integrated likelihood approximations and provide evidence that adaptive Gaussian quadrature is one of the optimal methods for obtaining suitable sentinel site, georeferenceable, diagnostically stratifiable, scaled-up, county-level, TB-related, transmission-related, hyper/hypoendemic, capture points employing a linear function approximated model function. As the integrand is the polynomial of degree $3\big(y(x) = 7x3 - 8x^2 - 3x + 3\big)$, (Weinstein 2002), here the 2-point Gaussian quadrature rule returned an exact result. A Gaussian quadrature employed nonlinear mixed models which advocated for a smooth nonparametric density for optimally, heuristically, optimizing, targeting, and prioritizing, the georeferenced, hyper/hypoendemic, aggregation/non-aggregation-oriented, county-level, capture point, sentinel site geolocation, geosampled, potentially eigen-orthogonalizable, eigen-spatial filterable, estimator determinants. These non-parametric procedures were employed for describing arbitrary regression distributions without the assumption that the form of the underlying estimator determinant densities is known. Conceptually the most simple and intuitive methods to estimate a PDF in an prognosticative, capture point, scaled-up sentinel site, georeferenceable, hyper/hypoendemic, aggregation/non-aggregation-oriented, epidemiological, TB scalable model is a histogram [42]. The range of each dimension xi of vector x was divisible into a fixed number m of intervals in the empirical, model output for optimally, optimizing, geosampled, hyper/hypoendemic, transmission-related, georeferenceable, aggregation/non-aggregation-oriented, county-level, hot/

cold spots and their potentially eigen-orthogonalizable, eigen-spatial filterable, estimator determinants employing a PDF.

Traditional approaches to fitting nonlinear mixed models involve Taylor series expansions, expanding around either zero, or the empirical best linear unbiased predictions of the random effects. The former is the basis for the well-known first-order method of [15, 16] and Sheiner and Beal (1985), and it is optionally available in Python. The latter is the basis for the estimation method of [60]. Here the closely related Laplacian approximation was an option,it was equivalent to adaptive Gaussian quadrature with only one quadrature point in the prognosticative, sentinel site, capture point, scaled-up, county-level, TB-related, stratified model.

The Laplacian approximation and its relationship to the Lindstrom-Bates as discussed by [17], Wolfinger (1993), Vonesh (1992, 1996), Vonesh and Chinchilli (1997), and Wolfinger and Lin (1997). provided an analytical expression for a posterior probability distribution in the county-level, TB, prognosticative model by fitting a capture point, Gaussian, sentinel site, stratified distribution with a mean equal to the MAP solution and a variance equal to the observed Fisher information. This approximation was justifiable by the Bernstein–von Mises theorem, which states that under regularity conditions the posterior converges to a Gaussian in a large sample.

We investigated the asymptotic normality of the posterior distribution in the discrete setting when the epidemiologic, prognosticative, stratified, TB model dimension increased with sample size artificially. We considered a PMF θ on N \{0} and a sequence of truncation levels (kn)n satisfying k 3n $\leq$ n infi $\leq$ knθ(i). We let ˆθ denote the ML estimate of (θ(i))i $\leq$ kn and let Δn(θ) denote the kn dimensional vector which in the empirical geosampled dataset of t diagnostic, stratified, TB model, potential eigen-orthogonalizable, eigen-spatial filterable, empirical forecasts was definable by $\sqrt{n}(\hat{}\theta n(i) - \theta(i))$ for 1 $\leq$ i $\leq$ kn. We checked to determine if under mild conditions on θ and on the sequence of prior probabilities on the kn-dimensional simplices. The variation distance between the posterior distribution and the kn-dimensional Gaussian distribution $N(\Delta n(\theta), I - 1(\theta))$ converged in probability to 0 after centering and rescaling. This theorem we assumed could prove the non-asymptotic normality of 'eigen-Bayesianizable' semi-parametric, Markovian, evidential, observational prognosticators generated from geosampled, hyper/hyperendemic, aggregation/non-aggregation-oriented, scaled-up, georeferenced, capture point, sentinel site, county-level, TB, stratified, clinical, sociodemographic and environmental, geosampled, eigen-orthogonalizable, eigen-spatial filterable, estimator determinants. The proofs would be based on concentration inequalities for centered and non-centered chi-square (Pearson) statistics. The latter allowed establishing robust posterior concentration rates with respect to Fisher measured distances in our county-level, scaled-up, capture point, sentinel site, TB model for optimally, heuristically, optimizing, targeting and prioritizing, georeferenced, hyper/hypo-endemic, aggregation/non-aggregation-oriented, eigenizable, estimator determinants rather than with respect to the Hellinger distance as it is commonplace in non-parametric Bayesian statistics employed in the literature for these prognosticative paradigms (Fig. 3).

Specifications	
Data Set	WORK.INFECTION
Dependent Variable	x
Distribution for Dependent Variable	Binomial
Random Effects	u
Distribution for Random Effects	Normal
Subject Variable	Clinic, sociodemiographic, environmental
Optimization Technique	Dual Quasi-Newton
Integration Method	Adaptive Gaussian Quadrature

Dimensions	
Observations Used	16
Observations Not Used	0
Total Observations	16
Subjects	8
Max Obs Per Subject	2
Parameters	3
Quadrature Points	5

Fig. 3 TB model information and dimensions for logistic-normal model of hillsborough couny

The "Specifications" table provided basic information about the county-level, georeferenced, nonlinear, mixed, scaled-up, potentially eigenizable, sentinel site, capture point model. For example, the distribution of the response variable, conditional on normally distributed random effects in the sentinel site, model for optimally, heuristically, optimizing, targeting and prioritizing, the capture point, hyper/hypo-endemic, stratified, TB, estimator determinants was binomial. The "Dimensions" table provided counts of various variables. We checked this table to make sure the dataset and model had been entered properly. Python selected five quadrature points to achieve the default accuracy in the likelihood calculation (Fig. 4).

The "Parameters" table listed the starting point of the optimization and the negative log likelihood at the earliest entered, geosampled, discrete, integer values (Fig. 5).

The "Iteration History" table indicated successful convergence in all iterations. The "Fit Statistics" table listed some useful statistics based on the maximized value of the log likelihood (Fig. 6).

The "Parameter Estimates" table indicated marginal significance of the fixed-effects in the scaled-up, capture point, sentinel site, county-level, TB-related stratified parameters. The positive value of the estimate of β_1 indicated that the uncertainty treatment significantly increased the chance for a favorable forecast of a geosampled,

Fig. 4 Starting values of parameter estimates

Parameters

beta0	beta1	s2u	NegLogLike
-1	1	2	37.5945925

Iteration History

Iter	Calls	NegLogLike	Diff	MaxGrad	Slope
1	2	37.3622692	0.232323	2.882077	-19.3762
2	3	37.1460375	0.216232	0.921926	-0.82852
3	5	37.0300936	0.115944	0.315897	-0.59175
4	6	37.0223017	0.007792	0.01906	-0.01615
5	7	37.0222472	0.000054	0.001743	-0.00011
6	9	37.0222466	6.57E-7	0.000091	-1.28E-6
7	11	37.0222466	5.38E-10	2.078E-6	-1.1E-9

NOTE: GCONV convergence criterion satisfied.

Fit Statistics

-2 Log Likelihood	74.0
AIC (smaller is better)	80.0
AICC (smaller is better)	82.0
BIC (smaller is better)	80.3

Fig. 5 Iteration history and fit statistics for logistic-normal model

Parameter Estimates

Parameter	Estimate	Standard Error	DF	t Value	Pr > |t|	Alpha	Lower	Upper	Gradient
beta0	-1.1974	0.5561	7	-2.15	0.0683	0.05	-2.5123	0.1175	-3.1E-7
beta1	0.7385	0.3004	7	2.46	0.0436	0.05	0.02806	1.4488	-2.08E-6
s2u	1.9591	1.1903	7	1.65	0.1438	0.05	-0.8554	4.7736	-2.48E-7

Fig. 6 Parameter Estimates for Logistic-Normal Model

county-level, scaled-up, capture point, sentinel site, hyper/hypo-endemic, stratified georeferenceable, estimator determinant (Fig. 7).

The "Additional Estimates" table displayed results from estimate statement in python. The estimate of $1/\beta_1$ equals $1/0.7385 = 1.3542$ and its standard error

Additional Estimates

Label	Estimate	Standard Error	DF	t Value	Pr > |t|	Alpha	Lower	Upper
1/beta1	1.3542	0.5509	7	2.46	0.0436	0.05	0.05146	2.6569

Fig. 7 Table of additional estimates

equaled $0.3004/0.7385^2 = 0.5509$ by the delta method Note that this approximation produced a_t statistic identical to that for the estimate of β_1.

We employed the regression line $(\mathbf{y}_i - \overline{\mathbf{y}}) = (\hat{\mathbf{y}}_i - \overline{\mathbf{y}}) + (\mathbf{y}_i - \hat{\mathbf{y}}_i)$ to generate a pseudo R^2 value where the first term was the total variation in the response y [county-level, CDC, archived, epidemiologic, empirical, geosampled, TB, prevalence data] and the second term was the variation in mean response based on the scaled-up, stratified, capture point, georeferenced, county-level, potentially eigen-orthogonalizable, eigen-spatial filterable, estimator determinants. The third term was the regressed, geosampled, stratified, clinical, environmental and/or sociodemographic, specified, transmission-related, hyper/hypoendemic, aggregation/non-aggregation-oriented, estimator determinants regressively derived, discrete, integer values as forecasted from the county-level, prognosticative, risk model. Squaring each of these terms and adding over all the stratified, observational, evidential prognosticators generated the equation $\sum (y_i - \overline{y})^2 = \sum (\hat{y}_i - \overline{y})^2 + \sum (y_i - \hat{y}_i)^2$. This equation was written as SST = SSM + SSE, where SS was notation for sum of squares and T, M, and E were the notation for total, uncertainty-related, vulnerability-oriented, stratified, model, error estimates. The square of the sample correlation was equal to the ratio of the estimates while the sum of squares was related to the total sum of squares: $r^2 = \mathrm{SSM/SST}41$. This formalized the interpretation of R^2 for elucidatively describing the fraction of variability in the epidemiologic, empirical, TB-related, time series, dependent, data, capture points and their stratifiable, aggregation/non-aggregation, hyper/hypoendemic, geosampled, scaled-up, capture point, sentinel site, estimator determinants. The sample variance s_y^2 was equal to $\sum \frac{(y_i - \overline{y})^2}{n-1}$, which in turn was equal to the SST/df, the total sum of squares divided by the total df. A regression equation was constructed by employing the mean square model (i.e., MSM) $= \sum \frac{(\hat{y}_i - \overline{y})^2}{l}$, which was equal to the SSM/df. The corresponding MSE was $\sum \frac{(y_i - \hat{y}_i)^2}{n-2}$ which we determined to be equal to SSE/df in the model explicatory estimate of the variance about the regression line (i.e., σ^2). The MSE is an estimate of σ^2 for determining whether or not the null hypothesis is true (Hosmer and Lemeshow 2000).

For parsimoniously quantitating the operationalizable, transmission-oriented, scaled-up, stratified, capture point, TB-related, eigen-orthogonalizable, eigen-spatial filterable, estimator determinant, prognosticative variables, (p), a DFM, was generated which we noted was equal to p and the error degrees of freedom (dfe)in the summary diagnostics. This product was also equal to $(n - p - 1)$, and the total degrees of freedom (dft) which was subsequently equal to $(n - 1)$ The sum of dfe and the dft was employable for conducting a non-asymptotic likelihood ratio test. The likelihood ratio test is one technique for detecting a shift in the mean of a sequence of independent, normal, random variables [38].

The likelihood ratio test here detected a shift in the mean of a sequence of independent, hyper/hypoendemic, transmission-related, potentially eigen-orthogonalizable, eigen-spatial filterable, estimator determinant, scaled-up, capture point, prognosticated, sentinel site, geosampled, stratified, TB, related random variables. We noted that if the time of the possible change of the geosampled estimator determinants is unknown, the non-asymptotic null distribution of the test statistic was an extreme value, rather than the usual chi-square distribution in the forecasted model output dataset. The non-asymptotic distribution was derivable here under the null hypothesis of no change. In our prognosticative paradigm the distribution of the likelihood ratio test was a mixture of chi square distributions with different df.

The relationship between the mean of the response variable (i.e., scaled-up, county-level, capture point, sentinel site, TB-related, prevalence count) and the level of the geosampled exogenous, potentially eigen-orthogonalizable, eigen-spatial filterable, prognosticators in the regression equation were assumed to be approximately linear (i.e., straight line). The corresponding table generated asymptotical, time-series, dependent, clinical, environmental and socio-demographic, stratifiable, estimator determinants in Python (see Table 2).

In the multiple regression analyses, the test statistic MSM/MSE had an $F(p, n-p-1)$ distribution. The null hypothesis was $\beta_1 = \beta_2 = \ldots = \beta_p = 0$, and the alternative hypothesis was deducible by encompassing the geosampled, county-level, quantifiable, capture point, hyper/hypoendemic, transmission-oriented, prognosticative, vulnerability-oriented, potentially, eigenizable, stratified parameters $\beta_j \neq 0,\ j = 1, 2, \ldots, p$.

The F test did not indicate which of the county-level, epidemiological, capture point, sentinel site, TB-related, stratified, estimator determinants $\beta_j \neq 0$ was not equal to zero only that at least one of them was linearly related to the response variable.

Homogeneity of variance–covariance matrices is the multivariate version of the univariate assumption of homogeneity of variance and the bivariate assumption of homoscedasticity (Hosmer and Lemeshow 2002). We evaluated the post "Homogeneity of variance" for determining equality of variances in the county-level, prognosticative, TB-related, stratified, hyper/hypoendemic paradigm. We wanted to determine if the homogeneity of variance–covariance matrices concerned the variance–covariance matrices of independently, geosampled, time series, dependent, scaled-up, capture point, sentinel site, county-level measures employing MANOVA for each stratified, discretely, regressively categorizable, TB-related, estimator determinant. Multivariate procedures such as MANOVA, Profile Analysis (GLM repeated models), Discriminant Function Analysis, Multivariate Regression (not Multiple Regression) can determine variances in predictive, epidemiologic, risk models [32]. Since the approximated mean was 10 dependent variables per county, the vulnerability analyses tested for approximately 10 correlations and 20 covariances for equality across each scaled-up, capture point, sentinel site, county-level,

Table 2 The time series regressed endemic TB regression-based model parameter estimate

Source	Sum of formula	Squares
Model	$\sum (\hat{\mathbf{y}}_i - \bar{\mathbf{y}})^2$	SSM/DFM
Error	$\sum (y_i - \hat{\mathbf{y}}_i)^2$	SSE/DFE
Total	$\beta_j \neq$	SST/DFT

georeferenced, stratified, sample group. The more dependent variables in a predictive equation, the higher the likelihood of non-equality of variances across the groups (Draper and Smith 1985).

Homogeneity of variance–covariance matrices for (e.g., MANOVA) was tested with the Box's M test which was non-significant. Box's M tests the statistical hypothesis that the variance–covariance matrices are equal [38]. In fact, we first checked the univariate homogeneity of variance with Levene's test for each individual, scaled-up, discretely, geosampled, stratifiable, hyper/hypoendemic, georeferenced, aggregation/non-aggregation-oriented, capture point, sentinel site, explanatory variable, which should have been non-significant for all the geosampled, stratified, clinical, environmental and sociodemographic, potential, eigen-orthogonalizable, eigen-spatial filterable, estimator determinants, or so we assumed. Levene's test is an inferential statistic used to assess the equality of variances for a variable calculated for two or more groups [38]. Some common statistical procedures assume that variances of the population of potential, epidemiologic, geo-spatiotemporal, stratified, TB-related, stratifiable, empirical, hot/cold spot sites from which different samples are drawn are equal [e.g. Verma et al. 2014]. Levene's tests here evaluated whether the null hypothesis that the population variances were equal (i.e., homogeneity of variance or homoscedasticity) in the county-level, hyper/hypoendemic, aggregation/non-aggregation-oriented, epiricial, epidemiologic, model forecasts. If a resulting p-value of Levene's test is less than some significance level (typically 0.05), the obtained differences in sample variances are unlikely to have occurred based on random sampling from stratifiable, TB, estimator determinant, hyper/hypoendemic, aggregation/non-aggregation-oriented, population, sampling groups with equal variances (Haase et al. 2007).

Here, the null hypothesis of equal variances was rejected, and it was concluded that there was a difference between the variances in the regressed, TB-related, county-level, geosampled, potentially, eigen-orthogonalizable, eigen-spatial filterable, estimator, determinant, population stratified, capture point, epidemiological, empirical, sentinel site data. However, Levene's test did not take account of covariances, so variance–covariance matrices were compared between groups using Box's M-test. Box's M test is a multivariate statistical test used to check the equality of multiple variance–covariance matrices [38]. Here the test was employed to determine the assumption of homogeneity of variances and covariances in MANOVA and for conducting a linear discriminant analysis of the CDC, archived, epidemiologic, geosampled dataset of vulnerability-oriented, hyper-hypoendemic, transmission related, scaled-up, capture point, sentinel sites for optimally heuristically, optimizing, targeting and prioritizing, georeferenceable, TB, diagnostically stratifiable, clinical, sociodemographic and environmental aggregation/non-aggregation-oriented, potentially eigen-orthogonalizable, eigen-spatial filterable, estimator determinants.

The non-asymptotic distribution of the LR statistic had a probability mass of one half at zero and one half - chi-square distribution with 1 df. In mathematics and statistics, an asymptotic distribution is a probability distribution that is in a sense the "limiting" distribution of a sequence of distributions (Draper and Smith 1985). One of the main uses of the idea of a non-asymptotic distribution is in providing approximations to the cumulative distribution functions of statistical estimators (Hosmer and Lemeshew 2000). The model data was \subsequently log-transformed. Summary of the regression estimation is provided in Table 3.

All derivatives from the model were subsequently evaluated in a spatial error (SE) model. Whereas the spatially lagged dependent variable model sees dependence as substance, in the sense that the yi is influenced by the value yj for ($j \neq i$), aspatial error models treat uncertainty correlations primarily as a nuisance, much like how statistical approaches often treat temporal serial correlation as something to be eliminated and solely as an estimation problem (Griffith 2003). An asymptotically unbiased, autoregressively, specified, county-level, TB, estimator, determinant risk model was constructed in Python which determined the geosampled, georeferenced, hyper/hypoendemic, aggregation/non-aggregation-oriented, transmission-related, potentially, eigen-orthogonalizable, eigen-spatial filterable, estimator determinant, significance levels. This process employed a variable Y as a function of nearby, scaled-up, capture point, geosampled, sentinel site, county-level, hot/cold spot, stratified, estimator determinant, Y values [i.e., an autoregressive response (AR) or spatial linear (SL) specification] and the error residuals of Y as a function of nearby Y residuals [i.e., an AR or SE specification].

First, we needed to import the necessary libraries:

```
import pandas as pd
import numpy as np
import matplotlib.pyplot as plt
from statsmodels.tsa.ar_model import AutoReg
```

Table 3 Summary of backward elimination of the sampled county-level TB paramterizable covariate coefficients

Step	Variable Removed	Number Vars In	Partial R-square	Model R-square	C(p)	F Value	Pr > F
1	Homeless	21	0	0.5154	21.002	0	0.968
2	Foreign born	20	0	0.5154	19.017	0.02	0.902
3	HIV positive	19	0	0.5153	17.036	0.02	0.889
4	Black	18	0.0025	0.5128	16.239	1.22	0.271
5	Long_Care	17	0.0027	0.5101	15.517	1.29	0.257
6	Age_25_44	16	0.0034	0.5067	15.146	1.65	0.201
7	Age_45_64	15	0.0043	0.5024	15.185	2.06	0.153
8	Age_65_More	14	0.0041	0.4983	15.147	1.97	0.162
9	Latin	13	0.0041	0.4942	15.12	1.97	0.162
10	NonInject_Drug	12	0.0046	0.4896	15.31	2.18	0.141
11	Alcohol	11	0.0049	0.4847	15.648	2.32	0.129
12	Jail	10	0.0049	0.4798	15.992	2.31	0.13

Next, we need to load the time-series data into python We used the CDC dataset, which contained the monthly number of TB hyper/hypo-endemic, aggregation/non-aggregtaion-oriented covariates from January 2015 to December 2023.

```
url = 'https://raw.githubusercontent.com/jbrownlee/Datasets/
master/airline-passengers.csv'
df = pd.read_csv(url, header=0, index_col=0, parse_dates
```

Distance between the georeferenced, evidential, observational prognosticators was defined in terms of an n-by-n geographic weights matrix, C, whose c_{ij} values were 1 if the scaled-up, capture point, georeferenced, sentinel site, county-level, epidemiological TB-related geolocations i and j were deemed nearby, and 0 otherwise. Adjusting this matrix by dividing each row entry by its row sum subsequently rendered the row sums given by C1 which then converted the C matrix to matrix W. The n-by-1 vector $x = [x_1 \cdots x_n]^T$ contained measurements of a quantitative, scaled-up, capture point, sentinel site, stratified, georeferenced, hot/cold spot, prognosticative explanatory variable for n spatial units and n-by-n spatial weighting matrix W. The formulation for the quantitating of the latent, erroneous coefficients employed was: $I(x) = \frac{n\sum_{(2)} w_{ij}(x_i-\overline{x})(x_j-\overline{x})}{\sum_{(2)} w_{ij}\sum_{i=1}^{n}(x_i-\overline{x})^2}$ where $\sum_{(2)} \sum_{i=1}^{n} \sum_{j=1}^{n}$ with $i \neq j$. The values w_{ij} were spatial weights stored in the symmetrical matrix $\mathbf{W}$ [i.e., $(w_{ij} = w_{ji})$] that had a null diagonal $(w_{ii} = 0)$. The matrix was generalizable to an asymmetrical matrix W. Matrix W can be generalized by a non-symmetric matrix W^* by using $W = (W^* + W^{*T})/2$ (Griffith 2003).

Subsequently, an autoregressive prognosticative model specification was employed to describe the stratified, hyper/hypoendemic, transmission-oriented, time–series, dependent, TB-related, signature, aggregation/non-aggregation-oriented model, variance, uncertainty estimates in eigenvector eigen-geospace. A non- asymptotical, prognosticative model specification was employed to describe both Gaussian and Poisson random variables. The resulting model specification took on the following form:

$$\mathbf{Y} = \mu(1-\rho)\mathbf{1} + \rho\mathbf{WY} + \varepsilon, \tag{4}$$

where μ was the scalar conditional mean of Y, and ε was an n-by-1 error vector, whose elements were statistically, independent randomized variates. We employed simple scalar mixing models coupled to a Lagrangian stochastic trajectory model in Python to calculate a range of conditional scalar statistics for multiple different, capture point model configurations. Among these sentinel site configurations were a uniform gradient source and a scalar mixing layer for qualitatively assessing the, geosampled, capture point, sentinel site, county-level, hyper/hypo-endemic, diagnostically stratifiable, clinical, sociodemographic and environmental, TB-related, aggregation/non-aggregation-oriented, hot/cold spot, potentially eigen-orthogonalizable, eigen-spatial filterable, estimator determinants.

The spatial covariance matrix for Eq. (4) incorporated the georeferenced, hyper/hypoendemic, TB, stratified, geosampled, diagnostic, estimator determinants in $E[(\mathbf{Y}-\mu\mathbf{I})'(\mathbf{Y}-\mu\mathbf{I})]=\sum=[(\mathbf{I}-\rho\mathbf{W}')(\mathbf{I}-\rho\mathbf{W})]^{-1}\sigma^2$, where $E(\bullet)$ denoted the calculus of expectations, I, which was quantifiable employing the n-by-n identity matrix. This was performed by denoting the matrix transpose operation where σ^2 was the error variance. A matrix for which all the column vectors are probability vectors is called transition or stochastic matrix (Afifi et al. 2004).

A stochastic matrix was employed for constructing an elucidative, robust, scaled-up, transmission-oriented, capture point, sentinel site, interpolative, operationizable, TB-related, epidemiologic, empirical, risk model for optimally, heuristically, optimizing, prioritizing and targeting, georeferenceable, hyper/hypoendemic, transmission-related, county-level, hot/cold spot, stratifiable, eigenizable, estimator determinants which was the transition matrix for a finite Markov chain (i.e., a Markov matrix). A Markov matrix is a square matrix used to describe the transitions of a Markov chain Markov chain. Markov process is a stochastic model describing a sequence of possible events in which the probability of each event depends only on the state attained in the previous event agniuc, (Gelman et al. 2013). PyMC (formerly PyMC3) is a Python package for Bayesian statistical modeling focusing on advanced Markov chain Monte Carlo (MCMC) and variational inference (VI) algorithms. Its flexibility and extensibility make it applicable to a large suite of problems.

```
# Creates the alpha and beta parameters
# Assumes a normal distribution
alpha = pm.Normal('alpha', mu=0.0, tau=0.05, testval=0.0)
beta = pm.Normal('beta', mu=0.0, tau=0.05, testval=0.0)
# The probability was modeled as a logistic function
p = pm.Deterministic('p', 1. / (1. + tt.exp(beta * time +
alpha)))
# Created the bernoulli parameter which uses observed data to
inform the algorithm
observed = pm.Bernoulli('obs', p, observed= TB _obs)
# Using Metropolis Hastings Sampling
step = pm.Metropolis()
# Draw the specified number of samples
sleep_trace = pm.sample(N_county-level TB Samples, step=step);
```

Each of our stratified, geosampled, potentially, eigen-orthogonalizable, eigen-spatial filterable, estimator determinant entries was a nonnegative discrete integer value representing a probability. The sum of the entries down any column was 1^4. In each row of the matrix the georeferenced, hyper/hypoendemic, aggregation/non-aggregation-oriented, geo-spatiotemporal, eigenizable, estimator determinant probabilities moved from the state represented by that row to other states. This matrix had elements that were statistically important variables [e.g., county geosampled, time series, dependent, georeferenced, stratified clinical, sociodemographic and environmental, geosampled, discrete, integer values with a 95% confidence interval (CI)] within a closed interval [0, 1]. Markov chain is collection of random variables where the index runs through 0, 1, ...) having the property that, given the present, the future may be predicted employing conditionally independent sentinel site, epidemiologic, data, capture points) (Gelman et al. 2013) [e.g., stratifiable, geo-spatiotemporal hot/cold spot, scaled-up, geosampled, TB-related, county-level, hyper/hypoendemic, potentially eigen-orthogonalizable, eigen-spatial filterable, aggregating/non-aggregating, non-heteroscedastic, uncertainty-oriented, iterative variables]. However, since a mixture of latent, positive, and negative, geo-spatiotemporal, eigen-autocorrelation coefficients was present in the georeferenced, county-level, scalable, capture point, sentinel site, geosampled, stratified, estimator, determinant empirical dataset, we assumed, a more explicit representation of both effects from the dataset in Python could generate a more accurate interpretation of the empirical results. Alternately, the excluded values were set to zero, although this was done assuming the mean and variance in the county-level, autoregressive, dependent, empirical, epidemiologic, diagnostically stratified, risk model would have to be adjusted.

Varying, georeferenceable, spatial, autoregressive parameters appeared in the covariance matrix, specifications which in the capture point, sentinel site, vulnerability-oriented, stratified, potentially eigen-orthogonalizable, eigen-spatial filterable, estimator, determinant, prognosticative, TB model output were quantifiable employing: $\Sigma = \left[(\mathbf{I}- <\rho> \mathrm{diag}\mathbf{W}')(\mathbf{I}- <\rho> \mathrm{diag}\mathbf{W})\right]^{-1}\sigma^2$. Here the diagonal matrix of the autoregressed parameters, $<\rho>_{diag}$, contained a minimum of two, geosampled, explanatorial, time series, dependent, hyper/hypoendemic, risk model TB-related, transmission-oriented, clinical georefernced, environmental and/or sociodemographic, estimator determinants ρ_+. These determinants were also employable to delineate specific, capture point, sentinel site, georeferenceable, hyper/hypoendemic, aggregation/non-aggregation-oriented, transmission-related, scalable, county-level geolocations and their eigenizable, estimator determinants. The pairs displayed positive spatial dependency (i.e., ρ, for those, capture point, sentinel site pairs displaying negative spatial dependency). For instance, by letting $\sigma^2 = 1$ and employing a 2-by-2 regular square tessellation rendered $\Sigma = \left[\begin{pmatrix}1&0&0&0\\0&1&0&0\\0&0&1&0\\0&0&0&1\end{pmatrix} - \begin{pmatrix}\rho_+&0&0&0\\0&\rho_+&0&0\\0&0&\rho_-&0\\0&0&0&\rho_-\end{pmatrix}\begin{pmatrix}0&\frac{1}{2}&\frac{1}{2}&0\\\frac{1}{2}&0&0&\frac{1}{2}\\\frac{1}{2}&0&0&\frac{1}{2}\\0&\frac{1}{2}&\frac{1}{2}&0\end{pmatrix}\right]^2$. For the vector $\begin{pmatrix}y_1\\y_2\\y_3\\y_4\end{pmatrix}$, this matrix enabled positing a positive relationship between the geosampled, georeferenced, TB-infected, hot/cold spot, county-level geolocations by adjusting the biased, uncertainty-oriented, estimator determinants y_1 and y_2. A negative relationship between the estimator determinants,y_3 and y_4, and, no relationship between the determinants, y_1 and y_3 and between y_2 and y_4 were noted. This covariance specification yielded:

$$\mathbf{Y} = \mu(\mathbf{I} - \rho_+ <\mathbf{I}_+>_{diag} - \rho_- <\mathbf{I}_->_{diag})\mathbf{1} + (\rho_+ <\mathbf{I}_+>_{diag} + \rho_- <\mathbf{I}_->_{diag})\mathbf{WY} + \varepsilon, \tag{5}$$

where $\mathbf{I}_+$ was a binary, 0–1, county-level, hyper/hypoendemic, capture point, sentinel site, geosampled, aggregation/non-aggregation-oriented, transmission-related, hot/cold spot, explanatory indicator variable. This equation also delineated those time-series, dependent, stratified, clinical, environmental and socio-demographic, potentially eigenizable, estimator determinants displaying positive spatial dependency where $\mathbf{I}_-$ was a binary 0–1 indicator variable denoting those scalable georeferenceable, capture point, sentinel site, county-level geolocations displaying negative spatial dependency employing $\mathbf{I}_+ + \mathbf{I}_- = 1$. Expressing the preceding 2-by-2 example in terms of Eq. (5) yielded:

$$\begin{pmatrix} y_1 \\ y_2 \\ y_3 \\ y_4 \end{pmatrix} = \mu \left[\begin{pmatrix} 1\,0\,0\,0 \\ 0\,1\,0\,0 \\ 0\,0\,1\,0 \\ 0\,0\,0\,1 \end{pmatrix} - \rho_+ \begin{pmatrix} 1\,0\,0\,0 \\ 0\,1\,0\,0 \\ 0\,0\,0\,0 \\ 0\,0\,0\,0 \end{pmatrix} - \rho_- \begin{pmatrix} 0\,0\,0\,0 \\ 0\,0\,0\,0 \\ 0\,0\,1\,0 \\ 0\,0\,0\,1 \end{pmatrix} \right]$$

$$\begin{pmatrix} 1 \\ 1 \\ 1 \\ 1 \end{pmatrix} + \left[\rho_+ \begin{pmatrix} 1\,0\,0\,0 \\ 0\,1\,0\,0 \\ 0\,0\,0\,0 \\ 0\,0\,0\,0 \end{pmatrix} + \rho_- \begin{pmatrix} 0\,0\,0\,0 \\ 0\,0\,0\,0 \\ 0\,0\,1\,0 \\ 0\,0\,0\,1 \end{pmatrix} \right]$$

$$\begin{pmatrix} 0 & \frac{1}{2} & \frac{1}{2} & 0 \\ \frac{1}{2} & 0 & 0 & \frac{1}{2} \\ \frac{1}{2} & 0 & 0 & \frac{1}{2} \\ 0 & \frac{1}{2} & \frac{1}{2} & 0 \end{pmatrix} \begin{pmatrix} y_1 \\ y_2 \\ y_3 \\ y_4 \end{pmatrix} + \begin{pmatrix} \varepsilon_1 \\ \varepsilon_2 \\ \varepsilon_3 \\ \varepsilon_4 \end{pmatrix}.$$

The Jacobian matrix was generated by $J_F(x_1,\ \ a_n)$ and $\frac{\partial(y_1,\ \ldots,\ y_m)}{\partial(x_1,\ \ldots,\ x_n)}$. The i-th row ($i = 1,\ m$) of this matrix was the gradient of the i-th component function $y_i\ :\ (\nabla y_i)$. In this experiment, p was a time-series, dependent, georeferenceable, hyper/hypoendemic, transmission-related, regressively forecastable, stratifiable, clinical, environmental or sociodemographic, aggregation/non-aggregation-oriented, hot/cold spot, potentially eigen-orthogonalizable, eigen-spatial filterable, estimator determinants in R^n when F was differentiable at p. A derivative was given by $J_F(p)$. The model described by $J_F(p)$ was the best linear approximation of F near the county-level, georeferenced, scaled-up, sentinel site, capture point p in the sense that: $F(x) = F(p) + J_F(p)\,(x - p) + o(\|x - p\|)$. The spatial structuring was achieved for constructing a linearized combination of a subset of the eigen-orthogonalizable, eigen-spatial filter eigenvectors of a modified geographic weights matrix employing $\left(\mathbf{I} - 11'/n\right)\mathbf{C}\left(\mathbf{I} - \mathbf{1}\mathbf{1}'/n\right)$ which we assumed would appear in the numerator of the Moran's coefficient, geo-spatiotemporal, latent, autocorrelation, post, eigen-Bayesian, eigen-algorithmic, semiparametric, MCMC, GARCH-related vulnerability-oriented, uncertainty analyses which here was indexed with a product moment correlation coefficient.

We defined a right eigenvector as a column vector $\mathbf{x}_R$ satisfying $A\mathbf{X}_R = \lambda_R\mathbf{X}_R$, where A was a matrix so $(\mathbf{A} - \lambda_R\ |)\mathbf{x}_R = \mathbf{0}$, which meant the right eigenvalues had zero determinant in our autoregressive, time-series, dependent, eigen-decomposition eigen-spatial filter, risk model. Similarly, we defined a left eigenvector as a row vector, X_L, satisfying $X_L A = \lambda_L X_L$. Taking the transpose of each side rendered $(X_L A)^T = \lambda_L X_L^T$ which was rewritten as $\mathbf{A}^T \mathbf{x}_L^T = \lambda_L \mathbf{x}_L^T$ We rearranged this equation once again to obtain $\left(\mathbf{A}^T - \lambda_L |\right) \mathbf{X}_L^T = \mathbf{0}$ which rendered $\det\left(\mathbf{A}^T - \lambda_L\ |\right) = \mathbf{0}$ The non-asymptotical, forecasting, equation, in turn, rendered $0 = \det\left(\mathbf{A}^T - \lambda_L\ |\right) = \det\left(\mathbf{A}^T - \lambda_L |^T\right)$, $\det(\mathbf{A} - \lambda_L\ |)^T$ and $\det(\mathbf{A} - \lambda_L\ |)$, Thereafter we derived the identity of $\det(A) = \det\left(A^T\right)$. We equated these equations to 0 for A and X, which required that in Python.

We subsequently let X_R be a matrix formed by the columns of the right eigenvectors X_L, which was a matrix formed by the rows of the left eigenvectors. We let $D \equiv \begin{bmatrix} \lambda_1 & \cdots & 0 \\ \vdots & \ddots & \vdots \\ 0 & \cdots & \lambda_n \end{bmatrix}$, and, as such, $\mathbf{AX}_R = \mathbf{X}_R D$ and $\mathbf{x}_L \mathbf{A} = D\mathbf{x}_L \mathbf{X}_L \mathbf{AX}_R = \mathbf{X}_L \mathbf{X}_R D$ while, $\mathbf{X}_L \mathbf{AX}_R = D\mathbf{X}_L \mathbf{X}_R$, so $\mathbf{X}_L \mathbf{X}_R D = D\mathbf{X}_L \mathbf{X}_R$. But this equation was of the form $CD = DC$, where D was a diagonal matrix, so, $\mathbf{C} \equiv \mathbf{X}_L \mathbf{X}_R$ was also diagonal in the epidemiologic model output. If A is a symmetric matrix, then the left and right eigenvectors are simply each other's transpose, and if A is a self-adjoint matrix (i.e., Hermitian), then the left and right eigenvectors are adjoint matrices (Griffth 2003). In an epidemiologic, prognosticative, autoregressive, aggregation/non-aggregation-oriented, hyper/hypoendemic, transmission-related, diagnostically stratifiable, TB-related, interpolative, hot/cold spot, time-series, dependent, vulnerability-oriented model, a Hermitian matrix can be designated as a square matrix with complex entries that is equal to its own conjugate transpose—that is, the element in the i-th row and j-th column is equal to the complex conjugate of the element in the j-th row and i-th column, for all indices i and $j\ :\ a_{i,\,j} = \overline{a_{j,\,i}}$. Employing the matrix with eigenvectors $\mathbf{x}_1, \mathbf{x}_2$, and $\mathbf{x}_3$ and corresponding eigenvalues λ_1, λ_2, and λ_3, an arbitrary vector, y here was written as $\mathbf{y} = b_1\mathbf{x}_1 + b_2\mathbf{x}_2 + b_3\mathbf{x}_3$.. Henceforth matrix A is employable for generating $\mathbf{Ay} = b_1\mathbf{Ax}_1 + b_2\mathbf{Ax}_2 + b_3\mathbf{Ax}_3 = \lambda_1\left(b_1\mathbf{x}_1 + \frac{\lambda_2}{\lambda_1}b_2\mathbf{x}_2 + \frac{\lambda_3}{\lambda_1}b_3\mathbf{x}_3\right)$; so $\mathbf{A}^n\mathbf{y} = \lambda_1^n\left|b_1\mathbf{x}_1 + \left(\frac{\lambda_2}{\lambda_1}\right)^n b_2\mathbf{x}_2 + \left(\frac{\lambda_3}{\lambda_1}\right)^n b_3\mathbf{x}_3\right|$ Hence since $\lambda_1 > \lambda_2,\ \lambda_3,\ \ldots,$ and $b_1 \neq 0$, (Griffth 2003) it follows that $\lim_{n\to\infty} \mathbf{A}^n\mathbf{y} = \lambda_1^n b_1\mathbf{x}_1$, so repeated application of the matrix to any vector in an eigendecomposed, capture point, sentinel site, scaled-up, county or district-level eigen-orthogonalizable, eigen-estimator, determinant, empirical, epidemiologic, geosampled dataset will result in a vector proportional to the eigenvector with largest robust, non multicollinear, homoscedastic, empirical eigenvalue. Unfortunately, non-zero latent, eigen-autocovariance quantification in epidemiologic, prognosticative, capture point, TB-stratified, sentinel site, vulnerability-oriented, scaled-up, district or county-level, hyper/hypo-endemic, transmission-related, hot/cold spot models appear sparasely in the literature.

In our experiment, the time-series, dependent, regression-related, prognosticated, eigendecomposed, eigen-spatial filters, optimally, heuristically, optimized, the geosampled, hyper/hypoendemic, aggregation/non-aggregation-oriented, county-level, TB-related, scaled-up, capture point, sentinel site, hot/cold spots and their estimator determinants. The epidemiologic, transmission-oriented, county-level, regressors were independent p-variate Gaussian variables which were drawn from a $N(0, \sum)$, distribution. If geosampled, hyper/hypoendemic, aggregation/non-aggregation-oriented, prognosticative, epidemiologic, scalable, observational, evidential prognosticators are independent p-variate Gaussian variables drawn, then the conditional distribution in a county-level, vulnerability, predictive TB model will have a distribution which is constructable employing the sample covariance matrix.

For a single sample setting with a discrepancy function $F = F(\Sigma(\Theta), \mu(\Theta); \mathbf{S}, \overline{\mathbf{x}})$, the gradient was definable in the vulnerability-oriented, scalable, hyper/hypoendemic, capture point, county-level, sentinel site, epidemiologic, TB-related, prognosticative model as the first partial derivatives of the discrepancy function with respect to the stratified, model parameters $\Theta : g(\Theta) = \frac{\partial}{\partial \Theta} F(\Theta)$ in Python. The Hessian was also definable as the second partial derivatives of the discrepancy function with respect to the sampled TB parameters $\Theta : H(\Theta) = \frac{\partial^2}{\partial \Theta \partial \Theta'} F(\Theta)$ The mean and covariance structures fit perfectly with $\Theta = \Theta_o$ in every sample county in Florida. The information matrix was definable as: $I(\Theta_o) = \frac{1}{2} \mathcal{E}(H(\Theta_o))$ where the expectation $\mathcal{E}(\cdot)$ took over the sampling space of $S, \overline{x}$.

The information matrix plays a significant role in statistical theory (Griffth 2003). Under certain regularity conditions, the inverse of the information matrix $I^{-1}(\Theta_o)$ in the prognosticative, epidemiologic, scalablTB model had the non-asymptotic covariance matrix for $\sqrt{N}\left(\widehat{\Theta} - \Theta_o\right)$, where N denoted the stratified, county sample size when $\widehat{\Theta}$ was an empirical geosampled, time series, dependent, hyper/hypoendemic, scalable, aggregation/non-aggregation-oriented, capture point, sentinel site, hot/cold spot, eigenizable, estimator determinant. In practice, Θ_o is never known and can only be approximated. Here, the information matrix was assessable by the so-called empirical information matrix: $I(\widehat{\Theta}) = \frac{1}{2} H(\widehat{\Theta})$ which was based on the eigendecomposed, eigen-spatial filtered estimator determinant, discrete, integer values derived from the sample empirical estimates $\widehat{\Theta}$. We noted that this information matrix, rather than the unknown $I(\Theta_o)$, was the "information matrix" displayed in the output. Taking the inverse of the empirical information matrix with sample size adjustment, PySAL assessed the estimated covariance matrix of $\widehat{\Theta}$ by $((N-1)I(\widehat{\Theta}))^{-1} = \left((N-1)\frac{1}{2}H(\widehat{\Theta})\right)^{-1} = \frac{2}{N-1} H^{-1}(\widehat{\Theta})$.

The Pearson product-moment correlation coefficient is a measure of the linear relationship between two questions/measures/variables, X and Y (Hosmer and Lemeshow 2000). The correlation value can range from +1 to −1. Here, a positive correlation (i.e., + 0.31) was renderable from the capture point, sentinel site, scaled-up, county-level, georeferenced, stratified, hyper/hypoendemic, aggregation/non-aggregation-oriented. potentially, eigenizable, evidential, observational prognosticators. There was a positive relationship between X and Y in the forecasted, TB, county-level, estimator determinant, model output.

Next, the georeferenced, aggregation/non-aggregation-oriented, autoregressable hyper/hypoendemic, epidemiologic coefficients derived from the geosampled, prognosticative, stratified, TB, estimator determinants were subsequently input into a hierarchical, inferential Bayesian estimation, matrix. Multiple MCMC chains were estimated for the intercept, which appeared to converge within the first 1,000 samples. The first 1,000 samples were discarded to allow the model to stabilize (i.e., "burn in"), and the next 10,000 samples were employable for deriving optimal, stratifiable, TB-related, empirical, epidemiologic, capture point, sentinel site, county-level, georeferenceable, scalable, hot/cold spots and their hyper/hypoendemic, aggregation/non-aggregation-oriented, estimator determinants based on statistical significance. The MCMC was able to numerically calculate multi-dimensional integrals.

PyMC3 has built in functions for assessing the quality of models, including trace and autocorrelation plots.

```
pm.traceplot(sleep_trace, ['alpha', 'beta'])
pm.autocorrplot(sleep_trace, ['alpha', 'beta'])
```

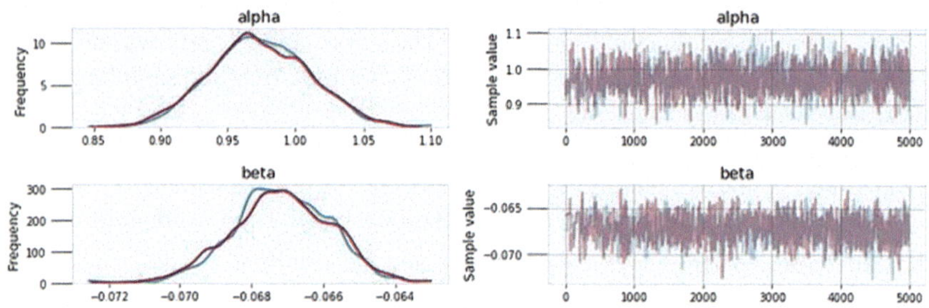

A Wishart distribution was generated from the geosampled, stratified, TB, estimator, determinant, georeferenced, epidemiologic, capture point, sentinel site data. The Wishart distribution is a multivariate extension of the Gamma distribution, although most statisticians use the Wishart distribution in the special case of integer df in which case it simplifies to a multivariate generalization of the $\chi 2$ distribution (Gelman et al. 2013). As the $\chi 2$ distribution describes the sums of squares of n draws from a univariate normal distribution, the Wishart distribution represents the sums of squares (and cross-products) of n draws from a multivariate normal distribution [26].

The calculation of moments of complex Wishart and complex inverse Wishart distributed random matrices in the geosampled, time-series, dependent, epidemiologic, TB-related, hyper/hypoendemic, transmission-oriented, operationalizable, hot/cold spot, predictive, stratified, empirical, risk, model dataset of clinical, sociodemographic and environmental, county-level, estimator determinants was subsequently addressed in python which returned a matrix that contained N random draws from the Wishart distribution with df. The Wishart distribution is a multivariate generalization of the gamma distribution (Gelman 2005). This is not the case with the Wishart distribution. A Wishart distribution is a probability distribution for nonnegative, definite matrix-valued, random variables. The complex inverse Wishart distribution is a matrix probability distribution defined on complex-valued positive-definite matrices and is the complex analog of the real inverse Wishart distribution (Gelman et al. 2013). It has greatest application in least squares optimization theory applied to complex valued data samples [26]. These distributions were employed to estimate covariance matrices in the prognosticative, non-frequentistic, simulation model.

We generated 1,000 samples from a Wishart distribution with 7 df and matrix parameter S. Each row of the returned matrix x represented a nonnegative definite matrix. In Scipy.Stats.Invwishart computed the sample mean and compared it with the expected value.

```
call randseed(1);
N = 1000;
DF = 7;
S = {1 1, 1 5};
x = RandWishart( N, DF, S );
Mean = DF * S;
SampleMean = shape(mean(x), 2, 2);
print SampleMean Mean;
Estimated Mean of Matrices
SampleMean Mean
7.0518633 7.2402925 7 7
7.2402925 36.056848 7 35
```

Our idea for optimizing Bayesian, modeling, complex Wishart and complex inverse Wishart distributed random matrices was to model the statistical properties of complex sample covariance matrices and complex inverse sample covariance matrices, respectively for simulating scalable, georeferenceable, capture point, sentinel site, county-level, hyper/hypoendemic, aggregation/non-aggregation-oriented, diagnostically stratifiable, TB-related, estimator determinants. Here we converted the covariance matrix to a correlation matrix in python by extracting the off-diagonal elements, as shown in the following.

Moments of random matrices were needed for determining the asymptotic properties of the georeferenced, clinical, environmental and sociodemographic, stratified, eigen-orthogonalizable, eigen-spatial filterable, estimator determinants. A derivation of the probability density function of complex inverse Wishart distributed random matrices was constructed. Strategies were outlined for the calculation of the moments of both complex Wishart and complex inverse Wishart distributed matrices. The distribution rendered from the stratified, clinical, environmental and sociodemographic, epidemiologic, hyper/hypoendemic, transmission-oriented, aggregation/non-aggregation-oriented, TB-related, stratified, estimator determinants had an inverse Wishart distribution $\mathrm{A} \sim \mathrm{W}^{-1}(\Psi, m)$.

We let x be an n-dimensional random vector with mean vector μ and covariance matrix Σ. The density was $pdf(x, \mu, \Sigma) = \frac{\exp\left(-\frac{1}{2}(x-\mu)^T \Sigma^{-1}(x-\mu)\right)}{\sqrt{(2\pi)^n|\Sigma|}}$ which here were the determinants of the covariance matrix Σ. The density function from the Wishart distribution in the model was: $pdf(x, \mu, \Sigma) = \frac{1}{C_n(\mu)}|\Sigma|^{-\frac{1}{2}}|x|^{-\frac{\mu-n-1}{2}} \exp\left(-\frac{1}{2}\mathrm{tr}\left(\Sigma^{-1}x\right)\right)$ with $\mu > n$, and the trace of the square matrices A was: $\mathrm{tr}(A) = \sum_i a_{ii}, C_n(\mu) = 2^{\frac{\mu n}{2}}\Gamma_n\left(\frac{\mu}{2}\right), \Gamma_n(z) = \pi^{\frac{n(n-1)}{4}} \prod_{i=1}^n \Gamma\left(z - \frac{i-1}{2}\right)$ Additionally, the density function from the inverse-Wishart distribution was $\text{pdf}(x, \mu, \Sigma) = \frac{1}{D_n(\mu)}|\Sigma|^{\frac{\mu-n-1}{2}}|x|^{-\frac{\mu}{2}}\exp\left(-\frac{1}{2}\text{tr}\left(\Sigma x^{-1}\right)\right)$ for $\mu > 2\text{n}$ and for $D_n(\mu) = 2^{\frac{(\mu-n-1)}{2}}\Gamma_n\left(\frac{\mu-n-1}{2}\right)$.

The marginal and conditional distributions from the inverse Wishart-distributed matrix were then further quantified employing $A \sim W^{-1}(\psi, \text{m})$. In the epidemiologic, non-frequentistic, simulation model ouput, we partitioned the matrices for determining if ψ was conformable with each other employing: $A = \begin{bmatrix} A_{11} & A_{22} \\ A_{21} & A_{22} \end{bmatrix}, \Psi = \begin{bmatrix} \Psi_{11} & \Psi_{22} \\ \Psi_{21} & \Psi_{22} \end{bmatrix}$ where A_{ij} and Ψ_{ij} were $p_i \times p_j$ matrices. We subsequently determined if A_{11} was independent of $\text{A}_{11}^{-1}\text{A}_{12}$ and $\text{A}_{22\cdot 1}$, when $A_{22\cdot 1} = A_{22} - A_{21}A_{11}^{-1}A_{12}$ was the Schur complement of A_{11} in $A_{11} \sim W^{-1}(\Psi_{11}, m - p_2)$ employing $A_\downarrow 11^\uparrow(-1)A_\downarrow 12|A_\downarrow(22 \cdot 1) \sim MNN_\downarrow(p_\downarrow 1 \times p_\downarrow 2)\,\psi_\downarrow 11^\top(-1)\psi_\downarrow 12, A_\downarrow(22 \cdot 1) \otimes \psi_\downarrow 11^\top(-1))$ in the capture point, sentinel site, county-level, stratified, TB, model forecasts where $\text{MN}_{\text{pxq}}(, \cdot)$ was a matrix normal distribution generatable from the empirical, geosampled, clinical, sociodemographic and environmental, stratified, hyper/hypoendemic, aggregation/non-aggregation-oriented, estimator determinants; and $\text{A}_{22\cdot 1} \sim \text{W}^{-1}\left(\Psi'_{11}, \text{m}\right)$. In linear algebra and the theory of matrices. the Schur complement of a matrix block (i.e., a submatrix within a larger matrix) is commonly defined using $p \times p, p \times q, q \times p$ and $q \times q$ matrices, where D is invertible and $M = \begin{bmatrix} A & B \\ C & D \end{bmatrix}$ so that M is a $(p+q) \times (p+q)$ matrix [26]. We assumed if the stratified, geosampled, georeferenced, county-level, estimator determinants $X = x_1, \ldots, x_n$ were independent p-variety Gaussian variables drawn from a distribution, then the conditional distribution, was $\text{W}^{-1}(\text{A}+\Psi, \text{n}+\text{m})$ distribution, where $\text{A} = \text{XX}^\text{T}$ w was quantifiable n times the sample covariance matrix. Because the prior and posterior distributions are the same family (Gelman 2005), the specified, potentially eigenizable, estimator determinant, inverse Wishart distribution in this experiment was the conjugate to the multivariate Gaussian.

The simulated, epidemiological, prognosticated, capture point, sentinel site, county-level, TB, prevalence distributions represented a family of probability distributions definable over an symmetrical, scalable, nonnegative-definite geosampled, empirical dataset of matrix-valued, random variables. These distributions were of great importance in this experiment, for the estimation of covariance matrices for parsimoniously interpolating the multivariate, georeferenced, CDC archived, hyper/hypoendemic, aggregation/non-aggregation-oriented, geosampled, epidemio logic data. The Wishart distribution arises as the distribution of the sample covariance matrix for a sample from a multivariate normal distribution (Hoff 2009). In statistics, the inverse Wishart distribution, also called the inverted Wishart distribution, is a probability distribution defined on real-valued positive-definite matrices [57]. The Wishart distribution is the conjugate prior of the inverse covariance-matrix of a multivariate-normal random-vector (Christian 2007). The probability density function of the inverse Wishart for the epidemiological, prognosticative, county-level, TB model was $\frac{|\Psi|^{\frac{v}{2}}}{2^{\frac{vp}{2}}\Gamma_p(\frac{v}{2})}|X|^{-\frac{v+p+1}{2}}e^{-\frac{1}{2}\mathrm{tr}(\Psi X^{-1})}$, where **X** and Ψ were $p \times p$ positive definite matrices, and $\Gamma_p(\cdot)$ was the multivariate Gamma function. A fit of a hierarchical Bayesian model revealed that the adjusted capture point, sentinel site, georeferenced, stratified, estimator determinant 'Personal Income' was statistically important to the prolific, geosampled, hyperendemic, TB-related, county-level, hot spot, foci (unadjusted improvement χ^2 was – 1.299, while the adjusted χ^2 was – 0.311). The Wishart distribution is a family of distributions for symmetric positive definite matrices. Let X1,..., Xn be independent Np(0, Σ) and form a $p \times n$ data matrix X = [X1, . . . , Xn]. The distribution of a $p \times p$ random matrix M = XX0 = Pni = 1XiX0i is said to have the Wishart distribution (Gelman et al. 2013).

The Markovian property was formulated as follows; $\mathrm{E}[f(X_t)|F_s] = \mathrm{E}[f(X_t)|\sigma X_s]$ for all $t \geq s \geq 0$ and $f : \mathbb{R}^n \rightarrow \mathbb{R}$ bounded and measurable capture point, scalable sentinel site, estimator determinants in the epidemiological, capture point, sentinel site model. In mathematics, specifically in the theory of Markovian stochastic processes in probability theory, the Chapman–Kolmogorov equation(CKE) is an identity relating the joint probability distributions of different sets of coordinates on a stochastic process (Gelman et al. 2013). In mathematics, specifically in the theory of Markovian stochastic processes, the CKE is an identity relating the joint probability distributions of different sets of coordinates on a stochastic process (Sheldon 2014). Here the CKE equation was given us P(n + 1) = P(n) P(1) = P(n) P. We proved that an epidemiologic, prognosticative, regression-related, county-level, TB-related, estimator, determinant model can be optimally, heuristically, optimized, for targeting and prioritizing, georeferenceable, hyper/hypoendemic, aggregation/non-aggregation-oriented, scalable, sentinel site, capture point, geolocations and their estimator determinants. by employing mathematical induction, where $P(n) = Pn$ for all $n \geq 0$.

The epidemiological, prognosticated, stratified, TB county-level vulnerability model, residual estimates revealed $\Pr(X_{n+1} = x|X_1 = x_1, X_2 = x_2, \ldots, X_n = x_n) = \Pr(X_{n+1} = x|X_n = x_n)$. where the probability of going from state i to state j in n time steps was $p_{ij}^{(n)} = \Pr(X_n = j|X_0 = i)$ and the single-step transition was $p_{ij} = \Pr(X_1 = j|X_0 = i)$ For quantifying the external values in the Markov chains we employed $p_{ij}^{(n)} = \Pr(X_{k+n} = j|X_k = i)$ and $p_{ij} = \Pr(X_{k+1} = j|X_k = i)$. The n-step transition probabilities satisfied the CKE, that for any k such that $0 < k < n$, $p_{ij}^{(n)} = \sum_{r\in S} p_{rj}^{(k)} p_{rj}^{(n-k)}$ in the prognosticated, non-frequentistic, model output. Here the CKE. delineated $p_{ij}^{(n)} = \sum_{r\in S} p_{rj}^{(k)} p_{rj}^{(n-k)}$ for $n, m \geq 0$ and for all i, j,. For a finite-state Markov chain, the summation does not go up to infinity and would have an upper limit (Gelman et al. 2013). The number P_{ij}^{n+m} was the probability that the chain was in state j after taking $n + m$ steps in the TB model output i. The following formulation was employed for quantifying the stratified derivatives:

$$
\begin{aligned}
P_{ij}^{n+m} &= P[X_{n+m} = j|X_0 = i] \\
&= \sum_{k=0}^{\infty} P[X_{n+m} = j, X_n = k|X_0 = i]
\end{aligned}
$$

$$= \sum_{k=0}^{\infty} \frac{P\left[X_{n+m} = j, X_n = k, X_0 = i\right]}{P[X_0 = i]}$$

$$= \sum_{k=0}^{\infty} \frac{P[X_n = k, X_0 = i]}{P[X_0 = i]} \times \frac{P\left[X_{n+m} = j, X_n = k, X_0 = i\right]}{P[X_n = k, X_0 = i]}$$

$$= \sum_{k=0}^{\infty} P[X_n = k|X_0 = i] \times P\left[X_{n+m} = j|X_n = k, X_0 = i\right]$$

$$= \sum_{k=0}^{\infty} P[X_n = k|X_0 = i] \times P\left[X_{n+m} = j|X_n = k\right]*$$

$$= \sum_{k=0}^{\infty} P_{ik}^{n} \times P_{kj}^{m}$$

The idea behind the derivation was to determine if the path from state i to state j in $n+m$ steps could be broken down into two paths, one from state i to an intermediate state k in the first n transitions, and the other from state k to state j in the remaining m transitions in the model output. Summing over all intermediate states in the capture point, sentinel site, epidemiological, TB simulation model k yielded the probability that the process will go from state i to state j in $n + m$ transitions. The entries in the matrix Q_n subsequently examined levels of geo-spatiotemporal dependence in the estimator determinants. By doing so, we noted that P_{kj}^{ni} varied in row Q_n (the row corresponding to the state i) and that P_{kj}^{n} varied in column Q_m [the column corresponding to the state j] The final matrix was $\begin{bmatrix} P_{i0}^{n} & P_{i1}^{n} & \cdots & P_{ik}^{n} & \cdots & P_{iw}^{n} \end{bmatrix} \begin{bmatrix} P_{0j}^{m} \\ P_{1j}^{m} \\ \vdots \\ P_{kj}^{m} \\ \vdots \\ P_{wj}^{m} \end{bmatrix}$.

The Markov chain is said to be time homogeneous if the transition probabilities from one state to another are independent of time index n. $p_{ij} = \Pr\{X_n = j|X_{n-1} = i\}$ (Gelman et al. 2013). The transition probability matrix, P, in the prognosticative, capture point, scalable, sentinel site, county-level model was the matrix consisting of the one-step transition probabilities, p_{ij}. We quantitated the m-step transition probability here employing the probability of transitioning from state i to state j in m steps employing $p_{ij}^{(m)} = \Pr\{X_{n+m} = j|X_n = i\}$ so as to encompass the heterogenous, uncertainty-oriented variables. The m-step transition matrix elements in the m-step transition probabilities $p_{ij}^{(m)}$ was expressible as $P^{(m)}$. The m-step transition probabilities for the epidemiologic, TB-related, potentially potential eigenizable, prognosticative, model output was found from the single-step transition probabilities as follows. To transition from i to j in m steps, the process we first transitioned from i to r in $m - k$ steps, and subsequently transitioned from r to j in k steps, where

$0 < k < m.p_{ij}^{(m)} = \sum_r p_{ir}^{m-k} p_{rj}^k$ In matrix form, this became: $P^{(m)} = P^{(m-k)}P^{(k)}$ Setting $k = m - 1$ yielded: $P^{(m)} = P \cdot P^{(m-1)}$ From this equation we derived $P^{(m-1)} = P \cdot P^{(m-2)}$ Substituting this back into the previous equation yielded: $P^{(m)} = P \cdot P \cdot P^{(m-2)}$ Continuing these substitutions, eventually the model rendered $P^{(m)} = P \cdot P \cdot P \cdots P = P^m$ Therefore, the m -step transition probability matrix for the georeferenced, scalable, capture point, sentinel site, prognosticative TB county level model was found by multiplying the single-step probability matrix by itself m times. The state vector at time m was found in terms of the transition probability matrix and the intial state vector.

We first observed that: $\pi_j(m) = \sum \pi_i(m-1)p_{ij}$ in the forecasted model results. In vector and matrix form, this became: $\Pi(m) = \Pi(m-1)P$ We also found that, through substitution: $\Pi(m-1) = \Pi(m-2)P$ or, $\Pi(m) = \Pi(m-2)P \cdot P$ Continuing the substitution yielded: $\Pi(m) = \Pi(0)P^m$ where $\Pi(0)$ was the vector containing the initial probabilities of being in each state at time 0. The probabilities of transition of a Markov chain ξ(t)) from a state i into a state j in the prognosticative, capture point, epidemiologic, TB model at a sample time interval was [s, t][pij(s, t) = P{ξ(t) = j | ξ(s) = i}, s < t. In view of the basic property of a Markov chain, for any states i, j ∈ S ∈ (where S was the set of all states of the chain) and any $s < t < u$ when pij(s, u) = ∑k ∈ Spik(s, t)pkj(t, u) (Gelman et al. 2013). The transition probabilities pij(t)) for a Markovian chain with discrete time were subsequently iteratively deductible by the values of pij(1) i, j ∈ S for any $t > 0 > \sum j \in$ Spij(t) = 1 which subsequently forecasted prevalence rates per county.

Subsequently, a simulation study was performed in which the impact of families of prior distributions for the covariance matrix of the multivariate dataset of the random effects, transmission-related, capture point, sentinel site, county-level, hot/cold spot, scalable, TB model, estimator determinants. Inferences about georeferenced, hyper/hypoendemic, cluster sizes were not particularly sensitive to prior choice, but the related covariance estimates were able to distinguish hot/cold spots at the county level. The multivariate, stratified, capture point, county-level, aggregation/non-aggregation-oriented, hyper/hypoendemic, scaled-up, sentinel site, forecast, model formulation was based on a random effects model which here was given by $\mathbf{Y}_{\rm i} = \theta + \mu_{\rm i} + \varepsilon_{\rm i};\ 1 \le {\rm i} \le {\rm n}$, where $\mathbf{Y}_{\rm i} = \left({\rm Y}_{\rm i1}, {\rm Y}_{\rm i2}, \ldots, {\rm Y}_{\rm ip}\right)^{\top}$. This was the vector of the stratified, hyper/hypoendemic, TB-related, estimator, determinant, cluster sizes, $\theta = \left(\theta_1, \theta_2, \ldots, \theta_{\rm p}\right)^{\top}$ which was based on the vector of a, scalable, capture point, sentinel site, county-level, georeferenceable county-level, cluster population, effect size, and $\mu_i = \left(\mu_{\rm i1}, \mu_{\rm i2}, \ldots, \mu_{\rm ip}\right)^{\top}$ and $\varepsilon_{\rm i} = \left(\varepsilon_{\rm i1}, \varepsilon_{\rm i2}, \ldots, \varepsilon_{\rm ip}\right)^{\top}$. These latter products were the vectors of the stratified random effects and sampling errors which we employed for optimally optimizing, targeting and prioritizing, the georeferenced, county-level, capture point, sentinel sites for delineating, hyper/hypoendemic, transmission related, aggregation/non-aggregation-oriented estimator determinants.

In our multivariate, stratified, vulnerability-oriented, epidemiological, hot/cold spot, prognosticative, TB model, we needed to estimate covariance matrix, **D**. It is often assumed that $\boldsymbol{\mu}_{\rm i}$ and $\boldsymbol{\varepsilon}_{\rm i}$ are independent, with E[$\mathbf{m}_{\rm i}$] = E[$\mathbf{+}_{\rm i}$] = $\mathbf{0}$ and $\text{Var}(\varepsilon_{\rm i}) = \boldsymbol{\Sigma}_{\rm i}$, where **D** and $\boldsymbol{\Sigma}_{\rm i}$ (assumed to be known) are the $p \times p$

between- and within covariance matrices (Griffth 2003). We assumed that ($\boldsymbol{\theta}$, $\mathbf{D}$) were random with a prior distribution of the form $\pi(\theta, \mathbf{D}) = \pi(\theta)\pi(\mathbf{D})$. Inference about ($\boldsymbol{\theta}$, $\mathbf{D}$) was based on the posterior distribution of ($\boldsymbol{\theta}$, $\mathbf{D}$), given the scalable, geosampled, capture point, sentinel site, county-level, data capture point ($\mathbf{y}_i$, Σ_i), which in the model forecasts was proportional to $\left[\exp\left\{-1/2 \sum \mathrm{i} = \ln\left[(\mathrm{yi} - \mu\mathrm{i})' \sum -1\mathrm{i}(\mathrm{yi} - \mu\mathrm{i}) + (\mu\mathrm{i} - \theta)'\mathrm{D} - 1(\mu\mathrm{i} - \theta)\right]\right\}\right]$ $\pi(\theta)\pi(\mathrm{D})$ [Eq. 4] Prior distributions for $\boldsymbol{\theta}$ under model 3.1 have been studied in the literature [19]. These priors included an improper prior of the form $\pi(\theta) \propto 1$, and also a normal prior $\mathrm{N_p}(0, \mathbf{C}_0)$, where $\mathbf{C}_0$ was a known $p \times p$ matrix. Here the variance covariance matrix for the prognosticative, TB-related, county-level model was $\mathbf{D}$ and $\theta \sim \mathrm{N_p}(0, \mathrm{C}_0)$, the conditional posterior densities of $\boldsymbol{\mu}_i$ and $\boldsymbol{\theta}$ on $\mathfrak{D}_i = \{\mathbf{y}_i, \Sigma_i, \mathbf{D}\}$ when $\mu\mathrm{i} \mid \theta, \mathrm{Di} \sim \mathrm{Np}\left(\mathrm{Vi}\left(\sum -1\mathrm{iyi} + \mathrm{D} - 1\theta\right), \mathrm{Vi}\right)$ and $\theta|\mu\mathrm{i}, \mathrm{Di} \sim \mathrm{Np}\left(\mathrm{V}\theta\mathrm{D} - 1 \sum \mathrm{i} = \ln \mu\mathrm{i}, \mathrm{V}\theta\right)$ and where $\mathrm{Vi} = \left(\sum -1\bar{\mathrm{i}} + \mathrm{D} - 1\right) - 1$ and $\mathrm{V}\theta = (\mathrm{nD} - 1 + \mathrm{C} - 10) - 1$.

The matrix $\mathbf{D}$ was the covariance matrix on the second level of the TB-related, estimator determinant, model as provided by Jacob et al. (2013). Here the matrix contributed to the marginal variability of $\mathbf{Y}_i$ since $\mathrm{Var}(\mathbf{Y}_i) = \mathbf{D} + \Sigma_i$. The diagonal elements of $\mathbf{D}$ represented the level of heterogeneity between each georeferenced, stratified, prognosticated, model outcome [i.e., aggregation/non-aggregation-oriented, hyper/hypo-endemic, county-level, geolocation, geosampled, estimator determinant] and its off-diagonal elements were the outcomes of the intra-cluster correlations. The correlations indicated how the scalable, hot/cold spot, population "true" effect sizes were correlated across the counties by the affected differences [i.e., hyper/hypo endemicity of the stratified, aggregation/non-aggregation, transmission-related, interpolated, GPS indexed, estimator, determinant, hot/cold spot, potentially, eigenizable, feature, attribute LULC characteristics]. One of the advantages of utilizing multivariate equations in eigen-Bayesian eigenvector eigengeospace for constructing epidemiologic, prognosticative, county-level, TB, prevalence models may be the robust estimation of effect cluster sizes with higher degree of precision. Here the eigenizable, estimator determinant iteration was aided by quantifying the stratified, georeferenceable, intra-cluster, TB-related, capture point, sentinel site, estimator determinant correlations of aggregation/non-aggregation, hyper/hypo-endemic propensities.

We considered the conjugate inverse Wishart prior parametrization under non-informative and informative set-ups; objective and reference priors for $\mathbf{D}$ that lead to a proper posterior distribution for the epidemiologic, geosampled, capture point, sentinel site, county-level, prognosticative hierarchical model; independent priors on the covariance components; and. finally, the constrained and mixture Wishart prior distribution families. We illustrated the parametrization of these priors assuming that $p = 2$ when $\mathbf{D}$ and $\boldsymbol{\Sigma}_{\mathrm{iwere}}$ were as follows: $\mathrm{D} = \left[\tau^2 1\tau 12\tau 12\tau^2 2\right], \sum \mathrm{i} = \left[\sigma^2 \mathrm{i}1\sigma\mathrm{i}12\sigma\mathrm{i}12\sigma^2\mathrm{i}2\right], \tau 12 = \tau 1\tau^2\rho\mathrm{b}, \sigma\mathrm{i}12 = \sigma\mathrm{il}\sigma\mathrm{i}^2\rho\mathrm{iw}$, where $\tau_2 > 0$, $\tau_2 > 0$, $-1 < \rho_b < 1$, $\sigma_{i1} > 0$, $\sigma_{i2} > 0$, and $-1 < \rho_{iw} < 1$ $\forall_i$. Here, $\tau^2 1, \tau^2 2$ and ρ_b described the variances and intra-correlations between the, stratified, TB, estimator determinants whereas σ^2i1, σ^2i2 and ρ_{iw} revealed the cluster variation. We also defined $\mathbf{\Sigma}_0$ as $\ln \sum \mathrm{ni} = 1\Sigma$ which was denotable by $\lambda_1 \geq \lambda_2$ the ordered eigenvalues of **D**.

A $p \times p$ matrix **X** has a Wishart distribution, $\mathrm{W_p(v, \mathbf{R})}$ if its associated PDF is proportional to $|\mathbf{X}|^{(v-p-1)/2} \exp\{\mathrm{tr}(-\mathbf{X}\mathbf{R}^{-1})/2\}, \mathbf{X} > 0$, where $v > 0$ is the df and **R** is a $p \times p$ symmetric scale matrix (Gelman 2005). An inverse Wishart prior distribution was generatable from the geosampled, stratified, prognosticative, capture point, sentinel site model, eigenizable, estimator determinants. We assumed that $\mathbf{D}^{-1}$ had a Wishart distribution. Specifically, assuming that $\mathbf{D}^{-1} \sim W_p(v, \mathbf{R}^{-1})$, the prior mean of the matrix $\mathbf{D}^{-1}$ was $\mathrm{v}\mathbf{R}^{-1}$, which corresponded to a prior mean of **D** equaled to $\mathbf{R}/(v-p-1)$. The condition $v > p+1$ here implied a finite prior mean for **D**, but the Wishart prior for $\mathbf{D}^{-1}$ was proper for $v > p{-}1$ in the prognosticated, stratifiable, TB, scalable, capture point, sentinel site, county-level, epidemiologic, georeferenceable, model output.

An inverse Wishart prior of the form $\mathrm{IW_p(p, \mathbf{I}_p)}$ is commonly used as a non-informative proper prior and its implementation is straightforward (Gelman et al. 2013). Unfortunately, the selection of v and **R** was not straightforward in our model. The matrix **R** is often set at the ML estimator of **D** for non-hierarchically forecasting models (Gelman 2015). Here we employed an empirical Bayes approach to estimate v in a non-hierarchical normalized model. In general, the specification of **R** and v are influential for constructing prognosticative models (Griffth 2003). The simulation study evaluated several priors and showed that inverse Wishart prior distributions (hierarchical prior and fixed v) estimated the covariance matrix. The model revealed high Bayes risk under the loss function $\mathrm{L}(\mathrm{D}^{\sim}, \mathbf{D}) = \mathrm{tr}(\mathrm{D}^{\sim}\mathbf{D}^{-1}) - \log|\mathrm{D}^{\sim}\mathbf{D}^{-1}| - \mathrm{p}.]$. Hence, the model had optimal performance since it had an output that exemplified a robustifiable, hierarchy of the geosampled, stratified, potentially eigenizable, hyper/hypoendemic, aggregation/non-aggregation-oriented, prognosticative, estimator determinants.

We considered the matrix variate normal model, $\mathrm{Xp} \times \mathrm{m} \sim \mathrm{Np} \times \mathrm{m}(\mu\mathrm{p} \times \mathrm{m}, \Phi\mathrm{p} \times \mathrm{p} \otimes \Sigma\, \mathrm{m} \times \mathrm{m})$ with density function given $\mathrm{byf(X)} = (2\pi) - \mathrm{mp}/2|\Phi| - \mathrm{m}/2|\Sigma| - \mathrm{p}/2\,\mathrm{etr}\{-12(\mathrm{X}-\mu)T\Phi - 1(\mathrm{X}-\mu)\Sigma - 1\}\mathrm{by\, f(X)} = (2\pi){-}\mathrm{mp}/2|\Phi|{-}\mathrm{m}/2|\Sigma|{-}\mathrm{p}/2\,\mathrm{etr}\{-12(\mathrm{X}{-}\mu)T\Phi{-}1(\mathrm{X}{-}\mu)\Sigma{-}1\}$ with Φ a known $p \times p$ matrix and μ and Σ unknown. Next, we considered an objective prior for μ, and the hypergeometric Wishart distribution which on our model was definable as $s = 0$, $t = 0$ and $\mathrm{h(x)} = \mathrm{vFw(a1, \ldots, av; b1, \ldots, bw; cx)}$ where the time series dependent, geosampled, stratified, clinical, sociodemographic and environmental, epidemiological, county-level TB estimator determinants $\Theta, a1, \ldots, av, b1, \ldots, bw, c$ and n were subjective priors for Σ. In this experiment these were denotable as $\Sigma \sim \mathrm{HWm}(\Theta, a1, \ldots, av, b1, \ldots, bw, c, n)$, such that the joint prior density function was $(11)\pi(\mu, \Sigma) = |\Theta| - \mathrm{n}/2|\Sigma|\mathrm{n}/2 - (\mathrm{m}+1)/2\Gamma\,\mathrm{m(n/2)v} + 1\mathrm{Fw(\, nm/2, a1, \ldots, av; b1, \ldots, bw; c)vFw\{a1, \ldots, av,\ b1, \ldots, bw; ctr}(\Theta - 1\Sigma)\}\,\mathrm{etr}(-\Theta - 1\Sigma)$ for $v = w = 0$ and $c = 1/2$. The prior density function of Σ was the density of a Wishart random matrix with parameter Θ and n df.

Based on a scalable, hyper/hypoendemic, capture point, TB-related, geosampled, epidemiological observation, X, the joint PDF was $(\mu, \Sigma) \propto |\Sigma|(\mathrm{n}-\mathrm{p})/2 - (\mathrm{m}+1)/2\,\mathrm{etr}\{-12(\mathrm{X}-\mu)\mathrm{T}\Phi - 1(\mathrm{X}-\mu)\Sigma - 1\}\mathrm{vFw\{a1, \ldots, av,\ b1, \ldots, bw; ctr}(\Theta -$

1Σ)} etr(−Θ − 1Σ) · T. The model margin PDF of μ was q(μ | x) ∝ ∫ Sm |Σ|(n − p)/2 − (m + 1)/2 etr{−12(x − μ)TΦ − 1(x − μ)Σ − 1 − Θ − 1Σ}vFw{a1, ..., av, b1, ..., bw; ctr(Θ − 1Σ)}dΣ ∝ EΣ ∗ [etr{−12(x − μ)TΦ − 1(x − μ)Σ ∗ −1} × vFw{a1, ..., av, b1, ..., bw; ctr(Θ − 1Σ∗)}], where Σ∗ ∼ Wm(Θ/2, n − p). Subsequently, q(μ | x) = CμEΣ ∗ [etr{−12(x − μ)TΦ − 1(x − μ)Σ ∗ −1} × vFw{a1, ..., av, b1, ..., bw; c tr(Θ − 1Σ∗)}], where Cμ − 1 = ∫ μEΣ ∗ [etr{−12(x − μ)TΦ − 1(x − μ)Σ ∗ −1} × vFw{a1, ..., av, b1, ..., bw; ctr(Θ − 1Σ∗)}]dμ was rewritten as Cμ − 1 = ∫ Sm 1Γ m{(n − p)/2}2 m(n − p)/2|Σ|(n − p)/2 − (m + 1)/2|12Θ| − (n − p)/2 etr(−Θ − 1Σ) ∫ μ etr{−12(x − μ)TΦ − 1(x − μ)Σ − 1}dμ × vFw{a1, ..., av, b1, ..., bw; ctr(Θ − 1Σ)}Dσ. Therefore, Cμ − 1 = (2π)mp/2|Φ|m/2Γm(n/2)|Θ|p/2Γm{(n − p)/2}EΣ1[vFw{a1, ..., av, b1, ... , bw; ctr(Θ − 1Σ∗)}], wherere Σ1 ∼ Wm(Θ/2, n). Henceforth, q(μ|x) = Γm{(n − p)/2}(2π) − mp/2|Φ| − m/2|Θ| − p/2Γm(n/2)EΣ1 [vFw{a1, ..., av, b1, ..., bw; ctr(Θ − 1Σ∗)}]EΣ ∗ [etr{−12(x − μ)TΦ − 1(x − μ)Σ ∗ −1} × vFw{a1, ..., av, b1, ..., bw; ctr(Θ − 1Σ∗)}] may be considered a scalable, capture point, sentinel site, county-level, iteratively, interpolatable epidemiological TB model margin posterior density where Σ∗ ∼ Wm(Θ/2, n − p) and Σ1 ∼ Wm(Θ/2, n).

The marginal posterior density function of Σ in our county-level, prognosticative, capture point, sentinel site, scalable TB model was given by q(Σ|x) = CΣ|Σ|(n − p)/2 − (m + 1)/2vFw{a1, ..., av, b1, ..., bw; ctr(Θ − 1Σ)}etr (−Θ − 1Σ) ∫ μetr{−12(x −μ)TΦ − 1(x − μ)Σ − 1}dμ∝ |Σ|n/2 − (m + 1)/2vFw{a1, ..., av, b1, ..., bw; ctr(Θ − 1Σ)} × etr(−Θ − 1Σ) with CΣ − 1 = 2 nm/2Γ m(n2)|Θ2|n/2EΣ1[vFw{a1, ..., av, b1, bw; ctr(Θ − 1Σ1)}], where Σ1 ∼ Wm(Θ/2, n). Hence, q(Σ | x) = |Θ| − n/2|Σ|n/2 − (m + 1)2Γ m(n/2)EΣ1[vFw{a1, ..., av, b1, ..., bw; ctr(Θ − 1Σ1)}]vFw{a1, ..., av, b1, ..., bw; ctr(Θ − 1Σ)} etr(−Θ − 1Σ) Note, q(μ | Σ) ∝ etr{−12(μ − x)TΦ − 1(μ − x)Σ − 1},i.e., μ | Σ, x ∼ N(x, Φp × p ⊗ Σ m × m) and q(Σ | μ) ∝ |Σ|(n − p)/2 − (m + 1)/2 etr{−12(X − μ)TΦ − 1(X − μ)Σ − 1 − Θ − 1Σ}vFw{a1, ..., av, b1, ..., bw;ctr(Θ − 1Σ)}.

Under the squared error loss function, the eigenizable, Bayes estimator in the prognosticative, hierarchical, generalizable, model stratifiable estimator determinants was the posterior mean values. Hence, the Bayes estimator of μ was (12)μ̂ = E(μ|X) since E(μ − X | X) = ∫ μ(μ − X)q(μ | X)dμ = Γm{(n − p)/2}(2π) − mp/2|Φ| − m/2|Θ| − p/2Γm(n/2)EΣ1[vFw{a1, ..., av, b1, ..., bw; ctr(Θ − 1Σ∗)}]EΣ ∗ [vFw{a1, ..., av, b1, ..., bw; ctr(Θ − 1Σ∗)}] ∫ μ(μ − X) etr{−12(X − μ)TΦ − 1(X − μ)Σ ∗ −1}dμ. As for the Bayes estimator of |Σ|r, it was given by E(|Σ|r | X) = v + 1Fw{(n/2 + r)m, a1, ..., av, b1, ..., bw; c}Γm(n/2 + r)Γm(n/2)EΣ1[vFw{a1, ..., av, b1, ..., bw; ctr(Θ − 1Σ1)}] from where Σ1 ∼ Wm(Θ/2, n).

The results of the preceding subsection were applied to the simulated, empirical, georeferenced, uncertainty-oriented dataset of the geosampled, time series, dependent, semi-parameterized estimators to illustrate the advantage of the hypergeometric Wishart distribution as a robust prior for the covariance matrix of the matrix variate normal model for teasing out error variance non-normality and other potential biases due to violations of regression assumption in eigen-Bayesian, eigen-vector. eigen-geospace. A sample of size n was simulated from a multivariate normal error distribution with a dimensionality of m, zero mean vector and an identity covariance matrix, [i.e., $X1 \times m \sim Nm(01 \times m, Im \times m]$. The assumed priors were the inverse Wishart, Wishart and hypergeometric Wishart distributions, [i.e., $\Sigma \sim IWm(\Theta 1, n1)$, $\Sigma \sim Wm(\Theta 1, n2)$ and $\Sigma \sim HWm(\Theta 1, a1,\ b1, c, n3]$, respectively, with $\Theta 1 = 4 \times Im$, $m = 3$, $nl = 9.5$, $n2 = 3$, $a1 = 1$, $b1 = 2$, $c = 1$, $n = 20$, $n3 = 3$.

Posterior samples of size 10,000 were simulated employing a Gibbs sampling scheme with an additional Metropolis–Hastings algorithm. In statistics, Gibbs sampling or a Gibbs sampler is a MCMC algorithm for obtaining a sequence of observations which are approximated from a specified multivariate probability distribution, when direct sampling is difficult (Gelman et al. 2013). The convergence of the Markov chain of matrices were evaluated graphically employing two measures, the estimator determinant, trace and largest eigenvalue of the simulated matrix. Gibbs sampling can be used to obtain draws from complicated joint distributions such as those generated by time series dependent, geosampled, potentially eigenizable, TB stratified, estimator determinants [1].

The Frobenius norm of the errors, were definable as $\|\Sigma - \Sigma\|F = \mathrm{tr}\{(\Sigma^2 - \Sigma)T(\Sigma - \Sigma)\}$ which was calculated for each geosampled, hyper/hypoendemic, clinical, sociodemographic and environmental, stratified, estimate given in Table 4. The Frobenius norm is sub-multiplicative and is very useful for numerical linear algebra (Hosmer and Lemeshow 2000).

The sub-multiplicativity of Frobenius norm in the scalable, capture point, county-level, vulnerability-oriented, prognosticative, epidemiological model output was proved employing Cauchy–Schwarz inequality [i.e. for any n-dimensional vectors u and v, and equality occurs if and only if $v = cu$]. The hypergeometric Wishart prior resulted in the smallest Frobenius norm of the error in the sampled, scalable, hyper/hypoendemic, capture point, sentinel site, estimator determinants. For further investigation, this sampling scheme was repeated 100 times to obtain 100 estimates under each prior as well as the ML for each simulated, county-level, scalable TB-stratified sample. The Frobenius norm of the eigenizable, estimator determinants and every repetition of it was calculable. It was clear that the hypergeometric Wishart prior resulted in the lowest Frobenius norm and therefore produced less error than the all the other estimator determinants.

Table 4 Error norms calculated from the simulated sample

Frobenius norm	Value
$\|\hat{}\Sigma MLE - \Sigma\|F$	0.4191
$\|\hat{}\Sigma IW - \Sigma\|F$	74.136
$\|\hat{}\Sigma W - \Sigma\|F$	0.2713
$\|\hat{}\Sigma HW - \Sigma\|F$	0.2433

Additionally, the ECDF of each set of Frobenius norms was calculable for, the capture point, sentinel site, county-level, TB, stratified, hyper/hypoendemic, transmission-related, aggregation/non-aggregation, hot/cold spot determinants which was quantified here using the ECDF function in Numpy.linalg.norm. This technique provided a simple way to compute and plot a "stairstep" ECDF for the estimator determinant, county-level, geosampled, epidemiologic data. In the simplest cases, this estimate made discrete jumps of 1/n at each georeferenced, epidemiologic, data capture point. [Fi, xi] = ecdf (x); stairs (xi, Fi, 'r'); xlim ([−2.5 5]); xlabel ('x'); ylabel ('F (x)'); The ECDF is regarded as the best uncertainty estimator since for a specific value of the error norm, a higher proportion of estimates from that particular prior results in less error [32]. To validate the graphical interpretation, a two-sample K-S test was subsequently performed, pairwise, on the varying different ECDF's and the p-value for all calculated (see Table 5).

From Table 5 it is clear that the ECDF of the errors under the hypergeometric Wishart prior was significantly different from the other priors. Henceforth, the assertion can be made that the hypergeometric Wishart prior structure renders an estimate with frequentist superiority when constructing a capture point, scaled up, sentinel site, epidemiological, TB-related, prognosticative, vulnerability-oriented, county-level, stratified model for optimally, heuristically, optimizing, targeting and prioritizing, georeferenceable, hyper/hypo-endemic, aggregation/non-aggregation-oriented, scalable, data capture points. The assumed priors here were $\Sigma \sim \mathrm{IWm}(\Theta\mathrm{IW}, \mathrm{nIW})$, $\Sigma \sim \mathrm{Wm}(\Theta\ \mathrm{W}, \mathrm{nW})$ and $\Sigma \sim \mathrm{HWm}(\Theta HW, \mathrm{a1}, \ \mathrm{b1}, \mathrm{c}, \mathrm{nHW})$, respectively. The hyperparameters were chosen as $\Theta\mathrm{IW} = 4\ \mathrm{S1}, \mathrm{nIW} = 9, \Theta\ \mathrm{W} = 0.25\ \mathrm{S1}, \mathrm{nW} = 3, \Theta HW = 0.5\ \mathrm{S1}, \mathrm{a1} = 1, \ \mathrm{b1} = 2, \mathrm{c} = 0.9, \mathrm{nHW} = 3$, with S1 equal to the sample covariance matrix. Based on these estimates of Σ, a multivariate empirical dataset of size 100 was simulated for each georeferenced, capture point, sentinel site, hyper/hypo-endemic, stratified, clinical, sociodemiographic, or environmental, prognosticated, geosampled, estimator determinant per county [e.g., $\mathrm{Yi}, 1 \times 6 \sim \mathrm{N6}\left(\mathrm{X}^{-}, \Sigma\mathrm{Q}\right), \mathrm{Q} = (\mathrm{MLE}, \mathrm{IW}, \ \mathrm{W}, \mathrm{HW})$].

We subsequently generated the inverse-Gamma distribution which was a univariate specialization of the inverse-Wishart distribution which summarized the geosampled, georeferenced, hyper/hypoendemic, transmission-oriented, aggregation/non-aggregation-oriented, scalable, estimator determinants. The PDF was $\frac{\beta^{\alpha}}{\Gamma(\alpha)} x^{-\alpha-1} \exp\left(\frac{-\beta}{x}\right)$ while the mean of the model was $\frac{\beta}{\alpha-1}$ for $\alpha > 1$. The variance was $\frac{\beta^2}{(\alpha-1)^2(\alpha-2)}$ for $\alpha > 2$. The skewness was $\frac{4\sqrt{\alpha-2}}{\alpha-3}$ for $\alpha > 3$, while the kurtosis was $\frac{30\alpha-66}{(\alpha-3)(\alpha-4)}$ for $\alpha > 4$. The moment generating function was $\frac{2(-\beta t)^{\frac{\alpha}{2}}}{\Gamma(\alpha)} K_{\alpha}\left(\sqrt{-4\beta t}\right)$,

Table 5 p-values of the K-S two-sample test based on samples of the Frobenius norms

Pairwise comparison	p-value
W and HW	<0.001
MLE and HW	<0.001

while the characteristic function was $\frac{2(-i\beta t)^{\frac{\alpha}{2}}}{\Gamma(\alpha)} K_\alpha\left(\sqrt{-4i\beta t}\right)$. In probability theory and statistics, the moment-generating function of a real-valued random variable is an alternative specification of its probability distribution [38]. Here the stratified, prognosticated, epidemiologic, capture point, sentinel site, stratified, TB, estimator. determinant, georeferenced, model output provided the basis of an alternative route to analytical results compared with working directly with the PDF.

The stratified model revealed that w $p = 1$ (i.e., univariate) and $\alpha = m/2, \dot{\beta} = \Psi/2$, and $x = B$ the PDF of the inverse-Wishart distribution was $p(x|\alpha, \beta) = \frac{\beta^\alpha x^{-\alpha-1} \exp(-\beta/x)}{\Gamma_1(\alpha)}$. The PDF of the Gamma distribution was $f(x) = x^{k-1} \frac{e^{-x/\theta}}{\theta^k \Gamma(k)}$. We optimally defined the transformation $Y = g(X) = \frac{1}{X}$, employing the resulting transformation where we found:$f_Y(y) = f_X\left(g^{-1}(y)\right)\left|\frac{d}{dy} g^{-1}(y)\right| = \frac{1}{\theta^k \Gamma(k)}\left(\frac{1}{y}\right)^{k-1} \exp\left(\frac{-1}{\theta y}\right)\frac{1}{y^2} = \frac{1}{\theta^k \Gamma(k)}\left(\frac{1}{y}\right)^{k+1} \exp\left(\frac{-1}{\theta y}\right) = \frac{1}{\theta^k \Gamma(k)} y^{-k-1} \exp\left(\frac{-1}{\theta y}\right)$.

Replacing k with α; θ^{-1} with β; and y with x resulted in the inverse-Gamma PDF. In probability theory and statistics, the inverse gamma distribution is a two-parameter family of continuous probability distributions on the positive real line, which is the distribution of the reciprocal of a variable distributed according to the Gamma distribution [38]. $f(x) = \frac{\beta^\alpha}{\Gamma(\alpha)} x^{-\alpha-1} \exp\left(\frac{-\beta}{x}\right)$. The inverse Gamma distribution's PDF was definable over the support $x > 0$ employing the equation $f(x; \alpha, \beta) = \frac{\beta^\alpha}{\Gamma(\alpha)}(x)^{-\alpha-1} \exp\left(-\frac{\beta}{x}\right)$ with shape parameter α and scale parameter β. The PDF was quantifiable employing the regularized Gamma function $F(x; \alpha, \beta) = \frac{\Gamma\left(\alpha, \frac{\beta}{x}\right)}{\Gamma(\alpha)} = Q\left(\alpha, \frac{\beta}{x}\right)$ where the numerator in the hyper/hypo-endemic, aggregation/nn-aggregation-oriented, county-level, scalable, capture point, sentinel site, geo-spatiotemporal, geosampled, epidemiologic, risk-related model was the upper incomplete Gamma function and the denominator was the Gamma function.

The PDF of a random, explanatory, TB-related, geosampled, capture point, sentinel site, county-level variable X was denotable by F (x), and was definable as $F(x) = \Pr(X \leq x)$. Using the probability of disjoint events, if X is a discrete random variable, we can write where xn is the largest possible value of X that is less than or equal to x (Casella 1990). The regularized Gamma functions were definable in the eigenized, hot/cold spot, transmission-oriented hyper/hypoendemic, estimator, determinant, capture point, county-level, prognosticative, risk model by $\frac{\gamma(a, z)}{\Gamma(a)}$ and $\frac{\Gamma(a, z)}{\Gamma(a)}$ where $\gamma(a, z)$ and $\Gamma(a, z)$ were incomplete Gamma functions, and $\Gamma(a)$ was a complete Gamma function.

We also employed methods in PyMC3 to calculate the multivariate Gamma function for geo-spatiotemporally quantitating the stratified, georeferenced, hyper/hypoendemic, transmission-oriented, prognosticative, epidemiological, vulnerability-oriented model interpolators and their uncertainty-related estimates. The function was constructed employing $\Gamma_p(a) = \int_{S>0} \exp(-\mathrm{trace}(S))\,|S|^{a-(p+1)/2} dS$ where $S > 0$ thus, S was positive-definite. We employed the Gamma function to determine the recursive relationships in the estimator determinants employing $\Gamma_p(a) = \pi^{(p-1)/2}\Gamma(a)\Gamma_{p-1}(a - \frac{1}{2}) = \pi^{(p-1)/2}\Gamma_{p-1}(a)\Gamma\left[a + (1 - p)/2\right]$.

Next, we quantified $\Gamma_1(a) = \Gamma(a)$, $\Gamma_2(a) = \pi^{1/2}\Gamma(a)\Gamma(a - 1/2)$, and $\Gamma_3(a) = \pi^{3/2}\Gamma(a)\Gamma(a - 1/2)\ \Gamma(a - 1)$. We defined the multivariate digamma function in the asymptotical, prognosticative, estimator determinants as $\psi_p(a) = \frac{\partial \log \Gamma_p(a)}{\partial a} = \sum_{i=1}^{p} \psi(a + (1 - i)/2)$ and the general polygamma function as $\psi_p^{(n)}(a) = \frac{\partial^n \log \Gamma_p(a)}{\partial a^n} = \sum_{i=1}^{p} \psi^{(n)}(a + (1 - i)/2)$ The digamma function in the epidemiological, risk model ouput was definable as the logarithmic derivative of the Gamma function:$\psi(x) = \frac{d}{dx} \ln \Gamma(x) = \frac{\Gamma'(x)}{\Gamma(x)}$ This equation calculated the digamma function which was expressable as $\frac{\partial \Gamma(a+(1-i)/2)}{\partial a} = \psi(a+(i-1)/2)\Gamma(a+(i-1)/2)$. In mathematics, the digamma function is defined as the logarithmic derivative o the gamma function (Abramowitz and Stegun 1972). The roots of the digamma function are the saddle points of the complex-valued gamma function (Weisstein 2002); thus, all the residually, forecasted, georeferencable, explanatory, time–series dependent, clinical, sociodemiographic and environmental, stratified, TB-related, optimally, heuristically, optimizable, hyper/hypoendemic, transmission-related, stratifiable, estimator, determinant, derivatives generated, lay on the real axis. The digamma function was expressable as $\psi_0(x)$, $\psi^0(x)$.

Subsequently, we generated the following expression: $\frac{\partial \Gamma_p(a)}{\partial a} = \pi^{p(p-1)/4} \prod_{j=1}^{p} \Gamma(a+(1-j)/2) \sum_{i=1}^{p} \psi(a+(1-i)/2) = \Gamma_p(a) \sum_{i=1}^{p} \psi(a+(1-i)/2)$ Since in this experiment $\Gamma_p(a) = \pi^{p(p-1)/4} \prod_{j=1}^{p} \Gamma(a + \frac{1-j}{2})$, was also an epidemiologic, prognosticatable, model, robust, ouput, it followed that $\frac{\partial \Gamma_p(a)}{\partial a} = \pi^{p(p-1)/4} \sum_{i=1}^{p} \frac{\partial \Gamma(a+\frac{1-i}{2})}{\partial a} \prod_{j=1, j\neq i}^{p} \Gamma(a + \frac{1-j}{2})$. The PDF of the inverse Wishart for the, explanatory, eigen decomposable, eigen-spatil filter, TB-related, scalable, capture point, county-level, stratified, model was found to be $\frac{|\Psi|^{\frac{\nu}{2}}}{2^{\frac{\nu p}{2}}\Gamma_p(\frac{\nu}{2})}|\mathbf{X}|^{-\frac{\nu+p+1}{2}} e^{-\frac{1}{2}\mathrm{tr}(\Psi\mathbf{X}^{-1})}$ when X and Ψ were $p \times p$ positive definite matrices, and $\Gamma_p(\cdot)$ was the multivariate Gamma function. The Wishart distribution W(Σ, d, n) is a probability distribution of random nonnegative-definite $d \times d$ matrices that is used to model random covariance matrices (Gelman et al. 2013). The parameter n is the number of df, and Σ is a nonnegative-definite symmetric $d \times d$ matrix that is called the scale matrix (Gelman et al. 2003).

A wide variety of options exist for checking the convergence of an MCMC chain. PyMC3 includes many of those options by default. For the sake of brevity, this experiment focused on producing one diagnostic which required a bit of extra effort in PyMC3, the Gelman-Rubin diagnostic test. The Gelman-Rubin diagnostic compares the results from two or more chains (Gelman et al. 2013). Once the chains converged, the principle drew from both chains which was virtually indistinguishable. As such, the Gelman-Rubin diagnostic compared the variance within one chain to the variance between the chains rendered from the scalable, capture point, sentinel site, stratified, TB, county-level, simulation model, geosampled, hyper/hypoendemic, aggregation/non-aggregations-oriented, estimator determinants. We noted that the ratio of the variances [i.e., the potential scale reduction factor] was close to 1, and, as such, the chains were indistinguishable, meaning they may have converged on the target distribution [i.e., georeferenced geolocation of an epidemiologic, capture point, sentinel site, transmission-related, hyper/hypo-endemic, county-level, scaled-up, TB, estimator determinant, stratified, hot/cold spot].

Finally, the deviance residual in the model output was definable as the square root of the contribution of the nth geosampled, scalable, capture point, sentinel site, county-level, stratifiable, TB-related, estimator determinant to the deviance, with the sign of the raw, residual, quantitated, summary, epidemiologic data in PyMC3. The adjusted Pearson, deviance and likelihood residuals were definable employing the models of [5], and Davison and Snell (1991). The residuals were useful for outlier detection and for assessing the influence of single, geosampled, scalable, capture point, sentinel site, TB-related, transmission-related, aggregation/non-aggregation-oriented, observational, evidential prognosticator on the fitted model. For our county-level, epidemiologic, operationalizable, time-series, dependent, iterative, interpolative, risk model, the variance of the nth individual, geosampled, TB-related, stratified, eigenizable, estimator determinants were simulatable when the dispersion parameter was a user-specified prior weight and when the mean, and the variance function was geo-spatiotemporally quantifiable.

We also constructed a robust, diagonal, georeferenceable matrix for denoting the nth diagonal element from the geosampled, empirical, epidemiological georeferenced dataset of stratified, estimator determinants. The weight matrix was employed in computing the expected information matrix. The Pearson residual was standardizable to have unit asymptotic variance. We requested Pearson residuals in our model output employing PyMC3. The likelihood residuals were definable. We iteratively quantitated the local DIC to assist in model selection to visualize the global and local impacts of adding the georeferenced, stratified, TB parameters to the eigen-Bayesian, estimation matrix. DIC statistics were generated to identify the best fitting model.

The deviance in the prognosticative, county-level, estimator determinant, TB model was definable as -2^* log (likelihood), where the 'likelihood' was approximately expressible as p(y | and theta). This included all the normalizing constants where y comprised all stochastic node values and theta stochastic parents of y in the autoregressive, georeferenced, time-series, dependent, hyper/hypo-endemic, aggregation/non-aggregation-oriented, transmission-oriented, capture point, sentinel site, county-level, scaled-up, geosampled, TB-related, risk model, stratified iteratively, interpolative, estimator determinants. The definition of deviance was -2^* log (likelihood) and the 'likelihood' was definable as p(y | theta),when y comprised all stochastic nodes given the geosampled, estimator, determinant, discrete, integer values (i.e., stratified hyper/hypoendemic, transmission-related, capture point, sentinel site, simulated, observational, evidential prognosticators), which was comprised of the of y – 'stochastic parents' when the stochastic nodes were dependent upon the distribution of y. 'Stochastic parents' are the stochastic nodes upon which the distribution of y depends upon when collapsing over all logical relationships [25]. In this experiment when $y^{\sim}$ Dnorm (mu, tau), then tau was a function of a parameter phi, the prior distribution in the prognosticative, transmission-oriented, capture point, epidemiological, sentinel site, risk model was definable. Thereafter, the likelihood was definable as a function of phi in the model. The expectation $\overline{D} = E^{\theta}[D(\theta)]$ was employable as a measure of model fitness based on the hyper/hypoendemic, aggregation/non-aggregation-related, regressed, interpolated, geosampled, estimator determinant, discrete, integer values. The effective number of the scalable, hot/cold spot, stratified, estimator determinants included in the model was computable employing $p_D = \overline{D} - D(\overline{\theta})$, where $\overline{\theta}$ was the expectation of θ.

The DIC generated the following conclusions: (1) the Dbar, was the posterior mean of the deviance, (2) the Dhat, was the point estimate of the deviance (i.e., –2* log (likelihood)) obtained by substituting the posterior means theta bar of theta which subsequently rendered Dhat = – 2* log (p(y — theta. bar); and 3) pD when the effective number of transmission-related, hyper/hypo-endemic, aggregation/non-aggregtation-oriented, estimator determinants was provided by pD = Dbar – Dha and pD using the posterior mean of the deviance minus the deviance of the posterior means. In normal hierarchical models, Pad = TR (H), where H is the 'hat' matrix maps the observed data to their fitted values (Afifi et al. 2004). The DIC was subsequently calculable as: $DIC = p_D + D$. The DIC value for the finalized, capture point, sentinel site, county-level, risk model, stratifiable, hot/cold spot, prognosticated, transmission-related, county-level, estimator determinants weas 947.6.

When the stochastic process under consideration is Markovian, the CKE is equivalent to an identity on transition densities (Spiegelhalter 2002). In our Markovian, county-level, capture point, sentinel site, prognosticative, eigen-spatial filter, simulation model we assumed that $i_1 < \ldots < i_n$. As suggested by Gelman et al. (2013) this was because of the Markovian property. Our capture point, georeferenced, sentinel site, county-level, capture point, sentinel site, TB risk model rendered, $p_{i_1,\ldots,i_n}(f_1,\ldots,f_n) = p_{i_1}(f_1)p_{i_2;i_1}(f_2|f_1)\ldots p_{i_n;i_{n-1}}(f_n|f_{n-1})$ where the conditional probability $p_{i;j}(f_i|f_j)$ was the transition probability between the sample times $i > j$.

In this experiment, CKE took the form $p_{i_3;i_1}(f_3|f_1) = \int_{-\infty}^{\infty} p_{i_3;i_2}(f_3|f_2)p_{i_2;i_1}(f_2|f_1)df_2$ where S was the state space of the Markov chain in the capture point, sentinel site, county-level, simulation model. We let $p_{i_1,\ldots,i_n}(f_1,\ldots,f_n)$ be the joint PDF. If the stratified, clinical, sociodemiographic and environmental, estimator, determinant, geosampled, discrete, integer values delineating the random variables f_1 to f_n then, the CKE equation generated by the random variables was $p_{i_1,\ldots,i_{n-1}}(f_1,\ldots,f_{n-1}) = \int_{-\infty}^{\infty} p_{i_1,\ldots,i_n}(f_1,\ldots,f_{n-1})df_n$ employing a straightforward marginalization over the nuisance variables. Note, that we did employ temporality for maintaining ordering of the geosampled, clinical, sociodemographic and environmental, stratified, capture point, sentinel site, random variables in the equation. The potentially eigenizable variables were applied equally to the marginalization of any geosampled, capture point, sentinel site, estimator determinant. When the stochastic process under consideration was Markovian, the CKE was equivalent to an identity on transition densities.

In the Markov chain setting, we assumed that $i_1 < \ldots < i_n$. Then, because of the Markov property, $p_{i_1,\ldots,i_n}(f_1,\ldots,f_n) = p_{i_1}(f_1)p_{i_2;i_1}(f_2|f_1)\ldots p_{i_n;p_{n-1}}(f_n|f_{n-1})$ the conditional probability $p_{i;j}(f_i|f_j)$ was the transition probability between the times $i > j$ in the model forecasts. The CKE generated employing the, scalable, county-level, capture point, hyper/hypoendemic, aggregation/non-aggregation-oriented, geosampled epidemiologic, TB stratfiable, transmission-related, data then took the form $p_{i_3;i_1}(f_3|f_1) = \int_{-\infty}^{\infty} p_{i_3;i_2}(f_3|f_2)p_{i_2;i_1}(f_2|f_1)df_2$. Our model revealed that when the probability distribution on the state space of a Markov chain is discrete, the chain is homogeneous. We expressed the CKEs in terms of (possibly infinite-dimensional) matrix multiplication, thus: $P(t+s) = P(t)\,P(s)$ where $P(t)$ was the transition matrix of jump t, (i.e., $P(t)$ is the matrix such that entry (i, j) contained the probability of the chain moving from state i to state j in t steps).

Additionally in the prognosticated, capture point, sentinel site, county-level, epidemiologic, risk analyses it followed that to calculate the transition matrices of jump t, it was sufficient to raise the transition matrix of jump one to the power of t, that is $P(t) = P^t$. The marginal distribution $\Pr(X_n = x)$ was the distribution over states at time n in the residuals. The initial distribution was $\Pr(X_0 = x)$. The evolution of the process through each step was described by

$$
\begin{aligned}
&\Pr(X_n = j)\\
&= \sum_{r \in S} p_{rj} \Pr(X_{n-1} = r)\\
&= \sum_{r \in S} {}_{p}^{(n)}\text{prj} \Pr(X_0 = r)
\end{aligned}
$$

Further, we extended the capture point, sentinel site, vulnerability-oriented, potentially eigenizable, county-level, estimator determinant prognosticative risk analyses to show that that distance between the nth step transition probability and the invariant probability measure in the scalable, vulnerability-oriented, capture point, sentinel site, epidemiological, prognosticative, diagnostically stratified, TB model output was bounded by p*[a + bg(x)] for the constant $a > b > 0$ and $p < 1$. The invariant was employed to obtain convergence rates to quantify the transition probabilities for quantifying the autoregressive processes in the epidemiological prognosticated model output employing a random walk on a half line. Here a random walk with reflecting zone on the nonnegative integers generated from the stratified, simulated, iterated, estimator determinants was a Markov chain whose transition probabilities $q(x, y)$ were those of a random walk [i.e., $q(x, y) = p(y - x)$] which was outside a finite set $\{0, 1, 2, \ldots, \text{K}\}$. As such, the distribution $q(x,)$ stochastically dominated $p(-x)$ for every $x \in \{0, 1, 2, \ldots, \text{K}\}$ Under mild hypothesis, we proved that if $\sum xp_x > 0$, occurs in an epidemiological capture point, georeferenced, prognosticated, TB-related, county-level, vulnerability-oriented, regression, model output, there estimator determinants transition probabilities can be easily quantitated by employing $q^n(x, y) \sim C_{xy}R^{-n_n} - \frac{3}{2}$ or $n \to \infty$ [in our case the forecasted model output revealed $\sum xp_x\rangle = 0$, $q^n(x, y) \sim C_{xy}n^{-\frac{1}{2}}$. We did extend and strengthen the stratified, TB model residuals in countable state space. Our inference for MCMC simulation techniques was based on weighted OLS proposals. We employed a weighting scheme for these proposals within the MCMC algorithm to improve statistical efficiency for optimally heuristically optimizing, targeting and prioritizing, georeferenced, geosampled, capture point, scalable, sentinel site, TB-related, hyper/hypoendemic, aggregation/non-aggregation-oriented, stratifiable, estimator determinants. These methods may be more powerful when combined with multi-proposal MCMC algorithms such as multiple-try Metropolis, which may efficiently exploit modern computer architectures with large numbers of cores. The locally weighted MCMC method may improve upon a partial parallelization of the Metropolis Hastings algorithm via Rao-Blackwellization for optimally heuristically optimizing, targeting and prioritizing, georeferenceable, hyper/hypo-endemic, aggregation/non-aggregation-oriented, stratifiable estimator determinants. We may be able to parsimoniously derive the effective sample size from the output of this algorithm and show how to approximate scalable, TB, estimator determinant, stratifiable, county-level, capture point, sentinel sites in future research efforts. This epidemiological TB-related, forecast modeling system may introduce a universal scheme that guarantees variance reductions also in Metropolis–Hastings-treated, scalable, county-level, hyper/hypoendemic, aggregation/non-aggregation-oriented, estimator determinants while keeping the computation cost under control. A TB researcher or epidemiologist may establish a central limit theorem for the improved estimator determinants and even illustrate their performances employing a probit model.

We let δ_{ij} denote the hyper/hypoendemic, aggregation/non-aggregation-oriented, stratified determinants for j for subject $i, i = 1\ldots, n$, and $j = 1, \ldots 4$. We assumed that δ_{ij} was determinable by a latent variable z_{ij}. Specifically, we assumed $\delta_{ij} = I(z_{ij} > 0), z_i = (z_{i1}, \ldots, z_{i4})^T \sim N(\mu_i, I_4)$ where I(A) was equal to 1 if A was true. The parameter $\mu_i = (\mu_{i1}, \mu_{i2}, \mu_{i3}, \mu_{i4})^T$ and I_4 was a 4 by 4 identity matrix. We decomposed μ_{ij} as spatial random effects $f(x_i, y_i)$ and subject effect η_i, i.e., $\mu_{ij} = f_i(x_i, y_i) + \eta_i, f_i \sim GP\left(f_0, \sigma_j^2 x\right), \eta_i \sim N\left(0, \sigma^2\right)$ where x_i and y_i represented the latitude and the longitude of each geosampled, capture point, sentinel site, hyper/hypoendemic, county-level, aggregation/non-aggregation-oriented, scalable, TB estimator determinant. The random effect f_i'were modeled using Gaussian Process. The prior distributions for the parameters.

Unadjusted effects			Adjusted effects	
	Improvement		Improvement	
Variable	df	χ^2	χ^2	df
Personal income	1	1	1.299	1

were: $\sigma^2 \sim \text{Inv} - \text{Gamma}(a_0, b_0), \sigma_j^2 \sim Inv - \text{Gamma}(a_0, b_0), f_0 \sim GP\left(0, \sigma_0^2 x\right)$ when $k(a, b) = \exp(-\|a - b\|/\rho)$ and $\rho \sim \text{Uniform}(c, d)$. We decomposed $\kappa(a, b) = \sum_{l=1}^{\infty} \lambda_l \phi_l(a)\phi_l(b)$ as $\sum_{l=1}^{L} \lambda_i \phi_l(a)\phi(b)$ and in doing a representation of f_i for $i = 1 \ldots 2$ was iterative.

The PDF of the inverse Wishart for the scalable, capture point, sentinel site, hyper/hypo-endemic, county-level, prognosticative, epidemiological model

$\frac{|\Psi|^{\frac{\nu}{2}}}{2^{\frac{\nu p}{2}}\Gamma_p(\frac{\nu}{2})}|X|^{-\frac{\nu+p+1}{2}}e^{-\frac{1}{2}\mathrm{tr}(\Psi X^{-1})}$, where **X** and Ψ were $p \times p$ positive definite matrices, and $\Gamma_p(\cdot)$ was the multivariate Gamma function. A fit of a generalizable, hierarchical, inferential, eigen-Bayesian, eigen-algorithmic MCMC model revealed that the adjusted covariate Personal Income was statistically important to prolific geosampled estimator determinants(adjusted improvement χ^2 was – 1.299) (see Table 6).

The simulated stratified, TB, iteration chains converged graphically. In the traceplot we noted that the chains took draws from the same target distribution which appeared to be indistinguishable. The potential scale reduction factor was approximated.

Median parameter values, as well as the 95% credibility intervals (2.5 percentile and 97.5 percentile values), were generatable from the autoregressive, interpolated, asymptotically normalized, hyper/hypoendemic, TB-related, estimator, determinant, geosampled, empirical dataset. As the sampling sites increased based on the explanatory, determinant ‘Personal income of < 20,000, the median log-count of TB increased. The adjusted model robustly quantized the independence amongst the geosampled, time-series, dependent, stratified, clinical, socio-demiographic, and environmental, hyper/hypo-endemic, transmission-oriented, aggregation/non-aggregation-related, stratified, hot/cold spot, estimator determinants representing the count data. We noted this model fit better that the model that adjusted for correlation within the study site based on the root mean squate error RMSE.

The (RMSE) and mean absolute error (MAE) are widely used metrics for evaluating models. The root-mean-square deviation (RMSD) or RMSE is a frequently used measure to quantify differences between values (sample or population values) predicted by a model or an estimator and the values observed (Draper and Smith 1985). Yet, there remains enduring confusion over their use, such that currently a standard practice is to present both, leaving it to the TB researcher or an epidemiologist to decide which is more relevant. In a recent reprise to the 200-year debate over their use, Willmott and Matsuura (2005) and Chai and Draxler (2014) give arguments for favoring one metric or the other. However, this comparison can present a false dichotomy in epidemiological, prognosticative, capture point, sentinel site, scalable,

Table 6 Estimation results using for the numpy and pandas libraries

Variable	DF	Estimate	Standard error	t Value	Approx Pr > |t|
Intercept	1	0.3867	0.08764	7.53	<0.0001
Variable	DF	Estimate	Standard error	t Value	Approx Pr > |t|
Intercept	1	0.3211	0.0494	13.12	<0.0001
ARCH0	1	02,213	0.0693	3.69	0.0003
ARCH1	1	0.3462	0.0224	6.18	<0.0001
GARCH1	1	0.8564	0.0118	17.45	<0.0001

vulnerability-oriented TB county-level modeling. Neither metric is inherently better: RMSE is optimal for normal (Gaussian) errors, and MAE is optimal for Laplacian errors rendered from these paradigms. When errors deviate from these distributions, other metrics may be superior especially for optimally, optimizing, georeferenced, aggregation/non-aggregation-oriented, hyper/hypoendemic, TB-related, capture point, scalable, transmission-oriented, geo-spatiotemporal, geosampled, stratifiable, clinical, sociodemiographic and environmental, eigenizable estimator determinants.

The geo-spatiotemporal, hyper/hypoendemic, aggregation/non-aggregation-oriented, transmission-related, TB, county-level, estimator determinants where subsequently analyzed with an eigen-spatial filtering eigen-algorithm employing a positive-definite covariance matrix. In doing so, we were able to transform all the geosampled, scalable, georeferenceable, interpolatable, hyper/hypo-endemic, TB-related, aggregation/non-aggregation-oriented, empirical, geosampled, epidemiological, data, capture point feature attributes containing dependence into estimator determinants free of dependence by partitioning the original, sentinel site, stratified data into two synthetic variates: (1) an eigen-spatial filter variate capturing latent spatial dependency and (2) a non-spatial variate that was free of spatial dependence. The geographic distribution of the county-level, georeferenced, epidemiologic, TB data exhibited PSA based on the autocorrelated, county-level, geosampled counts in all the county-level, capture point, hot/cold spot, geo-spatiotemporal, stratified geolocations which were aggregated in eigenvector eigen-geospace based on the sentinel site count data.

We subsequently partitioned the matrices A and Ψ employing $\mathbf{A} = \begin{bmatrix} \mathbf{A}_{11} & \mathbf{A}_{12} \\ \mathbf{A}_{21} & \mathbf{A}_{22} \end{bmatrix}$, $\Psi = \begin{bmatrix} \Psi_{11} & \Psi_{12} \\ \Psi_{21} & \Psi_{22} \end{bmatrix}$, where A_{ij} and Ψ_{ij} were $p_i a \times p_j$ matrices in PyMC3 i We generated the conjugate distribution of the georeferenced, scalable, hyper/hypo-endemic, county-level, transmission-oriented, geosampled, aggregation/non-aggregation-related, hot/cold spot, eigenized, estimator determinants by employing a covariance matrix $\sum$, whose prior $p(\sum)$ had a $W^{-1}(\Psi,\ m)$ distribution. Using PyMC3 can aid in testing for the stability of the stratified, regression coefficients and test for stationarity or unit roots in the time series (https://support.sas.com/).

The eigenvectors generated from county-level, TB estimator determinants were selected with a stepwise regression procedure in PyMC3. Because $(\mathbf{I} - 11'/n)\mathbf{C}(\mathbf{I} - 11'/n) = \mathbf{E} \wedge \mathbf{E}'$, where E is an n-by-n matrix of eigenvectors and Λ is an n-by-n diagonal matrix of the corresponding eigenvalues, (Griffith 2003) the resulting geosampled, scaled-up, capture point, sentinel site, time series, dependent, eigenized, estimator determinant, county-level, TB, risk model specification was given by: $\mathbf{Y} = \mu\mathbf{1} + \mathbf{E}_k\beta + \varepsilon$ In the unbiased, asymptotical, autoregressive, specified, hot/cold spot, vulnerability-oriented, prognosticative model, μ the scalar mean of Y, E_k was an n-by-k matrix containing the subset of $k <<_n$ eigenvectors selected with a stepwise regression technique where β was a k-by-1 vector of regression coefficients.

Several eigen-spatial filter eigenvectors were extractable from $(\mathbf{I} - \mathbf{11}'/n)\, C\, (\mathbf{I} - \mathbf{11}'/n)$ which we noted were affiliated with the georeferenced, geosampled, stratifiable, hyper/hypoendemic, epidemiological, stratified, geographic patterns of county-level, clinical, environmental and sociodemiographic, TB-related, interpolated, time series, dependent, aggregation/non-aggregation-oriented, estimator determinants. These determinants portrayed a negligible degree of latent geo-spatiotemporal autocorrelation. Consequently, only k of the n eigenvectors was of interest for generating a candidate set for conducting a stepwise regression procedure. Here candidate eigen-spatial filter eigenvectors represented a level of latent geo-spatiotemporal autocorrelation which we assumed could account for the redundant information in the epidemiologic, prognosticative, scaled-up, capture point, sentinel site, county-level, TB-related, stratified risk model for optimally, heuristically, optimizing, transmission-related, hot/cold spots and their time-series, dependent, hyper/hypo-endemic, georeferenceable, aggregation/non-aggregation-oriented, eigenized, estimator, determinant, propensity patterns.

Of note, the 2-by-2 square tessellation rendered a repeated eigenvalue in the prognosticated, TB-related, epidemiologic, risk, model output. To identify spatialized clusters of georeferenced, scaled-up, capture point, county-level, geosampled, high density, hyper/hypoendemic, transmission-related, aggregation/non-aggregation-oriented, stratifiable, estimator determinants, Thiessen polygon surface portioning was performed in python. In doing so, we generated a weighted neighborhood-matrix. Each grid cell had a unique identifier. This identifier was the county-level prevalence count. These matrices were employable in the autocorrelation, synthetically, eigenized, estimator, determinant, eigenvector risk analysis. Entries in matrix were 1, if two, stratified, georeferenced, hyper/hypoendemic, transmission-related, semi-parameterizable, estimator determinants shared a common Thiessen polygon boundary, and 0 otherwise.

Next, the linkage structure for each surface in the PyMC3 model was edited to remove unlikely geographic neighbors to identify pairs of georeferenceable, county-level, hyper/hypoendemic, estimator determinants sharing a common Thiessen polygon boundary. Attention was restricted to those georeferenceable, interpolative, iterative, capture point, sentinel site, hot/cold spot, stratifiable, risk-related, country level, map patterns associated with at least a minimum level of non-zero, latent, geo-spatiotemporal, autocorrelation coefficients which, for implementation purposes was definable here by $|MC_j/MC_{\max}| > 0.25$, where MC_j denoted the j-th value and $MC_{\max}$ was the maximum value of Moran's coefficient. This threshold value allowed two candidate sets of eigenvectors to be considered for quantitation of substantial PSA and NSA, respectively. These statistics indicated that the detected NSA may be considered statistically significant in an epidemiologic, prognosticative, TB-related, capture point, sentinel site, time-series, dependent, regressable, empirical. epidemiologic, dataset of scaled-up, county-level, georeferenceable, eigenizable, hyper/hypoendemic, transmission-oriented, diagnostically stratifiable, geosampled, clinical, environmental and sociodemiographic, synthetically, eigenized, estimator determinants based upon a randomization perspective.

Of note, the ratio of the PRESS (i.e., predicted error sum of squares) statistic to the sum of squared errors from the Moran coefficient scatterplot trend line was 1.24. Fortunately, this value fell within two standard deviations of the average standard prediction error value (roughly 1.17) for a prognosticatable, synthetic, georeferenceable, scaled-up, capture point, sentinel site, county-level, hyper/hypoendemic, transmission-related, aggregation/non-aggregation-oriented, time series, dependent, stratifiable, TB, estimator determinant. Because prevalence counts were being analyzed, a Poissonian spatial filter model specification was employed in PyMC3.

The epidemiological, prognosticative, transmission-oriented, hyper/hypoendemic, county level, risk model specification was written as follows: $LN(\mu) = \alpha\mathbf{1} + \mathbf{E}_k\beta,\ \sigma_i^{\angle} = \mu_i(1 - \eta\mu_i)$, in PyMC3 i where μ_i was the expected prevalence count for a georeferenced, county-level, geosampled, aggregation/non-aggregation-oriented, hot/cold spot, capture point, sentinel site, stratifiable geolocation I and where μ was an n-by-1 vector of the expected scalable count variables. LN denoted the natural logarithm (i.e., the GLM link function), α was an intercept term; and η was the negative binomial dispersion parameter. This log-linear equation had no error term. Estimation was parsimoniously executable assuming a negative binomial random variable.

Since adjacent, georeferenced, scalable, time-series, dependent, hyper/hypoendemic, transmission-oriented, prognosticatable, clinical, sociodemiographic and environmental, time-series, dependent, stratified, geosampled, epidemiologic, eigenized, estimator determinants were involved in the autoregressive function, we employed a first-order specification, with the autoregressive term being CY. An important matrix was optimally derivable from C1 which was the vector of the number of county-level, high density, hot/cold spot, scaled-up, hyper/hypoendemic, transmission-related, stratifiable, georefernceable, capture point sentinel site neighbors. The inverse of the elements of C1 were insertable into the diagonal of a matrix (i.e., $\mathbf{D}^{-1}$) rendering matrix W = $D^{-1}C$, which in this experiment became a stochastic matrix (i.e., each of its row sums equaled 1). One appealing feature of this matrix was the autoregressive term became WY, which generated averages, rather than sums of the neighboring, hyper/hypoendemic, aggregation/non-aggregation-oriented, geosampled, geo-spatiotemporally eigenizable, estimator, determinant, discrete, integer values. Because a covariance matrix for a robust, county-level, hyper/hypo-endemic, prognosticative, TB-related, transmission-related, dependent distribution must be symmetrical [42], we employed a matrix W specification with a conditional autoregressive model by articulating the individual, geosampled, estimator variance w, non- constantly employing $(\mathbf{I} - \rho\mathbf{D}^{-1}\mathbf{C})\mathbf{D}^{-1} = (\mathbf{D}^{-1} - \rho\mathbf{D}^{-1}\mathbf{C}\mathbf{D}^{-1})$ An appealing feature of this model version was that it restricted, risk-related, stratifiable, estimator, determinant, hyper/hypoendemic, aggregation/non-aggregation-oriented, geosampled, discrete, integer values of the autoregressive parameter to the more intuitively interpretable range of $0 \leq \hat{\rho} \leq 1$. The predictive, scalable, capture point, sentinel site, risk-related model furnished an alternative specification which here was also written in terms of matrix W. The spatial covariance was a function of the matrix $(\mathbf{I} - \rho\mathbf{C}\mathbf{D}^{-1})(\mathbf{I} - \rho\mathbf{D}^{-1}\mathbf{C}) = (\mathbf{I} - \rho\mathbf{W}^T)(\mathbf{I} - \rho\mathbf{W})$ where T denoted the matrix transpose. The resulting matrix was symmetric and was considered a second-order specification as it included the product of two spatial structure matrices (i.e., $W^T W$).

We generated a geographically weighted regression [GWR]-inspired, random effects. model specifications for the geosampled, stratified, TB-related, capture point, sentinel sites. GWR is an extension of OLS regression that models relationships as they vary across geospace by evaluating where locally weighted regression coefficients deviate from global coefficients [44] Griffith (2003). Here the GWR identified local non-stationarity in the geosampled, discrete integer values for delineating sentinel site, capture point, georeferenced, TB-related estimator determinant; stationarity referred to the situation where mean, variance, and dependency of location did not change through space. In effect, GWR moved a weighted window over the spatial TB, stratified epidemiological data and estimated unbiased slope and intercept parameters from "fit" scalable, georeferenced, hyper/hypoendemic, aggregation/non-aggregation-oriented, sentinel site, capture points which were then compared to global parameters estimated from a distance matrix of all the estimator determinants within the CDC archived, empirical, geosampled dataset. The size of the window (i.e., the bandwidth) were calibrated in various ways, and a kernel density function was employed to weigh data as a function of distance from the fit, stratified, TB, estimator, determinant, transmission-related, sentinel site, capture points. In this way, the regression equation was modifiable over geographic space to factor out variation due to changing autocorrelation structure across a Floridan county landscape.

One assumption about the GWR random effects term in our vulnerability-oriented, county-level, eigenvector, TB-related, capture point, sentinel site, hyper/hypoendemic, epidemiological, prognosticative risk model was that its frequency distribution was bell-shaped (i.e., normally/Gaussian distributed) with a mean of zero. The empirical estimate had a mean of 0.0191, with a Shapiro–Wilk normality probability of 0.0023. In other words, the mean was not exactly zero, although its t statistic was 0.07 which indicated it was not significantly different from zero, and the frequency distribution deviated from a bell-shaped one, but not dramatically. This random effect term accounted for roughly 41% of the variability in the observed probability of TB by county in Florida and yielded an under-dispersed binomial model.

The epidemiologic, prognosticative, hyper/hypoendemic, county-level, TB-related, stratified, transmission-related, scalable, capture point, sentinel site, aggregation/non-aggregation-oriented, estimator determinants generated from the second order eigenfunction, eigen-decomposition eigen-algorithm in python quantitated any latent, eigen-autocorrelation, erroneous, non-normal, zero, biased aggregation/non-aggregation-oriented, geosampled clinical, sociodemiographic and environmental, coefficients in the interpolated, time-series, dependent, variance estimates. Eigenvectors were extracted from the matrix $\left(\mathbf{I}-\mathbf{11}'/n\right) C \left(\mathbf{I}-\mathbf{11}'/n\right)$ employing the, georefernced, hot/cold spot, stratified determinants. It was noted that the autoregressive, estimators captured the latent, non-zero, geo-spatiotemporal, eigen-autocorrelation coefficients in the georeferenced, multivariate, hyper/hypoendemic, regressed, stratified, transmission-oriented, epidemiologic, geosampled, county-level, TB, epidemiological, risk model, residually, prognosticated, operationalizable, stratified, discrete derivatives. This quantification involved ρ, a conditional autoregressive covariance specification and the matrix $\left(\mathbf{I}^{-}\rho C\right)$, where I was an n-by-n identity matrix.

We noticed the stratified, asymptotical, semi-parameterized, hyper/hypoendemic, aggregation/non-aggregation-oriented, estimator, determinant, time series, dependent, clinical, environmental and sociodemiographic, stratified, eigen-orthogonalized, eigen-spatial filters had low discrete, integer, coefficient values and hence they were not suitable for defining spatial structures in the scalable, capture point, vulnerability-oriented, county-level, model forecasts. The diagonalization of the spatial weighting matrix rendered from the eigenized, estimator determinants instead consisted of finding the normalized vectors u_i stored as columns in the matrix $U = [u_1 \cdots u_n]$, for satisfying:$\Omega = HWH = U\Lambda U^T = \sum_{i=1}^{n} \lambda_i u_i u_i^T$, where $\Lambda = diag(\lambda_1 \cdots \lambda_n)$, $u_i^T u_i = \|u_i\|^2 = 1$ and $u_i^T u_j = 0$ when $i \neq j$. Note that double centering of Ω implied that the hyper/hypoendemic, transmission-related aggregation/non-aggregation-oriented, scalable, eigen-orthogonalized eigenvectors u_i generated from the interpolated, georeferenced, hot/cold spot, stratified, TB-related, scalable, sentinel site, empirical, county-level data, epidemiologic, data, capture points were centered whereby at least one eigenvalue was equal to zero. Introducing these eigenvectors in the original formulation of Moran's I led to

$$I(x) = \frac{n}{1^T W 1} \frac{x^T HWHx}{x^T Hx} = \frac{n}{1^T W 1} \frac{x^T U \Lambda U^T x}{x^T Hx} = \frac{n}{1^T W 1} \frac{\sum_{i=1}^{n} \lambda_i x^T u_i u_i^T x}{x^T Hx}. \quad (6)$$

Considering the centered vector $z = Hx$ and employing the properties of idempotence of H, Eq. (4) was calculable as being equivalent to the formulation:

$$I(x) = \frac{n}{1^T W 1} \frac{\sum_{i=1}^{n} \lambda_i z^T u_t u_i^T z}{z^T z} = \frac{n}{1^T W 1} \frac{\sum_{i=1}^{n} \lambda_i \left\| u_i^T z \right\|^2}{\|z\|^2}. \quad (7)$$

We subsequently transformed the geosampled, specified epidemiological, georeferenced dataset of county level, capture point, sentinel site, prognosticative indicators to define correlation, among the geosampled stratified, estimator determinants. The remote specified Bayesian eigenvectors u_i and vector z were subsequently centered, and the time-series, dependent, predictive, county-level, regression-related equation was optimally, heuristically, optimizable for targeting and prioritizing, georeferenceable, scaled-up, capture point, county-level, TB-related, epidemiologic, hot/cold spot, stratifiable, space–time, eigenized, estimator determinants which here was rewritten as:

$$I(x) = \frac{n}{1^T W 1} \frac{\sum_{i=1}^{n} \lambda_i cor^2(u_i, z) \mathrm{var}(z) n}{\mathrm{var}(z) n} = \frac{n}{1^T W 1} \sum_{i=1}^{n} \lambda_i cor^2(u_i, z). \quad (8)$$

where r was the number of null eigenvalues of $\Omega(r \geq 1)$. If an eigenvalue of A is zero, it means that the kernel (nullspace) of the matrix is nonzero (Griffith 2003). The null space of our geo-spatiotemporal, autocorrelation, regression matrix A consisted of all the vectors B such that AB = 0 when B was not zero [i.e., it was thought as the solution obtained from AB = 0 where A is known matrix of size m x n and B is matrix to be found of size n x k]. This meant that the matrix in our scalable, epidemiologic, prognosticative, capture point, sentinel site model for optimally, heuristically, optimizing, georeferenceable, hyper/hypo-endemic, transmission-related, aggregation/non-aggregation-oriented, county-level, TB-related, hot/spots and their respective estimator determinants was equal to zero. Our matrix however was not invertible.

In our prognosticative, epidemiological, eigenized, estimator, determinant model for optimally, heuristically, optimizing, targeting and prioritizing georeferenceable, stratifiable, TB-related, estimator, determinant, hyper/hypo-endemic, county-level, aggregation/non-aggregation-oriented, capture point, sentinel site, hot/cold spot geolocations Ax = 0 but had non-zero solutions (A had nonpivot columns).The null space always exists in an epidemiologic, regression-related, stratifiable, TB. time series dependent model [42]. In our model it reduced to 0, hence we had only trivial solutions.

Furthermore, the dimension of the null space in the scalable, epidemiologic, prognosticated, georeferenced capture point, sentinel site, estimator determinant, model derivatives was $k < n$, and $0 < k$ which revealed linearly independent vectors each of which yielded the 0 vector when the linear operator A was applied to it. We standardized the eigen-decomposition eigen-algorithm to quantitate 0 by employing an eigenvalue of geometric multiplicity k in the forecasted, county-level, stratified, vulnerability-oriented, model output. This implied that that the estimator, determinant, model algebraic multiplicity was greater than or equal to k. The characteristic equation subsequently had a factor x_n for $n \geq k$ rather than for "k eigenvalues" for the prognosticated values of k in the empirical, epidemiologic, geosampled dataset of eigendecomposed eigenvalues that happened to have the same discrete integer value. In this experiment these eigenvalues and corresponding eigenvectors were removed from Λ and U, respectively. The stratified epidemiological forecasts (i.e., county-level, hyper/hypoendemic, transmission-related epi-center georeferenced geolocations and their estimator determinants) were then strictly equivalent to $I(x) = \frac{n}{1^T W 1} \sum_{i=1}^{n-r} \lambda_i cor^2(u_i, z)$. Moreover, it was demonstrated that Moran's index for a given eigenvector u_i was equal to $I(u_i) = (n/ 1^T W 1)\lambda_i$, so the equation for the autoregressively forecasted, hot/cold spot, capture point, sentinel site, county-level, TB model was rewritten: $I(x) = \sum_{i=1}^{n-r} I(u_i)cor^2(u_i, z)$. The term $cor^2(u_i, z)$ represented the part of the variance of z that was explainable by u_i in the geosampled, time-series, dependent, transmission-oriented, epidemiologic, hyper/hypo-endemic, risk model [i.e., $z = \beta_i u_i + e_i$]. This quantity was equal to $\beta_i^2 / n\text{var}(z)$. By definition, the eigenvectors u_i were orthogonal and therefore the regression coefficients of the linear models $z = \beta_i u_i + e_i$ were those of the multiple regression model $z = U\beta + \varepsilon = \beta_i u_i + \ldots + \beta_{n-r} u_{n-r} + \varepsilon$.

Next, the distribution of the georeferenced, geosampled, eigenized, clinical, environmental and sociodemiographic, aggregation/non-aggregation-oriented, erroneous, non-normal, non-Gaussian hyper/hypoendemic, stratified, residuals in the county-level, transmission-oriented, autocovariance matrix revealed that the maximum value of I was obtainable by all of the variations of z, as explained by the eigenvector u_1, which here coincidentally corresponded to the highest eigenvalue λ_1 in the eigen-autocorrelation, diagnostic, error matrix.

Alternative coefficients were centered on their global constant coefficient counterparts (i.e., Moran eigenvectors having a mean of zero). Any multicollinearity was traceable to their common eigenvectors. The geographic visualizations of the capture point, hyper/hypo-endemic, TB-related, prognosticatable, capture point, sentinel site, epidemiologic, data tended to be fragmented rather than artificially smooth. Once constructed, interaction terms for a slope were selected by a stepwise forward selection procedure; the covariates simply needed to be factored from them, with the resulting linearizable combination of eigenvectors. One of the major advantages of eigen-spatial filtering is that it exploits the spatial eigen-autocorrelation, statistical distribution theory for linear regression residuals developed by [21] (Fig. 8).

We constructed a GARCH model. Many extensions of the simple GARCH model have been developed in the literature. For the implementation of the GARCH model, we need to import the autoregressive conditional heteroscedasticity (ARCH) module from Python. This library comes with some functions like archmodel() that help get the results. Along with this, the Numpy and Pandas libraries are also required for implementation.

The ARCH model is a statistical model for time series data that describes the variance of the current error term or innovation as a function of the actual sizes of the previous time periods' error terms; often the variance is related to the squares of the previous innovations (Hull and White 1987). The ARCH model is appropriate when the error variance in a time series follows an AR model; autoregressive moving, epidemiological, prognosticative, scalable, county-level, capture point, sentinel site, TB model for heuristically optimizing geolocating hot/cold spot geolocations and their capture point, sentinel site, scalable hyper/hypoendemic, aggregation/non-aggregation-oriented, estimator determinants. The general process for our epidemiologic, prognosticative, TB-related, capture point, sentinel site, county-level, GARCH model involved three steps. The first was to estimate a best-fitting

```
sensitivity = 0.90
specificity = 0.99
prior − 0.074
likelihood = sensitivity  # p(test|disease present)
marginal_likelihood = sensitivity * prior + (1 - specificity) * (1 - prior)
posterior = (likelihood * prior) / marginal_likelihood
posterior
0.8779330345373055
```

Fig. 8 The PyMC3 generated code to interpret posterior distributions in the epidemiological, eigen-Bayesian, semi-parametric, Markovian, TB county-level, non-Gaussian prognosticative, stratified model autoregressive model. The second was to compute determinant-related autocorrelations of the error term. The third step was to test for significance. In this experiment we illustrate the estimation of variants of GARCH models employing the Python procedures for optimally, heuristically, optimizing, the georeferenced, geosampled, TB-related, hyper/hypo-endemic, aggregation/non-aggregation-oriented, stratified, capture point, sentinel site, estimator determinants.

The data employed in this example were generatable with in step. The following code generated a simple, county-level, epidemiologic, prognosticative, TB-GARCH model with normally distributed georeferenced, hyper/hypoendemic, aggregation/non-aggregation-oriented, hot/cold spot residuals.

```
%let df = 7.5;
%let sig1 = 1;
%let sig2 = 0.1;
%let var2 = 2.5;
%let nobs = 1000;
%let nobs2 = 2000;
%let arch0 = 0.1;
%let arch1 = 0.2;
%let garch1 = 0.75;
%let intercept = 0.5;
data normal;
lu = &var2;
lh = &var2;
do i= -500 to &nobs ;
/* GARCH(1,1) with normally distributed residuals */
h = &arch0 + &arch1*lu**2 + &garch1*lh;
u = sqrt(h) * rannor(12345) ;
y = &intercept + u;
lu = u;
lh = h;
if i > 0 then output;
end;
run;
```

Subsequently a simple, capture point, sentinel site, TB GARCH(p,q) model was constructable as follows. We let $y_t = x_t\beta + u_t$ The residual u_t was modeled as $u_t = \sqrt{h_t} \cdot v_t$ where v_t was i.i.d. with zero mean and unit variance, and where h_t was expressable as $\kappa + \delta_1 h_{t-1} + \delta_2 h_{t-2} + \cdots + \delta_p h_{t-p} +$ and In a standard GARCH model, v_t is normally distributed [29]. To estimate the simple, scaled-up, georeferenced, capture point, sentinel site, county-level, prognosticative, eigen-Bayesian, geo-spatiotemporal, epidemiologic, GARCH model, we employed the MCMC algorithmic procedure. We employed the GARCH = option to specify the model, and the (P = , Q =) sub option to specify the orders of the model.

```
proc autoreg data = normal;
/* Estimate GARCH(1,1) with normally distributed TB residuals with Python */
model y = / garch = ( q = 1,p = 1);
run;
quit;
```

The python procedure produced the following output given in Table 6 for a GARCH model with normally distributed errors.

We employed the Python procedure to estimate a simple, stratified, epidemiologic, estimator, determinant, georeferenced, capture point, sentinel site, eigen-Bayesian, geo-spatiotemporal, epidemiologic, eigen-algorithmic, MCMC, TB, estimator determinant, capture point, sentinel site, scalable, GARCH model. We first specified the georeferenced, hyper/hypo-endemic, aggregation/non-aggregation-oriented, hot/cold spot, stratified, estimator determinants. Then we specified the mean model and the variance model. The XLAG function returned the lag of the first argument. If the lag of the first argument is missing then the second argument is returned (Engles 1962). The function made it easy to specify the lag initialization for the GARCH processing of the epidemiologic, geosampled, TB-related, empirical data. The mse.y variable contained the value of the MSE for y at each iteration. These similated discrete integer values were obtained automatically from the prognosticated, epidemiolgtic, capture point, sentinel site, eigen-Bayesian, geo- algorithmic, MCMC, TB, stratfiable, GARCH model estimates, and were employed to specify lagged values in the approximation.

We fit a GARCH(1,1) model using a point mass estimate zMLE.

```
:
model = pf.GARCH(returns,p=1,q=1)
x = model.fit()
x.summary()
GARCH(1,1)
Dependent Variable: JPM Returns Method: MLE
Start Date: 2006-01-05 00:00:00 Log Likelihood: 6594.7911
End Date: 2016-03-10 00:00:00 AIC: -13181.5822
Number of observations: 2562 BIC: -13158.188
Latent Variable Estimate Std Error z P>|z| 95% C.I.
Vol Constant 0.0
q(1) 0.0933
p(1) 0.9013
Returns Constant 0.0009 0.0065 0.1359 0.8919 (-0.0119 |
0.0137)
```

The standard errors are not shown for transformed variables. You can pass through a transformed = False argument to summary to obtain this information for untransformed variables. We can plotted the GARCH latent variables with plot_z():model.plot_z(figsize = (15,5)).

The capture point, sentinel site, epidemiologic, county-level, TB-model weighted the past squared, geosampled, capture point, sentinel, site residuals which revealed declining weights [i.e., clinical, sociodemiographic and environmental hyper/hypoendemic, aggregation/non-aggregation-oriented, stratified georeferenceable, scaled-up, capture point, sentinel site, estimator determinants] which never went back completely to zero. The most widely used GARCH specification, asserts that the best period is a weighted average of the long run average variance, the variance predicted for this period and the new information this period which is the most recent squared residual [29].

Figure 9 provides eigen-Bayesian algorithmic MCMC, capture point, sentinel site, scalable, GARCH simulations to evaluate the finite sample performance of the estimation and inference procedures for the georeferenced, capture point, sentinel site, county-level, hyper/hypo-endemic, aggregation/non-aggregtation, hot/cold spot, epidemiologic, prognosticative model. We applied the proposed methods to study the roles of hyper/hypoendemic, aggregation/non-aggregation-oriented, TB-related, estimator determinants. We document strong heterogeneity in returns to the conditional distribution in these regressors (Fig. 10).

Finally, the empirical, CDC-archived, georeferenced, epidemiological dataset of time series, dependent, geosampled, stratified, hyper/hypoendemic, TB-related, regressively, prognosticatable estimator determinants revealed capture point, sentinel site ins eigenvector eigen-Bayesian eigen-geospace and time, allowing estimation of a State level forecast model. The model output aided in optimally predicting the probability of TB by county throughout Florida. The random effects term in the model was comprised of spatially structured and stochastic effect (i.e., spatially unstructured) terms. These terms are substituted for hyper/hypoendemic, aggregation/non-aggregation-oriented, (i.e., hot/cold spot), determinants in the model. By employing random effects term in the model, we achieved a frequency distribution that was bell-shaped (i.e., normally/Gaussian distributed) with a mean of zero. An eigen-spatial filter, eigenfunction eigen-decomposition was performed on the CDC archived, stratified, estimator, determinant, geosampled dataset. We noted that the frequency distribution deviated from a bell-shaped one. The random effects term accounted for roughly a mean of 47% of the heterogeneity in the observed probability of TB by county in the CDC, estimator determinant. We were able to generate an underdispersed binomial model distribution. A description of the random effect term revealed, eigen-spatial filter eigenvectors portraying noticeable, latent, positive, non-zero, geo-spatiotemporal eigen-autocorrelation. The eigen-decomposition eigen-algorithm also revealed eigenvectors portraying noticeable residual negative geo-spatiotemporal eigen-autocorrelation. The spatially unstructured random effects component accounted for the non-homogeneity in the CDC archived, estimator determinant, epidemiologic dataset. The final model revealed that from 2025 to 2030 Duval, Orange and Broward counties would require immediate intervention to prevent TB transmission. The model also revealed that from 2035 to 2050 Hillsborough and Palm Beach counties could become hyperendemic without implementation of control strategies (Table 7).

The final model revealed that from 2025 to 2030 Duval, Orange and Broward counties would require immediate intervention to prevent TB transmission. The model also revealed that from 2035 to 2050 Hillsborough and Palm Beach counties could become hyperendemic without implementation of control strategies.

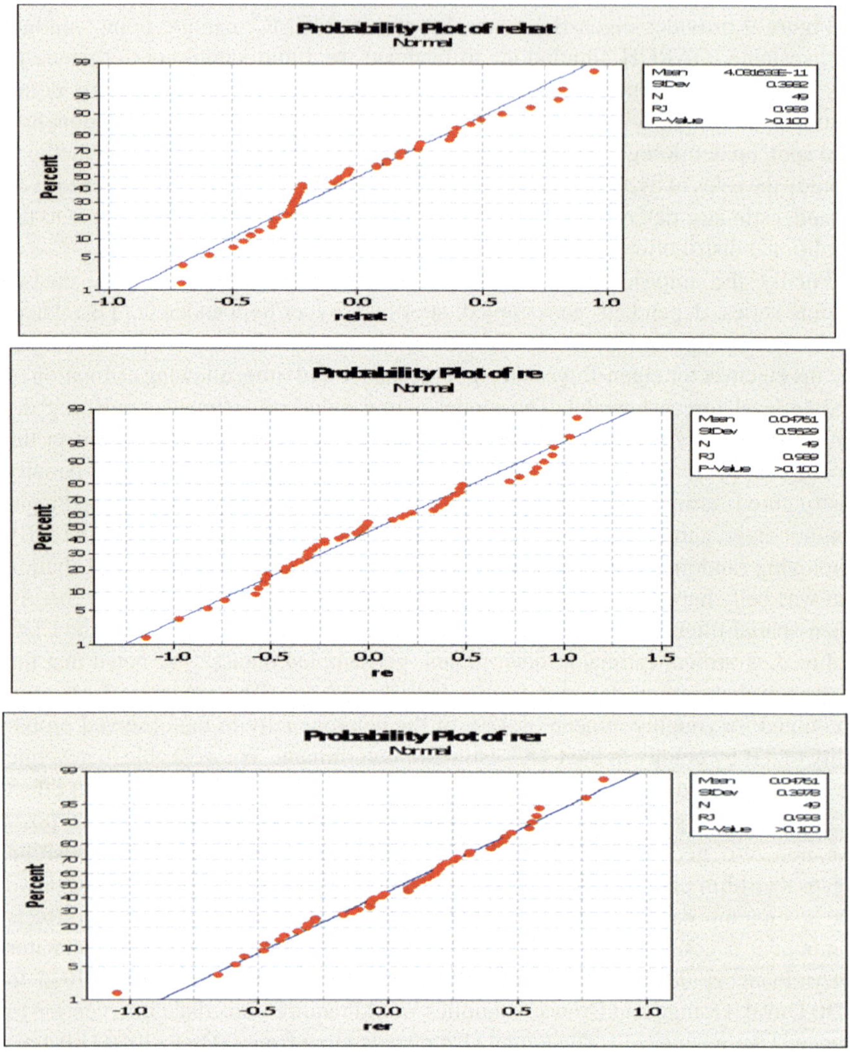

Fig. 9 Plotted residual variance error in the scaled up county-level autoregressive TB model. **a**: normal quantile plot for the random effects term. **b**: normal quantile plot for the spatially structure random effects term. **c**: normal quantile plot for hc spatially unstructured random effects term

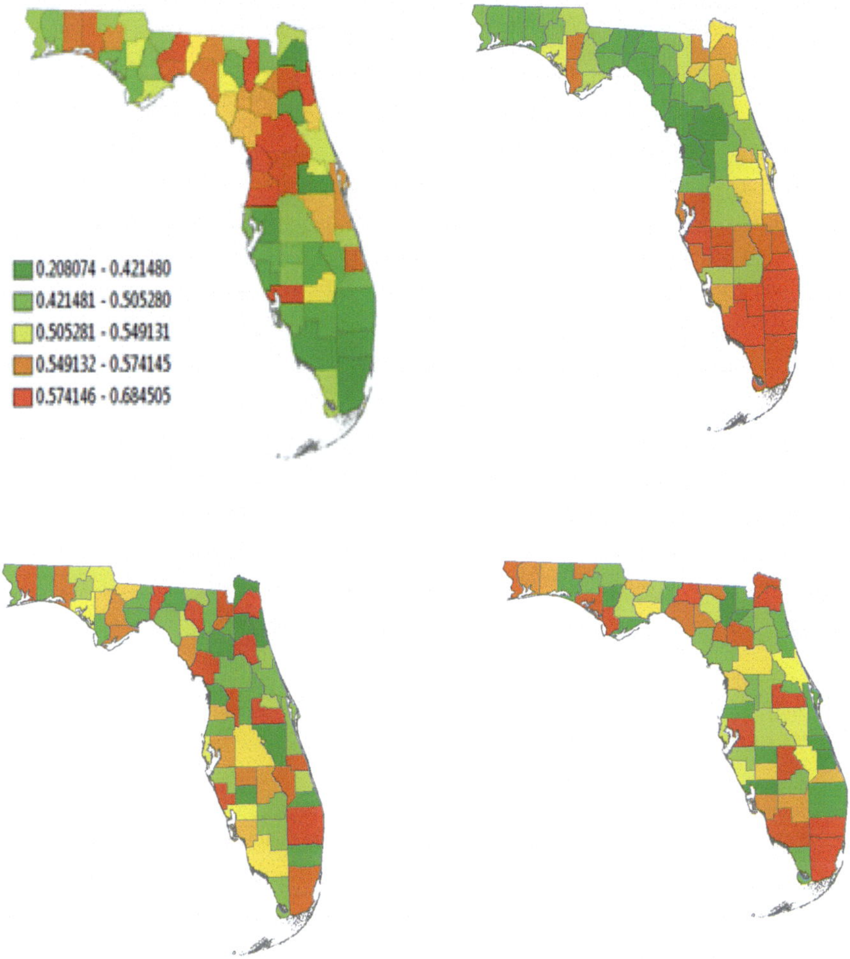

Fig. 10 Top left **a**: random effects term. Top right **b**: positive spatial autocorrelation eigenvector spatial filter. Bottom left **c**: negative positive spatial autocorrelation eigenvector eigen-spatial filter. Bottom right **d**: spatially unstructured random effects term

Table 7 The final predicted model forecast of county-level TB prevalence in Florida employing the eigen-Bayesianized, conditional variances

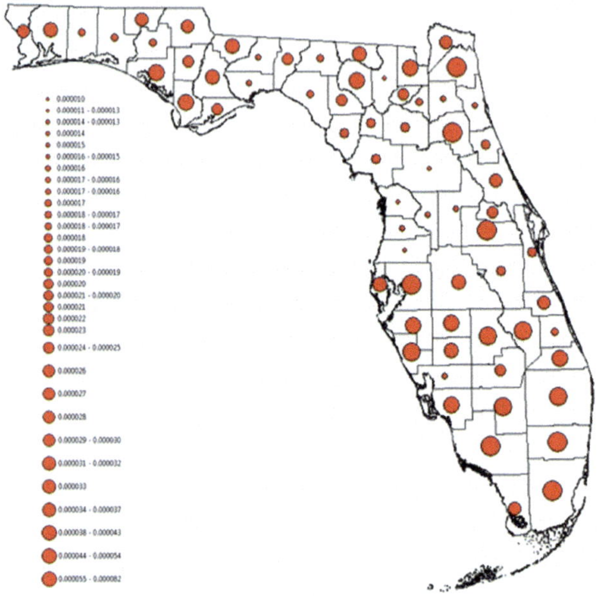

References

1. Abramowitz M (1972) 6.3 psi (Digamma) function. In: Stegun IA (ed) Handbook of mathematical functions with formulas, graphs, and mathematical tables, 10th ed, pp 258–259. New York, Dover
2. Abramowitz M (1983) Chapter 9. In: Stegun IA (ed) Handbook of mathematical functions with formulas, graphs, and mathematical tables, 1st ed., Applied Mathematics Series, vol 55, 9th reprint with additional corrections of 10th original printing with corrections [December 1972], pp.355, 435). Washington D.C., New York, United States Department of Commerce, National Bureau of Standards. Dover Publications
3. Addison AW (1967) A series representation for Euler's constant. Am Math Monthly 74:823–824. https://doi.org/10.2307/2315803
4. Afiti A, Clark VA, May S (2004) Computer aided multivariate analysis, 4th edn. Chapman and Hall, London
5. Agresti A (1987) An introduction to categorical data analysis. Wiley, New York
6. Antoine D, French CE, Jones J, Watson M (2001) Tuberculosis treatment outcome monitoring in England, Wales and Northern Ireland for cases reported in 2001. J Epidemiol Commun Health 61(4):307–307. https://doi.org/10.1136/jech.2005.044404
7. Apostol TM (1965) Mathematical analysis: a modern approach to advanced calculus. Addison-Wesley, London
8. Apostol TM (1974) Mathematical analysis. Addison-Wesley
9. Avriel M (1976) Nonlinear programming: analysis and methods. Prentice Hall, pp 220–221
10. Ayres F (1962) Schaum's outline of theory and problems of matrices. Schaum, New York, NY, p 134
11. Baltagi BH, Li Q (1995) Testing AR (1) against MA (1) disturbances in an error component model. J Econom 68(1):133–151. https://doi.org/10.1016/0304-4076(94)01646-H

12. Baltagi BH, Song SH, Koh W (2003) Testing panel data regression models with spatial error correlation. J Econom 117(1):123–150. https://doi.org/10.1016/S0304-4076(03)00120-9
13. Banerjee S, Roy A (2014) Linear algebra and matrix analysis for statistics, tests in statistical science (1st ed). Chapman and Hall
14. Bareiss EH (1969) Numerical solution of linear equations with Toeplitz and vector Toeplitz matrices. Numer Math 13(5):404–424
15. Beal SL, Sheiner LB (1982) Estimating population kinetics. Crit Rev Biomed Eng 8(3):195–222
16. Beal SL, Sheiner LB (1988) Heteroscedastic nonlinear regression. Technometrics 30(3):327–338
17. Beal SL, Sheiner LB (1992) NONMEN user guide–part VII. University of California, Conditional estimation methods. San Francisco
18. Behr MA, Warren SA, Salamon H, Hopewell PC, Ponce de Leon A, Daley CL, Small PM (1999) Transmission of Mycobacterium tuberculosis from patients smear-negative for acid-fast bacilli. The Lancet 353(9151):444–449. https://doi.org/10.1016/S0140-6736(98)03406-0
19. Berger JO, Strawderman W, Tang D (2005) Posterior propriety and admissibility of hyperpriors in normal hierarchical models. Ann Stat 33(2):606–646. https://doi.org/10.1214/009053605000000075
20. Billingsley P (1986) Probability and measure. Wiley
21. Cliff AC, Ord JK (1973) Spatial autocorrelation. Pion Limited, London
22. Cover TM, Thomas JA (1991) Elements of information theory. Wiley, p 22
23. Cox C (1998) Delta method. In: Amritage P, Colton T (eds) Encyclopedia of biostatistics. New York. John Wiley & Sons, pp 1125–1127
24. Cox DR, Hinkley DV (1974) Theoretical statistics (1st ed). Taylor & Francis Group, New York. https://doi.org/10.1201/b14832
25. Cressie N (1993) Statistics for spatial data. Wiley, New York, NY
26. Cressie N (2015) Statistics for spatial data. Hoboken, John Wiley & Sons
27. Crouch EAC, Spiegelman D (1990) The evaluation of integrals of the form 3 f(t) exp(−t 2)dt: application to logistic normal models. J Am Stat Assoc 85:464–469
28. Davidian M, Giltinan DM (1995) Nonlinear models for repeated measurement data. Chapman and Hall, London
29. Engle RF (1982) Autoregressive conditional heteroscedasticity with estimates of the variance of United Kingdom inflation. Econometrica 50(4):987–1007
30. Eskow E, Schnabel B (1991) Algorithm 695: software for a new modified Cholesky factorization. AMC Trans Math Softw 17:306–312. https://doi.org/10.1145/114697.116806
31. Evans L (1998) Partial differential equations, 1st ed. American Mathematical Society
32. Everitt BS (1998) The Cambridge dictionary of statistics. Cambridge University Press
33. Finch SR (2003) Mathematical constants. In: Encyclopedia of mathematics and its applications, vol 94. Cambridge University Press, Cambridge
34. Fisher RA (1925) Statistical methods for research workers. Oliver and Boyd, Edinburgh, Scotland
35. Floyd K, Blanc L, Raviglione M, Lee J (2002) Resource required for global tuberculosis control. Science 295(5562):2040–2041. https://doi.org/10.1126/science.1069771
36. Ford K (2002) Vinogradov's integral and bounds for the Riemann zeta function. Proc Lond Math Soc 85(3):565–633. https://doi.org/10.1112/S0024611502013655
37. Fraleigh KB, Katz VJ (2003) A first course in abstract algebra. Addison-Wesley
38. Freedman DA, Pisani R, Purves RA (2007) Statistics (4th ed.). W. W. Norton & Company, New York, NY
39. Freedman DA (2009) Statistical models: theory and practice. Cambridge University Press
40. Jacob BG, Krapp F, Ponce M, Gottuzzo E, Griffith DA, Novak RJ (2010) Accounting for autocorrelation in multi-drug resistant tuberculosis predictors using a set of parsimonious orthogonal eigenvectors aggregated in geographic space. Geospat Health 4(2):201–217. https://doi.org/10.4081/gh.2010.201

41. Jacob BG, Krapp F, Ponce M, Zhang N, Caliskan S, Griffith DA, Gotuzzo E, Novak RJ (2013) A Bayesian Poisson specification with a conditionally autoregressive prior and a residual Moran's coefficient minimization criterion for quantitating leptokurtic distributions in regression-based multi-drug resistant tuberculosis treatment protocols. J Pub Health Epidemiol 5(3):122–143
42. Jacob BG, Mendoza DM, Ponce M, Caliskan S, Moradi M, Gotuzzo E, Griffith DA, Novak RJ (2014) Pseudo R^2 probablity measures, Durbin Watson diagnostic statistics and Einstein summations for deriving unbiased frequentistic inferences and geoparametrizing non-zero first-order lag autocovariate error in regressed multi-drug resistant tuberculosis time series estimators. Am J Appl Math Stat 2(5):252–301. https://doi.org/10.12691/ajams-2-5-1
43. Jacob BG, Shafer S, Alinda P, Loun D, McKinnon A, Munu D, Katabarwa MN, Lakwo T, Habomugish P, Unnasch TR (2017) Lexicographically, cartesian-ordered, differential calculi in canonically extrac' in-situ near infra-red fluorescence quantum spectroscopic sub-surface continuous geodesic fluxions for metaheuristic chlorophyll-a translucent emissivity mapping intermittently canopied immature narrow riverine tributary Simulism damnosum s.l. oviposition sites for bio-optically delineating multivariate normalized Gaussian processes elucidative administrated by prior covariances and a spline within a reproducing non-frequentist simultaneous diagonalization of amalgamized positive definite kernels in Hilbert space: Implementation of a 'Slash and Clear' control intervention in two eco-georeferenceable agro-village complexes in Northern Uganda. J Geophys Remote Sens 4(5):26–221
44. Jacob BG, Izureta R, Bell J, Parikh J, Loum D, Casonova J, Gates T, Murray K, White L, Aceng JR (2023) Approximating non-asymptoticalness, skew heteroscedascity and geo-spatiotemporal multicollinearity in posterior probabilities in Bayesian eigenvector eigengeospace for optimizing hierarchical diffusion-orientated COVID-19 random effect specifications geosampled in Uganda. Am J Math Stat 13(1):1–43. https://doi.org/10.5923/j.ajms.20231301.01
45. Jaynes ET (2003) Probability theory: the logic of science. IN: Bretthorst LG (ed). Cambridge University Press, Cambridge, UK
46. Johnson RA, Wichern DW (2007) Applied multivariate statistical analysis. Person Prentice Hall
47. Kapoor M, Kelejian HH, Prucha IR (2007) Panel data models with spatially correlated error components. J Econom 140(1):97–130
48. Katz VJ (1979) The history of Stokes' theorem. Math Mag 52(3):146–156
49. Katz S, Balakrishnan N, Johnson NL (2000) Continuous multivariate distributions volue 1: models and applications, 2nd ed. Wiley
50. Kazemzadeh S, Yu J, Jamshy S, Pilgrim R, Nabulsi Z, Chen C, Beladia N, Lau C, McKinney SM, Hughes T, Kiraly AP, Kalidindi SR, Muyoyeta M, Malemela J, Shih T, Corrado GS, Peng L, Chou K, Chen PC, Prabhakara S (2022) Deep learning detection of active pulmonary tuberculosis at chest radiography matched the clinical performance of radiologists. Radiology 306(1):124–137. https://doi.org/10.1148/radiol.212213
51. Kelejian HH, Prucha IR, Yuzefovich Y (2005) Estimation problems in models with spatial weighting matrices which have blocks of equal elements. J Reg Sci 46(3):507–515. https://doi.org/10.1111/j.1467-9787.2006.00449.x
52. Kolmogorov A (1933) Sulla determinazione empírica di una legge di distribuzione. Giornale dell'Istituto Italiano degli Attuari 4:83–91
53. Kruschke J (2014) Doing Bayesian analysis: a tutorial with R, JAGS, and Stan (2nd ed.). Academic Press
54. Kullback S, Leibler RA (1951) On information and sufficiency. Ann Math Stat 22(1):79–86
55. Lane T, Russo DP, Zorn KM, Clark AM, Korotcov A, Tkachenko V, Reynolds RC, Perryman AL, Freundlich JS, Ekins S (2018) Comparing and validating machine learning models for mycobacterium tuberculosis drug discovery. Mol Pharm 15(10):4346–4360. https://doi.org/10.1021/acs.molpharmaceut.8b00083
56. Lee JHH, King M (1993) A locally most mean powerful based score test for ARCH and GARCH regression disturbances. J Bus Econ Stat 11:17–27. https://doi.org/10.1080/07350015.1993.10509930

57. Lee PM (2012) Bayesian statistics: an introduction, 4th ed. Wiley
58. Leonov GA, Kuznestov NV (2007) Time-varying linearization and the Perron effects. Int J Bifurc Chaos 17(4):1079–1107. https://doi.org/10.1142/S0218127407017732
59. Lin X, Breslow NE (1996) Bias correction in generalized linear mixed models with multiple components of dispersion. J Am Stat Assoc 91(435):1007–1016. https://doi.org/10.2307/2291720
60. Lindstrom ML, Bates DM (1990) Nonlinear mixed effects models for repeated measures data. Biometrics 46(3):673–687
61. Littell RC, Milliken GA, Stroup WW, Wolfinger RD, Schabenberger O (2006). Chapter 14 and Chapter 15. SAS for mixed models, 2nd ed, pp 369–422. SAS Institute Inc
62. Liu Q, Pierce DA (1994) A note on gauss-hermite quadrature. Biometrika 81:624–629
63. Longford N (1994) Logistic regression with random coefficients. Comput Stat Data Anal 17(1):1–15. https://doi.org/10.1016/0167-9473(92)00062-V
64. Mullahy J (1986) Specification and testing of some modified count data models. Hournal Econom 33(3):341–365. https://doi.org/10.1016/0304-4076(86)90002-3
65. Nocedal J, Wright SJ (1999) Numerical optimization. Springer-Verlag
66. Nocedal J, Wright SJ (2000) Numerical optimization. Springer-Verlag
67. Spivak M (1965) Calculus on manifolds: a modern approach to classical theorems of advanced calculus. New York
68. Stein EM, Shakarchi R (2005) Real analysis, 1st ed, vol 3. Princton Lectures in Analysis, p 91
69. Stein ML (1999) Interpolation of spatial data: some theory for kriging (Springer Series in Statistics, pp 40). Springer. modified matrices, memorandum report 42. Statistical Research Group, Princeton University, Princeton, NJ
70. Xie M, He B, Goh TN (2001) Zero-inflated poisson model in statistical process control. Comput Stat Data Anal 38(2):191–201. https://doi.org/10.1016/S0167-9473(01)00033-0
71. Zeng MR, Wong TST (2015) A note on the modified likelihood ration test for homogeneity in beta mixture models. Commun Stat Simul Comput 44(10), 2500–2512. https://doi.org/10.1080/03610918.2013.800208

Interpolating an Eigen-Decomposed *Aedes Aegypti* Swimming Pool Habitat for Implementing "Seek and Destroy" Larval Source Management in Los Angeles County, California

Sasha Marie Mosich, Ricardo Izurieta, Ismael Hoare, Namit Choudhari, Heather McDonald, Anthony Masys, and Benjamin Jacob

Abstract The notable increase in cases and the emergence of local transmission in Los Angeles County underscore the pressing need to understand and effectively manage dengue and *Aedes aegypti* mosquitoes. This study aims to develop accurate predictive models for identifying previously unknown *Ae. aegypti* swimming pool habitats, focusing on factors such as land classifications, vegetation cover, and discontinuous elevation. The ultimate objective is to implement targeted larval source management using the Seek & Destroy approach to control *Ae. aegypti* populations in Los Angeles County. To achieve this, a land use/land cover, normalized difference vegetation index, and digital elevation models were constructed in ArcGIS to use for analysis. Through interpolation of swimming pool habitats, identification of clusters of hot and cold spots was achieved. The results showed a Global Moran's Index of 0.762, a z-score of 7.18, and a p-value of less than 0.05, indicating a significant result. This p-value indicated rejection of the null hypothesis of random autocorrelation, thereby accepting the alternative hypothesis. Hot spots were predominantly clustered in low elevation areas, aligning with local transmission patterns observed in Los Angeles County. Conversely, cold spots were primarily located in mountainous, high elevation regions. These results contribute valuable insights for targeted mosquito control strategies in the region.

Keywords Ae. Aegypti · Dengue · Swimming pool · Bayesian · Los Angeles county · California

---

S. M. Mosich · R. Izurieta · I. Hoare · H. McDonald · A. Masys · B. Jacob (✉)
College of Public Health, University of South Florida, Tampa, USA
e-mail: bjacob1@usf.edu

N. Choudhari
School of Geosciences, University of South Florida, Tampa, USA

B. Jacob et al. (eds.), *Tech Enabled Global Health Security*, Advanced Sciences and Technologies for Security Applications, https://doi.org/10.1007/978-3-031-86997-6_7

1 Introduction

In 2023, the California Department of Public Health (CDPH) reported that Los Angeles County, California, had experienced 78 dengue cases [2]. Two were locally acquired, marking the first instance of local transmission in California's history. This development highlights the increasing threat *Aedes aegypti* mosquitoes pose and the need for robust surveillance and control measures. The 78 reported cases, including the new local transmission, demonstrate an urgency in understanding and managing mosquito-borne diseases in urban areas such as Los Angeles County, where environmental factors and population density can facilitate the spread of dengue. The year 2023 also marked the highest caseload since 2019, with 55 cases. Lastly, Los Angeles County had the most abundant cases in California, with Orange County coming in second with 18 cases, none of which were locally acquired [2]. Figure 1 illustrates a time series map from 2016 to 2023 depicting dengue cases across all California counties, indicating a rising trend.

Dengue, an arbovirus transmitted by *Ae. aegypti*, contains four serotypes: DENV-1, DENV-2, DENV-3, and DENV-4 [15]. All four serotypes have been documented in the United States [6]. Within these four serotypes, there are 47 identifiable strains [15]. The virus thrives in warm and tropical climates, with cases peaking during and after rainy periods [15]. Dengue is spread through a blood meal by female *Ae. aegypti* mosquitoes [15]. Human-to-human transmission is impossible, except for

Fig. 1 Time-series map of dengue cases in California counties

rare instances such as a blood transfusion from an infected individual to a non-infected individual. Another method of dengue transmission is when the disease is passed from mother to fetus, which can occur when dengue IgG antibodies cross the placental barrier [12, 18]. When a susceptible mosquito obtains a blood meal from a viremic individual, it can acquire and replicate the virus, which can then be transmitted to a new host [5].

The incubation period for dengue ranges from 3 to 14 days. While most cases are asymptomatic or mild with symptoms such as nausea, fatigue, musculoskeletal pain, joint pain, rash, and fever, severe dengue can occur, leading to organ damage and hemorrhaging [15]. Currently, there is no specific treatment for dengue; instead, care focuses on symptom management and fluid administration [22].

In the USA, 96% of the 378 locally acquired cases from 2010 to 2017 were DENV-1 [6]. Among the 5009 travel-associated cases, 45% were DENV-1. In this same timeframe, the US experienced eighteen fatalities caused by dengue, resulting in a 0.3% fatality rate. While the incidence is generally low in the US, it is rising. There have been 7895 cases overall reported between 2010 and 2021 in the US [6]. A few licensed vaccines are available [22]. The World Health Organization (WHO) recommends that those nine years and older receive three doses of Dengvaxia [12]. Despite the availability of licensed vaccines like Dengvaxia, developing an effective one-dose tetravalent vaccine remains a goal due to the complex nature of dengue immunity [12, 26]. When an individual is infected, they gain lifelong immunity to the serotype they were infected with [12]. The challenge comes with a secondary infection with dengue of a different serotype, which can lead to severe dengue [12].

Historically, *Aedes* mosquitoes transmitted dengue via sylvatic cycles, meaning it originated from forests [15]. Nowadays, dengue is transmitted through sylvatic cycles and vertical transmission. The life cycle of *Ae. aegypti* primarily revolves around transmission between mosquitoes and humans [15]. Occasionally, incidental hosts may serve as reservoirs. *Ae. aegypti*, a holometabolous mosquito, undergoes metamorphosis, progressing through four growth phases from egg to adult [15]. Female mosquitoes can lay 10 to 100 eggs four to five times during their two to four-week lifespan [15]. *Ae. aegypti* can live in temperatures above 4 °C but primarily thrives between 15 and 37 °C [15]. Breeding sites for *Ae. aegypti* include artificial water containers such as flower vases, waste tires, ornamental bird baths, etc. [15, 26]. Additionally, these mosquitoes can breed in untreated swimming pools [7]. Environmental factors significantly influence mosquito breeding and dengue transmission. Suitable oviposition breeding sites such as the artificial water containers mentioned provide ideal conditions for *Ae. aegypti* to thrive and multiply. Furthermore, *Ae. aegypti*'s adaptability to a wide range of temperatures contributes to its efficiency as a vector. Understanding these ecological dynamics is vital in designing effective vector control strategies and mitigating the risk of dengue outbreaks.

Mosquito biting patterns are not homogenous and will vary significantly depending on geospatial and temporal factors. Additionally, different mosquitoes exhibit different clustering patterns (i.e. hot and cold spots at the zip-code level), influencing empirical data interpretation. No consistent or reliable method currently exists in the literature to assess a population's disease susceptibility based on ArcGIS and/or Python interpolated spectral signatures for swimming pool habitats that are vulnerable to dengue. This lack of a dependable measure makes it challenging to predict outbreaks and plan and execute effective LSM strategies.

In this study, spatial statistics and Sentinel-2 visible and near-infrared (NIR) data was employed to signaturize and identify the swimming pool habitats of *Ae. aegypti* species and their occurrence, abundance, and distribution across Los Angeles County. Using proxy geospatial and temporal data, models for Land Use/Land Cover (LULC), Normalized Difference Vegetation Index (NVDI), and Digital Elevation Model (DEM) were constructed. These models helped us analyze habitat patterns and clustering of these swimming pool foci employing Moran's eigenvector spatial filtering autocorrelation indices for identifying habitat clustering and non-clustering propensities (i.e. hot and cold spots).

The primary objective of this research is to create accurate predictive models for identifying unknown *Ae. aegypti* swimming pool habitats. These models, based on factors such as land classifications, vegetation cover, and discontinuous elevation, aim to enhance understanding and management of mosquito populations in the region for implementing real-time larval control. Additionally, this research seeks to evaluate the effectiveness of these predictive models in guiding targeted interventions and optimizing resource allocation for real-time vector larval control efforts. By integrating advanced geospatial analysis techniques, this study aims to contribute to the development of proactive and data-driven strategies for mitigating dengue risk in Los Angeles County.

2 Literature Review

A study by Ranasinghe and Amerasinghe identified 22 microbiota species in *Ae. aegypti* breeding sites in Sri Lanka, with water ponds having the highest concentration [24]. Of the microbiota identified, chlorophyta (green algae) and cyanobacteria (blue-green algae) are especially notable. Cyanobacteria was also found in the gut of *Ae. aegypti* [24]. This finding is crucial because green and blue-green algae commonly grow in untreated pools, as seen in Fig. 2 [1, 16]. This suggests that algae growth in untreated pools can serve as an attractant for *Ae. aegypti* as a source of food. The presence of microbiota species in oviposition *Ae. aegypti* breeding sites and gut highlight the ecological and entomological dynamics that influence mosquito behavior and distribution. Alternatively, the presence of chlorine in treated swimming pools may have a repellent effect on *Ae. aegypti*, reducing their attraction to stagnant swimming pools [21]. Therefore, regular cleaning and maintenance of swimming pools to prevent algae buildup are essential in vector larval control efforts. Swimming pools should be cleaned weekly to avoid algae buildup [19]. Understanding these interactions between mosquitoes and their environment can inform targeted interventions to reduce *Ae. aegypti* breeding and dengue transmission.

In addition to dengue, *Ae. aegypti* can transmit yellow fever, chikungunya, and Zika viruses which pose a significant concern as an infestation of *Ae. aegypti*

Fig. 2 Google Earth image of green pool and blue pool

increases the risk of not only dengue but also other mosquito-borne viruses [23]. Furthermore, California's local *Ae. aegypti* populations may resist pyrethroids and organophosphates, highlighting the challenge faced in effectively managing mosquito populations and controlling dengue transmission [30].

Before 2013, there were introductions of *Ae. aegypti* in California, but they either failed to establish or were eradicated [20]. Initially, it was believed that California's climate, characterized by a lack of summer rain, low humidity, and lethal winter temperatures, would prevent the establishment of these mosquitoes. However, this perception changed in 2013 when *Aedes* was discovered infesting the city of Madera following a complaint about mosquitoes filed with the Madera County Vector Control District. The subsequent year revealed that *Ae. aegypti* had survived the winter, leading to its establishment in California. By 2015, *Aedes* had spread to more counties and cities, increasing from seven counties to twelve and 14 cities to 55. Surveillance efforts were initiated after the establishment of *Aedes*. Although attempts were made to eradicate *Ae. aegypti*, it has since become well-established and widespread. Consequently, the focus shifted to small-scale treatments, as large-scale eradication efforts were deemed unsustainable, costly, and labor-intensive [20].

In addition to *Ae. aegypti*, California also hosts other disease-carrying mosquitoes, including, but not limited to, *Culex tarsalis*, *Cx. pipiens*, *Cx. quinquefasciatus*, *Ae. albopictus*, *Ae. sierrensis*, and *Anopheles freeborni* [4]. These vectors contribute to the state's complexity of mosquito-borne disease surveillance and control efforts. Furthermore, with climate change, rising temperatures increase dengue risk in Los Angeles [27]. Changes in the urban environment can alter the distribution and epidemiology of *Aedes*, and further urbanization combined with climate change can impact California and Los Angeles County by increasing the incidence of cases [13]. Therefore, comprehensive mosquito surveillance, integrated vector management strategies (i.e. larval source management [LSM] tactics such as emptying artificial water containers biweekly, targeted drone larviciding, and others associated with Seek and Destroy [S&D]), and public health interventions are crucial to address the evolving challenges posed by mosquito vectors and the changing environmental conditions in California [18, 26].

The Greater Los Angeles County Vector Control District primarily employs *Bacillus thuringiensis* and *Bacillus sphaericus* to control larvae and pupae in standing water [9]. Additionally, they utilize methoprene, spinosad, surface film oil, and mosquitofish. To combat adult mosquitoes, they use pyrethroid, pyrethrin, and organophosphates [9]. However, as previously mentioned, resistance to these insecticides is developing, necessitating a multifaceted approach to vector control. The CDPH has released a comprehensive guide for surveillance and response to *Aedes* mosquitoes, which forms the basis of effective control strategies. The plan centers around surveillance and education, emphasizing the critical role of community engagement in mosquito control efforts [3]. This includes notifying health departments of *Ae. aegypti* presence, distributing educational materials to the public, enhancing surveillance through trapping and monitoring, implementing control measures such as larvicide applications and adult mosquito treatments using vehicle-mounted sprayers or aerial applications, and sending mosquitoes for testing, among other measures [3].

However, there is no indication in the guide regarding using geographical information systems (GIS) and Python to determine *Ae. aegypti* larval habitat areas and direct specific treatment to identified areas, which could enhance the precision and effectiveness of vector larval control efforts. Additionally, the method of aerial application remains unclear regarding the dispersal of larvicide or insecticide, highlighting the need for clear guidelines and protocols to ensure optimal distribution and efficacy. Furthermore, the guide's approach to handling *Aedes* mosquitoes appears to be more of a reactive approach, simply waiting for a mosquito from a trap to be positive for dengue or a human case of dengue to occur before acting. Instead, vector control districts should take a proactive and preventative approach by actively seeking out potential habitats and implementing targeted larval control measures to break the dengue transmission cycle at its source.

An alternative approach proposed by Jacob et al. involves implementing a real-time system for LSM. In their study, they developed an interactive web-configurable, convolutional neural network (CNN) smartphone app to identify potential malaria *Anopheles* mosquito breeding sites, in Northern Uganda [18]. This real-time app utilizes satellite signature interpolation to precisely locate unknown habitats, which can then be targeted for control measures by local vector control authorities. Known as the S&D strategy, this method could potentially be adapted for use in Los Angeles County by employing real-time drone technology to deliver insecticides to geostatistically and cartographically identified swimming pool habitats [18]. Integrating such innovative technology into mosquito control efforts in Los Angeles County could revolutionize the approach to vector management for dengue, allowing for more precise targeting of oviposition breeding sites of *Ae. aegypti* and efficient deployment of control measures.

While studies such as Hill et al. have employed remote sensing for mapping *Aedes* habitats or Ritchie et al. have mapped *Aedes* habitats in bird baths in Hillsborough

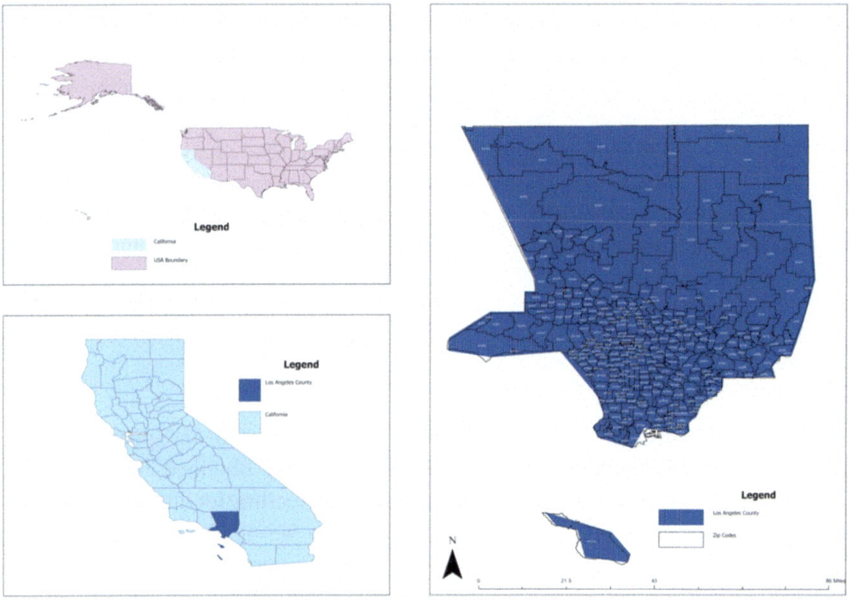

Fig. 3 Study area map of Los Angeles County

County, Florida, the current literature lacks discussion on prognosticative entomological signature interpolation for mapping specific habitats in Los Angeles County. This gap in research highlights the need for further investigation into the application of advanced predictive algorithmic techniques in characterizing and monitoring mosquito habitats, especially in predominantly urban environments like Los Angeles County, where precise mapping can significantly impact vector control strategies (Fig. 3).

3 Study Site and Data Source

3.1 Study Site

Los Angeles County is located on the west coast of the United States, specifically in the southwestern region of California [29]. As of the 2020 census, its population stood at 10,014,009, making it the most populous county in California. However, recent estimates for 2023 suggest a decrease in population, projected to dwindle to 9,663,345. The county's largest city is Los Angeles, the most populous city in California and the second most populous in the country. According to data from the US Census Bureau, the county spans an area of 4,751 square miles (12,310 km^2), of which 4,058 square miles (10,510 km^2) is land and 693 square miles (1790 km^2) is water. Remarkably, Los Angeles County boasts 70 miles (110 km) of coastline [29]. Considering its diverse demographics, expansive geographical features, and significant population density, Los Angeles County serves as an ideal setting for this study's research goals. The dynamic interplay between urban infrastructure, environmental factors, and human activities in the region presents unique challenges and opportunities for studying *Ae. aegypti* populations and implementing S&D. By focusing research efforts on Los Angeles County, this study aims to generate valuable insights that can inform proactive vector larval control management in urban environments.

3.2 Data Source

This study utilized detailed satellite imagery from the European Space Agency's (ESA) Sentinel-2B multispectral instrument, specifically obtained for July 2023. This imagery is notable for being cloud-free and correcting for factors like atmosphere and lighting, ensuring accuracy. This imagery was sourced from the ESA's Copernicus Data Space Ecosystem (https://dataspace.copernicus.eu/). In addition to the satellite imagery, elevation data was obtained from the US Geological Survey (USGS) to create an elevation model (https://earthexplorer.usgs.gov/). These datasets were crucial for creating geostatistical models and furthering the objective of this study. Furthermore, this study leveraged advanced remote sensing technologies and open-access data sources to enhance the precision and reliability of the predictive models. By integrating satellite imagery geostatistics and non-homogenous elevation data, this study was able to capture environmental variations and identify potential mosquito oviposition aquatic breeding habitats with greater accuracy. The utilization of these tools and datasets underscores this study's commitment to employing innovative methodologies in mosquito population analysis and control strategies, ultimately contributing to more effective and targeted vector larval management efforts in Los Angeles County.

4 Methodology

In this paper, a Moran's eigenvector semiparametric eigen-spatial filtering statistical approach is proposed for allowing local vector larval control mosquito abatement officers and other research collaborators to deal explicitly with (a) spatially lagged autoregressive models and (b) simultaneous autoregressive models. These paradigms were employed for quantitating potential georeferenced, hyper/hypoendemic, aggregation/non-aggregation-oriented, signature propensities among potential, surface-colored swimming pool *Ae. aegypti* habitats, wavelength signature reflectance, of various land covers (i.e., urban residential, water, etc.) regions in Los Angeles County, California. As in one non-parametric eigen-spatial filtering approach, a specific subset of eigenvectors from a transformed spatial link matrix in Python was employed to capture dependencies among the disturbances of a spatial regression-related surface-colored *Ae. aegypti* habitat signature model. However, it was assumed that the optimal subset in the proposed filtering pool habitat signature model could be identifiable more intuitively by an objective function that minimizes spatial autocorrelation rather than maximizing model fit. The proposed objective function in a spatial autocorrelation model can lead to a robust and smaller subset of selected eigenvectors [10]. Furthermore, it was assumed that the application of the proposed spatial eigenvector filtering approach on georeferenced multiple mapped surface-colored swimming pool *Ae. aegypti* habitats in Los Angeles County could demonstrate its feasibility, flexibility, and simplicity for prioritizing the breeding site oviposition aquatic foci in other concentrated georefrenceable geolocations.

This study employed the Spatial Autocorrelation Moran's *I* tool in Python (PySAL) to measure residual, non-zero autocorrelation in the empirical, georeferenced, eigen-decomposed pool signature-related estimator determinant empirical dataset. Spatial autocorrelation refers to the correlation among values of a single variable strictly attributable to their relatively close local positions on a two-dimensional surface, deviating from the independent observations assumption of classical statistics [10]. Using a dataset stratified by zip code with pool signature-related diagnostic features, it was evaluated whether eigenvectors derived from weighted aggregation/non-aggregation-oriented, prognosticative, model were clustered, dispersed, or random. The Moran's *I* value was calculated, along with a z-score (i.e., standard deviations) and p-value, to assess the significance of the eigen-spatial filter, aggregation/non-aggregation-oriented, zip code-stratified, pool-related signature estimator determinants.

Moran's Indices (*I*) statistics were employed in the analyses to estimate non-zero, probabilistic, latent autocorrelation and spatial distribution in the multivariate geosampled, zip code-level stratified, pool-interpolated signature covariates. Moran's *I* is a measure of geospatial autocorrelation, characterized by a correlation in a signal amongst nearby georeferenced locations in eigenvector eigen-geospace [10]. Geo-spatiotemporal autocorrelation can also occur as systematic patterns and in values of other exogenous observational predictors across geolocations on a georeferenced, scalable, capture point sentinel site map due to underlying common entomological cofactors causing health risk factors in a zip code stratified geospatial cluster [17]. It was assumed that latent geo-spatiotemporal autocorrelation in the pool signature-related determinant dataset was more complex than one-dimensional (d) autocorrelation since spatial correlation is multi-dimensional and multi-directional [11]. One research hypothesis was that Moran's *I* can remove residual non-Gaussian zero autocorrelation, multicollinearity, and/or skew heteroscedasticity in an eigen-decomposed, empirical, georeferenced dataset of interpolated distribution patterns in eigen-Bayesian eigenvector eigen-geospace.

In the context of predictive entomological regression-related vulnerability analysis, several methods are employable to control for statistical effects of non-normal propagation (e.g., multicollinearity, skew heteroscedasticity, zero autocorrelation, non-asymptoticalness). Maximum likelihood or other Bayesian interpolative approaches can account for geo-spatiotemporal classifiable dependencies in a semiparametric Markovian framework, whereas recent Bayesian eigenvector spatial filtering approaches focus on semi-parametrically removing autocovariance [10]. In this experiment, a semiparametric eigen-Bayesian Markovian, eigen-spatial filtering approach was employed to explicitly deal with (a) lagged autoregressive models and (b) simultaneous autoregressive signature models for heuristically optimizing forecasting empirically dependent potential zip code-stratifiable, georeferenceable, *Ae. aegypti* swimming pool habitats. This model's estimator determinants were used to delineate hyper/hypoendemic, clustering/non-clustering, propensities of the aquatic foci in eigen-Bayesian eigenvector geospace. This iterative quantitative trend in the empirical, geosampled, georeferenceable, dataset, *Ae. aegypti* swimming pool habitat estimator determinants in eigen-Bayesian semiparametric Markovian eigenvector eigen-geospace were identified.

It was assumed that the proposed iterative signature eigen-Bayesian semiparametric model could recursively update the posterior distribution of the georeferenced, zip code-stratified, oviposition *Ae. aegypti* habitat model estimators. Furthermore, it was assumed that simultaneously aggressively quantifying the process with new empirical predictive samples (i.e., evidential likelihoods) drawn from a proposal density in an empirical dataset of potential swimming pool interpolated signature capture point sentinel sites in eigenvector eigen-geospace would identify new unsampled foci. The assumption was that over iterations, the proposal density would progressively become localized near the posterior modes, allowing the definition of georeferenceable, hot/cold spot (e.g., clustering/non-clustering), *Ae. aegypti* swimming pool habitats. This model would also employ asymptotically normalized (e.g., non-multicollinear, non-skew, non-heteroscedastic), georeferenced, geosampled, zip code-stratified, geolocations. In the *Ae. aegypti* swimming pool habitat model, the posterior mean and posterior mode represented the mean and mode of the posterior distribution of Θ.

In this experiment, the Dirichlet process Gaussian mixture was trained with sparse eigenvalues derived from the previous iteration in Python to update the proposal density in the georeferenced, multivariate, zip code-stratified, clustering/non-clustering, *Ae. aegypti* swimming pool habitat model. This study employed the Dirichlet Process Gaussian-mixture model (a fully non-parametric Bayesian methodology) to estimate probability density functions (PDF) with a flexible set of assumptions. The PDF is a statistical expression that defines a probability distribution (the likelihood of an outcome) for a discrete signature random variable instead of a continuous random variable [14]. It was assumed that a novel algorithm, based on the dependent Dirichlet process mixture model (DDPMM), could optimally capture batch-sequential, empirical, clustering/non-clustering, georeferenceable, *Ae. aegypti* oviposition swimming pool habitats and their zip code-stratified, signature estimator determinants. The algorithm was derived via a low variance asymptotic analysis of the Markovian semiparametric sampling algorithm for the DDPMM, which in this experiment provided hard clustering with convergence guarantees like those of the k-means algorithm (i.e., a method of vector quantization). This method was originally aimed for signal processing to partition n observations into k clusters in which each observation (i.e., zip code stratified, *Ae. aegypti* swimming pool habitat, model estimator determinants) was delineated.

In this experiment, one assumption was that the empirical results from an eigen-synthetic test with moving Gaussian clusters and a test for unbiasing non-normality in a regressed dataset of zip code-stratifiable, georeferenceable, geosampled, clustering/non-clustering, *Ae. aegypti* swimming pool habitats would be reliable. It was assumed that the georeferenced, entomological capture points, in eigen-Bayesian eigenvector eigen-geospace would demonstrate that the algorithm requires orders of magnitude less computational time than contemporary clustering uncertainty algorithms discussed in the literature. Hence, multiple varying non-Gaussian heterogeneities such as multicollinearity, non-asymptoticness, and other uncertainties due to violations of regression assumptions (i.e. unequal variances) would be addressable. In doing so, the eigenspatial filter algorithm would provide higher estimation accuracy of the semi-parameterized, hierarchical, autocorrelated, georeferenced, *Ae. aegypti* swimming pool habitat, estimator determinants derived from the eigen-Bayesian Markovian model output.

It was assumed that an eigendecomposed, eigen-Bayesianized, asymptotic, prognosticative, estimator model may heuristically optimize predictive targeting and prioritizing hierarchical, diffusion-related, potential hyper/hypoendemic, georeferenecable, aggregation/non-aggregation-oriented, *Ae. aegypti* swimming pool habitats, georeferenced, geolocations by generating disturbance-free estimator determinants. Asymptotic regression is appropriate when a contiguous region on the right or left (or both) of the input is expected to deviate systematically from the regression model in a finite fashion [10]. Such regions (non-asymptotic eigen-Bayesianized semi-parametric regressed regions) often occur in vulnerability-oriented entomological risk model analysis [8, 17]. It was assumed that the local asymptotic normality property may be non-frequentistically quantifiable in eigenvector eigen-geospace for optimally regressing an empirical, entomological, georeferenceable, dataset of semi-parameterizable, *Ae. aegypti* swimming pool habitats. Interpolated, spectral signature, estimator determinants may be usable for constructing robust, prognosticative, vulnerability-oriented, regression, capture point sentinel site models. These paradigms may also be employable for heuristically optimizing targeting and prioritizing sub-county, potential georeferenceable, aggregation/non-aggregation-oriented, swimming pool habitat sites at the zip code level in Los Angeles County for implementing S&D.

4.1 Land Use/Land Cover (LULC)

A Land Use/Land Cover (LULC) model was constructed in Google Earth Pro using satellite imagery from 2022 and 2023. The LULC was mapped using supervised classification. Based on prior knowledge of the study site, five distinct LULC categories were identified: mountain, desert, water, urban commercial, and urban residential. This classification aimed to analyze the impact of different land covers on swimming

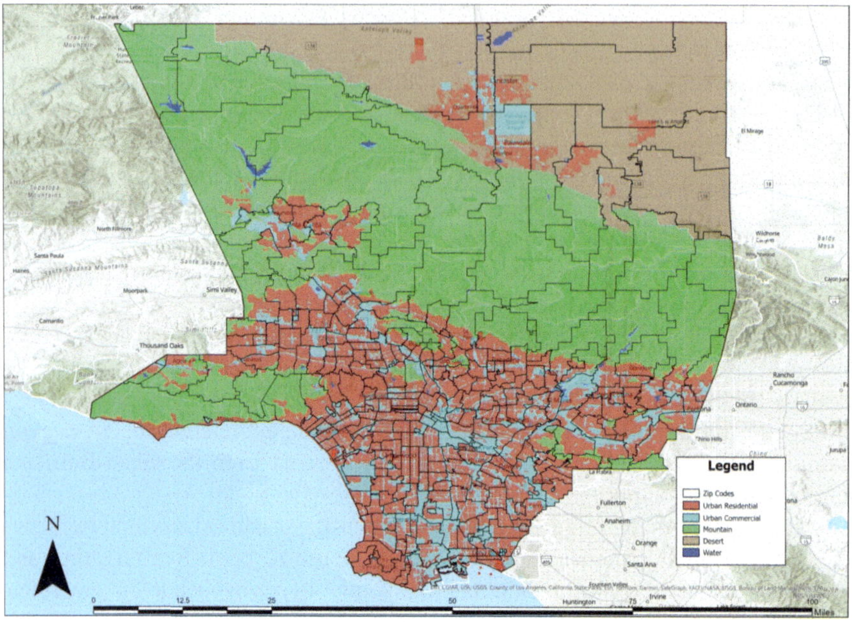

Fig. 4 Land use/land cover map

pool habitats. One key advantage of utilizing Google Earth Pro for LULC mapping is its capability to provide high-resolution satellite imagery of less than 5 m, compared to Sentinel-2's 10-m data. This finer spatial resolution allowed for precise and accurate classification of the land covers, enabling us to capture subtle variations in the landscape that may have an influence on *Ae.* aegypti breeding site oviposition preferences. It was hypothesized that the habitats would predominantly cluster in urban residential areas due to the presence of artificial water sources like swimming pools (Fig. 4).

4.2 Normalized Difference Vegetation Index (NDVI)

A Normalized Difference Vegetation Index (NVDI) was generated using ArcGIS Pro to assess the impact of vegetation on *Ae. aegypti* swimming pool habitats. The NDVI output ranges from positive to negative, with positive values indicating dense vegetation and negative values indicating sparse vegetation, water bodies, or urban areas [28]. The NDVI formula, $NDVI = \frac{NIR-Red}{NIR+Red}$, underpins this analysis [28]. It was hypothesized that the swimming pool habitats would predominantly occur in less densely vegetated areas, indicative of an urban environment (Fig. 5).

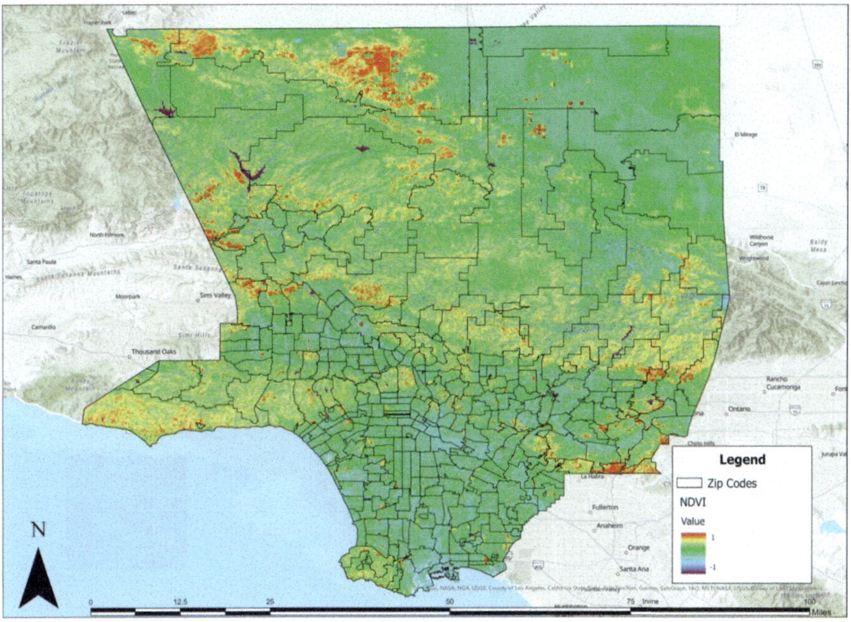

Fig. 5 Normalized difference vegetation index map

4.3 Digital Elevation Model (DEM)

A Digital Elevation Model (DEM) was generated in ArcGIS Pro using data from the USGS (https://earthexplorer.usgs.gov/). This step was undertaken to analyze the influence of discontinuous elevation on *Ae. aegypti* swimming pool habitats. It was hypothesized that these habitats would predominantly be in low-elevation areas, associating them with the urban environment of Los Angeles County. Across all three models—LULC, NDVI, and DEM—it was anticipated that the swimming pool habitats would primarily be in urban residential areas characterized by low vegetation and elevation (Figs. 6 and 7).

4.4 Autocorrelation Model

An autocorrelation model specification was used to describe *n* random autocorrelated *Ae. aegypti* swimming pool oviposition habitat explanatory variables. The resulting model specification took on the following form:

$$Y = \mu(1 - \rho)1 + \rho WY + \varepsilon \quad (1)$$

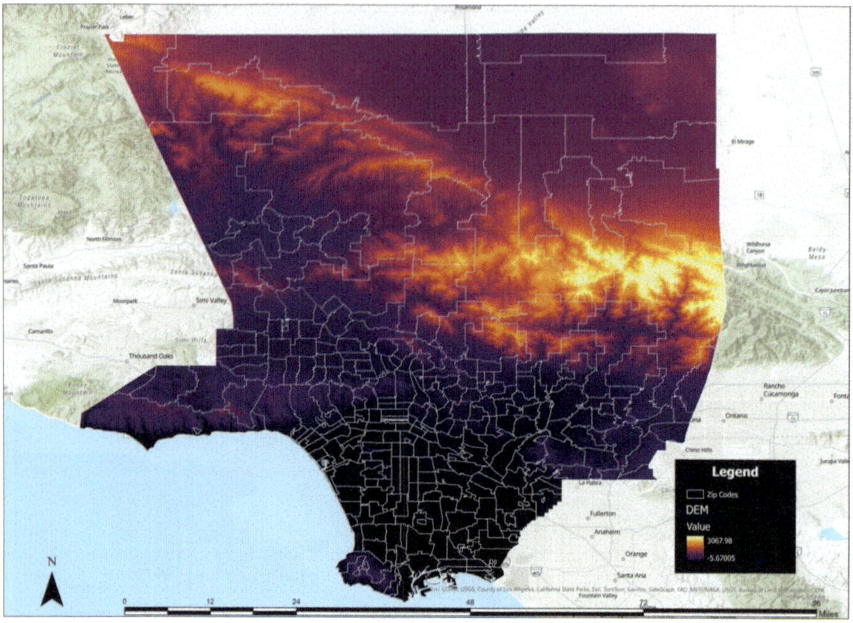

Fig. 6 Digital elevation map

where μ was the scalar conditional mean of Y, and ε was an n-by-1 error vector whose elements were statistically independent and identically distributed (i.i.d.) normalized random variates. The spatial covariance matrix for Eq. 1 employed the sampled *Ae. aegypti* swimming pool habitat signature covariates, which in this experiment was expressable as $E\left[(Y - \mu l)'(Y - \mu l)\right] = \Sigma = \left[(I - \rho W')(I - \rho W)\right]^{-1}\sigma^2$, in Python where E (●) denoted the calculus of expectations, I was the n-by-n identity matrix delineating the matrix transpose operation, and σ^2 was the error variance.

An assumption was made that a mixture of positive spatial autocorrelation (PSA) and NSA latent spatial autocorrelation may be present in the *Ae. aegypti* swimming pool habitat model, which, in turn, it was assumed that would provide an even more explicit representation of the eigen-synthesized results. Varying spatial autoregressive parameters appeared in the covariance matrix, which, for the signature model specification, was

$$\sum = 1- < \rho >_{\text{diag}} W')\left(\text{I}- < \rho >_{\text{diag}} \text{W})\right]^{-1}\sigma^2 \quad (2)$$

where the diagonal matrix of autoregressive parameters, $< \rho >_{\text{diag}}$, contained two sampled parameters: ρ_+ for those *Ae. aegypti* swimming pool habitat covariates displaying positive spatial dependency, and ρ_- for those displaying negative spatial dependency. For example, by letting $\sigma^2 = 1$ and employing a 2-by-2 regular square tessellation, Python revealed the following in the autocorrelation

Spatial Autocorrelation Report

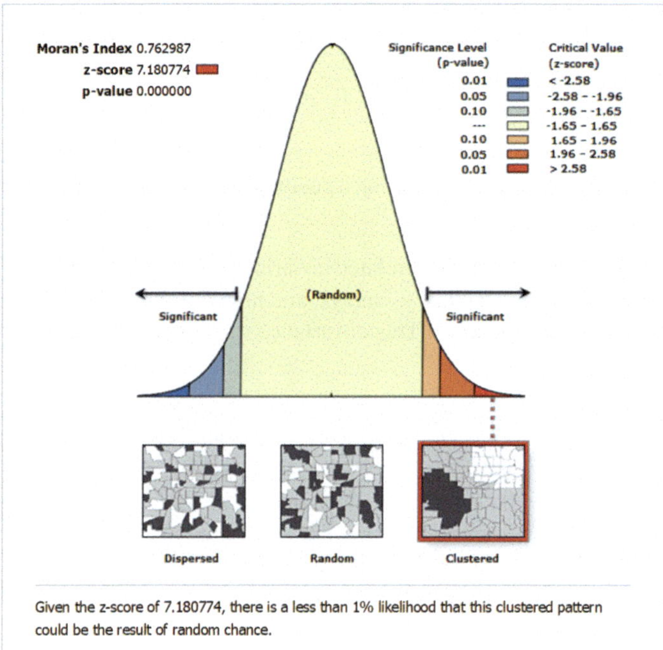

Global Moran's I Summary

Moran's Index	0.762987
Expected Index	-0.035714
Variance	0.012372
z-score	7.180774
p-value	0.000000

Dataset Information

Input Feature Class:	Extract_Random32
Input Field:	RASTERVALU
Conceptualization:	FIXED_DISTANCE
Distance Method:	EUCLIDEAN

Fig. 7 Spatial autocorrelation report

model: $\Sigma = \left[\begin{pmatrix} 1&0&0&0\\0&1&0&0\\0&0&1&0\\0&0&0&1 \end{pmatrix} - \begin{pmatrix} \rho_+ & 0 & 0 & 0\\ 0 & \rho_+ & 0 & 0\\ 0 & 0 & \rho_- & 0\\ 0 & 0 & 0 & \rho_- \end{pmatrix}\begin{pmatrix} 0 & \frac{1}{2} & \frac{1}{2} & 0\\ \frac{1}{2} & 0 & 0 & \frac{1}{2}\\ \frac{1}{2} & 0 & 0 & \frac{1}{2}\\ 0 & \frac{1}{2} & \frac{1}{2} & 0 \end{pmatrix}\right]^2$ in the autocorrelation model [i.e., $\begin{pmatrix} y_1\\y_2\\y_3\\y_4 \end{pmatrix}$] enabled positing a positive relationship between the sampled *Ae. aegypti* swimming pool habitat-related covariates, y_1 and y_2, a negative relationship between signature covariates, y_3 and y_4, and no relationship between covariates y_1 and y_3 and between y_2 and y_4. This covariance specification yielded:

$$\begin{aligned} \mathrm{Y} &= \mu\left(\mathrm{I} - \rho_+ < I_+ >_{\mathrm{diag}} - \rho_- < I_- >_{\mathrm{diag}}\right)1, \\ &\quad + \left(\rho_+ < I_+ >_{\mathrm{diag}} + \rho_- < I_- >_{\mathrm{diag}}\right)\mathrm{WY} + \varepsilon \end{aligned} \tag{3}$$

where I_+ was a binary 0–1 explanatory indicator variable, denoting *Ae. aegypti* signature oviposition swimming pool habitat covariates displaying positive spatial dependency, and I_- was a binary 0–1 indicator variable denoting sampled covariates displaying negative spatial dependency, using $\mathrm{I}_+ + \mathrm{I}_- = 1$. Expressing the preceding 2-by-2 example in terms of Eq. 3 yielded:

$$\begin{aligned} \begin{pmatrix} y_1\\y_2\\y_3\\y_4 \end{pmatrix} &= \mu\left[\begin{pmatrix} 1&0&0&0\\0&1&0&0\\0&0&1&0\\0&0&0&1 \end{pmatrix} - \rho_+\begin{pmatrix} 1&0&0&0\\0&1&0&0\\0&0&0&0\\0&0&0&0 \end{pmatrix} - \rho_-\begin{pmatrix} 0&0&0&0\\0&0&0&0\\0&0&1&0\\0&0&0&1 \end{pmatrix}\right]\begin{pmatrix} 1\\1\\1\\1 \end{pmatrix} \\ &\quad + \left[\rho_+\begin{pmatrix} 1&0&0&0\\0&1&0&0\\0&0&0&0\\0&0&0&0 \end{pmatrix} + \rho_-\begin{pmatrix} 0&0&0&0\\0&0&0&0\\0&0&1&0\\0&0&0&1 \end{pmatrix}\right]\begin{pmatrix} 0 & \frac{1}{2} & \frac{1}{2} & 0\\ \frac{1}{2} & 0 & 0 & \frac{1}{2}\\ \frac{1}{2} & 0 & 0 & \frac{1}{2}\\ 0 & \frac{1}{2} & \frac{1}{2} & 0 \end{pmatrix}\begin{pmatrix} y_1\\y_2\\y_3\\y_4 \end{pmatrix} \\ &\quad + \begin{pmatrix} \varepsilon_1\\ \varepsilon_2\\ \varepsilon_3\\ \varepsilon_4 \end{pmatrix} \end{aligned}$$

If either $\rho_+ = 0$ (and hence $\mathrm{I}_+ = 0$ and $\mathrm{I}_- = \mathrm{I}$) or $\rho_- = 0$ (and hence $\mathrm{I}_- = 0$ and $\mathrm{I}_+ = \mathrm{I}$), then Eq. 3 is reduced to Eq. 1. It was assumed that this indicator variable classification could be constructed in Python by the quadrants of the corresponding Moran scatterplot. This matrix was generated employing the zip code-stratified, *Ae. aegypti* swimming pool habitat, signature covariates interpolated in the Los Angeles entomological intervention study site.

To identify georeferenced *Ae. aegypti* swimming pool habitat diagnostically stratified clusters, Thiessen polygon surface partitioning was generated in Python to construct geographic neighbor matrices, which were then used in the autocorrelation analysis. Entries in the matrix were set to one if two sampled signature interpolated covariates shared a common Thiessen polygon boundary and zero otherwise. Next, the linkage structure for each surface was edited to remove unlikely geographic neighbors and identify pairs of zip code stratified, *Ae. aegypti* swimming pool habitat, signature, covariates sharing a common Thiessen polygon boundary. Attention was restricted to georeferenced map patterns associated with at least a minimum level of spatial autocorrelation, defined for implementation purposes as $|\mathrm{MC_j}/\mathrm{MC_{max}}| > 0.25$, where $\mathrm{MC_j}$ denoted the j th value and $\mathrm{MC_{max}}$, the maximum value of MC. It was assumed that this threshold value would allow candidate sets of eigen-orthogonalized, *Ae. aegypti* swimming pool habitat, signature, eigen-spatial filter, eigenvectors to reveal substantial PSA and NSA, respectively. Additionally, the ratio of the PRESS (i.e., predicted error sum of squares) statistic to the sum of squared errors from Moran's scatterplot trend line was evaluated in eigenvector eigen-Bayesian eigen-geospace.

5 Results

The upper and lower bounds for a spatial matrix generated using Moran's indices (I) in Python were given by $\lambda_{\max}\left(\mathrm{n}/1^T W 1\right)$ and $\lambda_{\min}\left(\mathrm{n}/1^T W 1\right)$, where $\lambda_{\max}$ and $\lambda_{\min}$ representing the extreme eigenvalues of $\Omega = HWH$. Therefore, in this experiment, the eigenvectors of Ω were vectors with unit norm that maximized Moran's I. These eigenvalues of the matrix were equivalent to Moran's I coefficients of spatial autocorrelation post-multiplied by a constant. Eigenvectors associated with high positive (or negative) eigenvalues exhibit high positive (or negative) autocorrelation [10]. The eigen-orthogonalized eigenvectors associated with the derived signature of *Ae. aegypti* swimming pool habitat, parameter, estimator-related, eigenvalues revealed extremely small absolute values, corresponding to low spatial autocorrelation for defining structures in the empirical, sampled, dataset. The diagonalization of the spatial weighting matrix generated from the covariates involved finding the normalized vectors u_i, stored as columns in the matrix $U = [u_1 \ldots u_n]$ for satisfying: $\Omega = HWH = U\Lambda U^{\mathrm{T}} = \sum_{i=1}^{n} \lambda_i u_i u_i^T$ where $\Lambda = diag(\lambda_1 \cdots \lambda_n)$, $u_i^T u_i = ||u_i||^2 = 1$ and $u_i^T u_j = 0$ for $i \neq j$. The double centering of Ω implied that the eigenvectors u_i generated from the interpolated, signature, *Ae. aegypti* swimming pool habitat, parameter, estimator, stratified covariates were centered, and at least one eigenvalue was equal to zero. Introducing these eigenvectors in the original formulation of Moran's index led to the following expression:

$$I(\mathrm{x}) = \frac{n}{1^T W 1}\frac{x^T HWHx}{x^T Hx} = \frac{n}{1^T W 1}\frac{x^T U \Lambda U^T x}{x^T Hx} = \frac{n}{1^T W 1}\frac{\sum_{i=1}^{n} \lambda_i x^T u_i u_i^T x}{x^T Hx} \tag{4}$$

Considering the centered vector $z = Hx$ and using the properties of idempotence of H, Eq. 4 was equivalent to

$$I(\mathrm{x}) = \frac{n}{1^T W 1} \frac{\sum_{i=1}^{n} \lambda_i z^T u_t u_i^T z}{z^T z} = \frac{n}{1^T W 1} \frac{\sum_{i=1}^{n} \lambda_i \left|\left|u_i^T z\right|\right|^2}{||z||^2} \quad (5)$$

Since both the eigenvectors u_i and the vector z were centered, Eq. 5 was rewritten in Python as:

$$I(\mathrm{x}) = \frac{n}{1^T W 1} \frac{\sum_{i=1}^{n} \lambda_i cor^2(u_i, z) var(z) n}{var(z) n} = \frac{n}{1^T W 1} \sum_{i=1}^{n} \lambda_i cor^2(u_i, z) \quad (6)$$

In this experiment, r represented the number of null eigenvalues of Ω ($r \geq 1$). These eigenvalues and their corresponding eigenvectors were subsequently removed from Λ and U, respectively. Equation 6 was then strictly equivalent to

$$I(\mathrm{x}) = \frac{n}{1^T W 1} \sum_{i=1}^{n-r} \lambda_i cor^2(u_i, z) \quad (7)$$

Moreover, in this experiment, it was demonstrated that Moran's index for a given habitat interpolated eigenvector u_i was equal to $I(u_i) = \left(n/1^T W 1\right)\lambda_i$, so the equation was rewritten: $I(\mathrm{x}) = \sum_{i=1}^{n-r} I(u_i) cor^2(u_i, z)$. The term $cor^2(u_i, z)$ represented the part of the variance of z that was explainable by u_i in the signature *Ae. aegypti* swimming pool habitat model $z = \beta_i u_i + e_i$. This quantity was equal to β_i^2/n var(z). The eigenvectors u_i were orthogonal, and therefore, regression coefficients of the linear models $z = \beta_i u_i + e_l$ were generated by the multiple regression model $z = U\beta + \varepsilon = \beta_i u_i + \cdots + \beta_{n-r} u_{n-r} + \varepsilon$.

Subsequently the distribution of the error residuals in the signature, *Ae. aegypti* swimming pool habitat, model in the autocovariance matrix was quantifiable using Python. The maximum value of I was obtainable for all variations of z, as explained by the eigenvector u_1, which corresponded to the highest eigenvalue λ_1 in the spatial autocorrelation error matrix. In this experiment, $cor^2(u_i, z) = 1\left(\text{and}\, cor^2(u_i, z) = 0 \text{ for} i \neq 1\right)$, and the maximum value of I was deduced for Eq. 7, which equaled $I_{\max} = \lambda_1\left(n/1^T W 1\right)$. The minimum value of I in the error matrix was parsimoniously obtained as all variations of z, which were explained by the eigenvector u_n corresponding to the lowest eigenvalue $\lambda_{n\text{-}r}$ generated in the signature, *Ae. aegypti* swimming pool habitat autocorrelation model. This minimum value equaled $I_{\min} = \lambda_{n-r}\left(n/1^T W 1\right)$. If the sampled explanatory predictor variables were not spatialized, the part of the variance explained by each eigen-orthogonalized eigenvector equaled, on average, $cor^2(u_i, z) = 1/n - 1$. Because the sampled signature, *Ae. aegypti* swimming pool habitat, parameter, estimator, variables in z were randomly permuted, it was assumed that this study would obtain this result. In this experiment, the set of $n!$ random permutations revealed that $E_R(I) = \frac{n}{1^T W 1(n-1)} \sum_{i=1}^{n} \lambda_i = \frac{n}{1^T W 1(n-1)} trace(\Omega)$. It was easily demonstrated that $trace(\Omega) = -\frac{1^T W 1}{n}$, and it followed that $E_R(I) = -\frac{1}{n-1}$ in the prognosticated spectral signature model output, revealing that the Euclidian distance measurement variable between *Ae. aegypti* habitats was the most influential prognosticative eigen-estimator.

An eigen-spatial filter algorithmic analyses selected eigenvectors as regressors, resulting in autocorrelation being filtered out of the residuals of the signature, *Ae. aegypti* swimming pool habitat, parameter, estimator data. The eigen-spatial filtering analyses transformed all the variables containing spatial dependence into covariates free of spatial dependence by partitioning the original georeferenced explanatory attribute variables within a generalizable linear model framework into two synthetic variates: (1) an eigen-spatial filter signature variate capturing latent spatial dependency that would otherwise have remained in the response residuals, and (2) a nonspatial variate that was free of spatial dependence. The eigenfunction, eigen-spatial filter, derived from Moran's indices determined the mean, variance, and statistical distribution characterizations and descriptions of the sampled covariates. The spatially adjusted models identified the clustering patterns of the sampled entomological data in the regressed datasets while accounting for conditional heteroscedastic error terms in the model output. Autocorrelation indices can enable significance testing zip code-stratifiable, georeferenceable, signature, semi-parameterizable, estimator determinants, models using field and remote sampled explanatory variables, which can be very useful for improving resource allocation for implementing control strategies (e.g., building a social media platform on X for prevention messaging especially in senior citizen homes, local schools, homeless shelters, food refugee camps, about removing stagnant water from artificial water containers such as residential bird baths).

6 Discussion

Second-order spatial interpolation was conducted to identify clusters of *Ae. aegypti* mosquito habitats in this experiment. The results showed a Global Moran's Index of 0.762, a z-Score of 7.18, and a p-value of less than 0.05, indicating a significant result. This p-value indicated rejection of the null hypothesis of random autocorrelation, thereby accepting the alternative hypothesis. Consequently, clustering propensity was revealed at a 95% confidence interval. The eigenfunction/eigen decomposition algorithm was employed to identify hot and cold spots using the fixed distance method, revealing hot spots in zip codes 90,731, 90,274, 90,505, 90,503, 90,806, 90,755, 90,813, 90,804, 90,220, 90,221, 90,808, 90,815, 90,638, 90,631 and 90,605. These areas are primarily characterized by low elevation, medium vegetation, and urban residential areas. Cold spots were identified in zip codes 91,390, 93,510, 91,042, 91,605, 91,345, 91,344, and 91,381, characterized primarily by medium elevation, higher vegetation, and mountainous regions. The mean NDVI value for the cold spots was 0.2845 and the mean elevation was 867.56 m. The mean NDVI value for the hot spots was 0.2296 and the mean elevation was 139.55 m. The fixed distance method was selected for identifying hot and cold spots due to its equal spatial weightage for all interpolated points across the study area; the inverse distance method lacks this advantage. Consequently, significant hot and cold spots were revealed. These findings have significant implications for targeted mosquito larval control strategies, particularly in urban residential areas with a high clustering propensity of mosquito habitats. Understanding the spatial distribution of hot and cold spots allows for a more efficient allocation of resources and interventions to mitigate mosquito-borne disease transmission in these identified areas.

Several hot spots clustered in the city of Long Beach, where local transmission occurred in 2023 [2]. Additionally, cold spots were found in mountainous regions or areas where the valley begins, indicating that higher elevation areas may have a protective effect. These findings suggest that geographical factors play a crucial role in the distribution of mosquito habitats and disease transmission. Ongoing monitoring and surveillance in both hot and cold spots are essential for effective vector control and disease prevention efforts.

The findings of this study could also have implications for *Culex quinquefasciatus* larval control, potentially breaking the transmission cycle of West Nile virus (WNV). In 2008, Reisen noted a rising *Culex* population in California, particularly attracted to abandoned green swimming pools with algal growth, contributing to an increase in WNV cases [25]. Similarly, *Ae. aegypti* is drawn to algal growth in untreated swimming pools. Implementing S&D in Los Angeles County could not only help control diseases like dengue, chikungunya, Zika, and yellow fever carried by *Ae. aegypti* but also mitigate the risk of WNV transmission by targeting oviposition capture point *Culex* breeding sites. By focusing on eliminating stagnant water sources and treating potential habitats, such as untreated swimming pools, public health authorities in Los Angeles County can address multiple vector-borne diseases simultaneously, enhancing overall disease control efforts.

However, this study faced significant limitations due to the spatial resolution of the satellite imagery obtained, with Sentinel-2 only offering up to 10 m resolution. It would have greatly benefited from submeter resolution data (e.g. WorldView4 with 31 cm visible and NIR spatial resolution data), which could have provided more detailed and precise information about specific pool classes and their correlation to dengue transmission. For instance, finer resolution data could have allowed for distinguishing between blue, green, and brown pools, each potentially associated with varying degrees of mosquito breeding and disease risk. This study had attempted to create a signature with 10 m resolution but unfortunately, due to Sentinel-2's resolution limitations, this level of differentiation among pool classes was not feasible, and all pool types had to be combined for analysis. This suggests that red, green, and blue (RGB) signatures using 10 m resolution cannot differentiate colors successfully. This aggregation may have obscured potential nuances in oviposition mosquito breeding preferences and disease transmission dynamics associated with different pool types.

Future studies should consider replicating this research using improved satellite resolution to overcome the limitations faced in this study. Exploring the three pool classes separately - blue, green, and brown - could provide valuable insights into the different impacts of these pool types on *Ae. aegypti* breeding and dengue transmission. Additionally, comparing Orange County and San Diego County could yield further insights into the spatial epidemiology of dengue in Southern California. Despite having similar population sizes, Orange County reported more cases in 2023 than San Diego County, indicating potential differences in mosquito habitat oviposition preferences, environmental factors, or public health interventions that warrant investigation [2]. Furthermore, future research could focus on developing a comprehensive California model for mosquito-borne disease risk assessment. This model could integrate various data sources, including high-resolution satellite imagery, meteorological data, land use patterns, and seasonal mosquito surveillance data, to create a real-time mosquito-biting risk index app. Such an app could be utilized by residents, public health agencies, and vector control authorities to enhance mosquito larval control efforts (e.g. implementing S&D at the county level), raise public awareness, and mitigate the risk of mosquito-borne diseases like dengue.

7 Conclusions

In conclusion, a RGB spectral signature of a swimming pool habitat can be interpolated in Python to identify other unknown *Ae. aegypti* swimming pool habitats. This capability empowers local vector control districts to adopt a more targeted approach to LSM, effectively breaking the cycle of transmission. By leveraging advanced computational techniques, such as machine learning algorithms, these vector control districts can optimize resource allocation, reduce environmental impact, and improve public health outcomes.

Author Contributions **Sasha Marie Mosich:** Conceptualization, Data curation, Formal Analysis, Investigation, Methodology, Project administration, Software, Supervision, Validation, Visualization, Writing—original draft, Writing—review and editing.

Ricardo Izurieta: Supervision.

Ismael Hoare: Supervision.

Namit Choudhari: Software, Validation, Methodology, Formal Analysis.

Heather McDonald: Software, Validation, Methodology, Writing- Review & Editing.

Anthony Masys: Supervision.

Benjamin Jacob: Methodology, Project administration, Resources, Supervision, Writing—original draft, Writing—review and editing.

Conflicts of Interest The authors declare no conflicts of interest.

Abbreviations

California Department of Public Health	CDPH
Geographical Information Systems	GIS
Larval Source Management	LSM
Seek and Destroy	S&D
Convolutional Neural Network	CNN
Near-Infrared	NIR
Land Use/Land Cover	LULC
Normalized Difference Vegetation Index	NDVI
Digital Elevation Model	DEM
European Space Agency	ESA
US Geological Survey	USGS
Probability Density Functions	PDF
Dirichlet Process Mixture Model	DDPMM
Positive Spatial Autocorrelation	PSA
Negative Spatial Autocorrelation	NSA
Predicted Error Sum of Square	PRESS
West Nile Virus	WNV
Red, Green, and Blue	RGB

References

1. Black AP, Kinman RN, Keirn MA, Smith JJ, Harlan WE (1970) The disinfection of swimming pool waters. Am J Public Health Nations Health 60(3):535–545. https://doi.org/10.2105/ajph.60.3.535
2. California Department of Public Health (2024a) CDPH monthly update on number of dengue infections in California. https://www.cdph.ca.gov/Programs/CID/DCDC/CDPH%20Document%20Library/TravelAssociatedCasesofDengueVirusinCA.pdf. Accessed 30 Mar 2024
3. California Department of Public Health (2024b) Guidance for surveillance of and response to invasive aedes mosquitoes and dengue, Chikungunya, and Zika in California. https://www.cdph.ca.gov/Programs/CID/DCDC/CDPH%20Document%20Library/InvasiveAedesSurveillanceandResponseinCA.pdf. Accessed 30 Mar 2024
4. California Department of Public Health (2016) Guide to important mosquitoes in california. https://www.cdph.ca.gov/. Accessed 30 Mar 2024
5. Carrington LB, Simmons CP (2014) Human to mosquito transmission of dengue viruses. Front Immunol 5:290. https://doi.org/10.3389/fimmu.2014.00290
6. Chen LH, Marti C, Diaz Perez C, Jackson BM, Simon AM, Lu M (2023) Epidemiology and burden of dengue fever in the United States: a systematic review. J Travel Med 30(7). https://doi.org/10.1093/jtm/taad127
7. Cunha HS, Sclauser BS, Wildemberg PF, Fernandes EA, dos Santos JA, Lage Mde, Lorenz C, Barbosa GL, Quintanilha JA, Chiaravalloti-Neto F (2021) Water tank and swimming pool detection based on remote sensing and deep learning: Relationship with socioeconomic level and applications in dengue control. Plos One 16(12). https://doi.org/10.1371/journal.pone.0258681

8. Getis A, Griffith DA (2002) Comparative spatial filtering in regression analysis. Geogr Anal 34(2):130–140. https://doi.org/10.1111/j.1538-4632.2002.tb01080.x
9. Greater Los Angeles County Vector Control District (n.d.) Public health pesticides. https://www.glamosquito.org/public-health-pesticides. Accessed 30 Mar 2024
10. Griffith DA (2003) Spatial autocorrelation and spatial filtering: gaining understanding through theory and scientific visualization. In: Advances in spatial science. Springer, Berlin, Heidelberg. https://doi.org/10.1007/978-3-54024806-4
11. Griffith DA (2005) Effective geographic sample size in the presence of spatial autocorrelation. Ann Assoc Am Geogr 95(4):740–760. https://doi.org/10.1111/j.1467-8306.2005.00484x
12. Halstead SB (2024) Three dengue vaccines—what now? N Engl J Med 390(5):464–465. https://doi.org/10.1056/nejme2314240
13. Hill M, Wilke ABB, Fuller DO, Kummer AG, Ventura PC, Beier JC, Kluh S, Vetrone S, Litvinova M, Ajelli M (2023) Environmental and socio-economic determinants of mosquito vector species presence and abundance in Los Angeles County, California. https://doi.org/10.21203/rs.3.rs-3338430/v1
14. Hosmer DW, Lemeshow S (2000) Applied logistic regression. Wiley. https://doi.org/10.1002/0471722146
15. Islam MT, Quispe C, Herrera-Bravo J, Sarkar C, Sharma R, Garg N, Fredes LI, Martorell M, Alshehri MM, Sharifi-Rad J, Daştan SD, Calina D, Alsafi R (2021) Transmission, pathogenesis, and control of dengue virus: a literature-based undivided perspective. Biomed Res Int. https://doi.org/10.1155/2021/4224816
16. Fitzgerald GP (1959) Bactericidal and algicidal properties of some algicides for swimming pools. Appl Microbiol 7(4):205–211. https://doi.org/10.1128/am.7.4.205-211.1959
17. Jacob BG, Izureta R, Bell J, Parikh J, Loum D, Casonova J, Gates T, Murray K, White L, Aceng JR (2023) Approximating non-asymptoticalness, skew heteroscedasticity and geospatiotemporal multicollinearity in posterior probabilities in Bayesian eigenvector Eigengeospace for optimizing hierarchical diffusion-oriented covid-19 random effect specifications geosampled in Uganda. Am J Math Stat 13(1):1–43. https://doi.org/10.5923/j.ajms.20231301.01
18. Jacob BG, Loum D, Kaddumukasa M, Kamgno J, Djeunga HN, Domche A, Nwane P, Mwangangi J, Hernandez Bojorge S, Parikh J, Casanova J, Izurieta R, Michael E, Mason T, Mubangizi A (2021) Geospatial artificial intelligence infused into a smartphone drone application for implementing 'Seek and Destroy' in Uganda. Am J Entomol 5(4):921–109. https://doi.org/10.11648/j.aje.20210504.11
19. Lowe's. (2021) How to maintain a pool. https://www.lowes.com/n/how-to/swimming-pool-maintenance. Accessed 30 Mar 2024
20. Metzger ME, Hardstone Yoshimizu M, Padgett KA, Hu R, Kramer VL (2017) Detection and establishment of *Aedes aegypti* and *Aedes albopictus* (Diptera: Culicidae) mosquitoes in California, 2011–2015. J Med Entomol 54(3):533–543. https://doi.org/10.1093/jme/tjw237
21. Navarro DMAF, De Oliveira PES, Potting RPJ, Brito AC, Fital SJF, Sant'Ana AEG (2003) The potential attractant or repellent effects of different water types on oviposition in *Aedes aegypti* L. (Dipt., Culicidae). J Appl Entomol 127(1):46–50. https://doi.org/10.1046/j.1439-0418.2003.00690.x
22. Palanichamy Kala M, St John AL, Rathore APS (2023) Dengue: update on clinically relevant therapeutic strategies and vaccines. Curr Treat Options Infect Dis 15(2):27–52. https://doi.org/10.1007/s40506-023-00263-w
23. Powell JR (2018) Mosquito-borne human viral diseases: Why *Aedes aegypti?*. Am J Trop Med Hyg 98(6):1563–1565. https://doi.org/10.4269/ajtmh.17-0866
24. Ranasinghe HA, Amarasinghe LD (2020) Naturally occurring microbiota in dengue vector mosquito breeding habitats and their use as diet organisms by developing larvae in the Kandy District, Sri Lanka. Biomed Res Int 2020:1–12. https://doi.org/10.1155/2020/5830604
25. Reisen WK, Takahashi RM, Carroll BD, Quiring R (2008) Delinquent mortgages, neglected swimming pools, and West Nile virus, California. Emerg Infect Dis 14(11):1747–1749. https://doi.org/10.3201/eid1411.080719

26. Ritchie KK, Izurieta R, Hoare I, Choudhari N, Murray L, Yost B, Fiess D, Pecora P, Masys A, Casanova J, Jacob BG (2024) Mapping Ae. aegypti bird bath habitats for implementing "Seek and Destroy" larval source management in Hillsborough County, FL, USA. Am. J. Entomol. 8(1):1–17. https://doi.org/10.11648/j.aje.20240801.11
27. Trejo I, Barnard M, Spencer JA, Keithley J, Martinez KM, Crooker I, Hengartner N, Romero-Severson EO, Manore C (2023) Changing temperature profiles and the risk of dengue outbreaks. Plos Clim 2(2). https://doi.org/10.1371/journal.pclm.0000115
28. Tucker CJ (1979) Red and photographic infrared linear combinations for monitoring vegetation. Remote Sens Environ 8(2):127–150. https://doi.org/10.1016/0034-4257(79)90013-0
29. Wikimedia Foundation (2024) Los Angeles County, California. https://en.wikipedia.org/wiki/Los_Angeles_County%2C_California. Accessed 30 Mar 2024
30. Yang F, Schildhauer S, Billeter SA, Hardstone Yoshimizu M, Payne R, Pakingan MJ, Metzger ME, Liebman KA, Hu R, Kramer V, Padgett KA (2020) Insecticide resistance status of *Aedes aegypti* (Diptera: Culicidae) in California by biochemical assays. J Med Entomol 57(4):1176–1183. https://doi.org/10.1093/jme/tjaa03

Diagnosing Space and Time Overdispersion Due to Heteroskedasticity

Nathanael B. Stanley, Namit R. Choudhari, and Benjamin G. Jacob

Abstract Introduction: Geotemporal modeling with count-based data requires diagnostic tests to ensure the statistical reliability of results, especially when assumptions like equidispersion and independence of observations are violated. Overdispersion, heteroskedasticity, and spatial or temporal autocorrelation are common in spatial data and require assessment to prevent distortion in model predictions. We demonstrate a methodology for diagnosing spatial and temporal errors using zipcode tabulation area (ZCTA) level data on uninsured populations within a 23-county area in central-southern Florida, focusing on addressing overdispersion, heteroskedasticity, and testing critical model assumptions. Methods: We used five American Community Survey (ACS) 5-year datasets (2014–2022) to diagnose model stability of uninsured populations over time and space. Techniques included multivariate regression, Poisson and negative binomial regression, spatial autocorrelation (Moran's I), cluster mapping (Getis-Ord Gi*), and GARCH modeling. R and ArcPro 2.6 were employed for analysis. Results: Model diagnostics and automated decision-making based on variable significance and model selection criteria showed systematic assessment of model stability in error-prone secondary data. Unemployment and disability status were consistent predictors across all year groups, while spatial analysis revealed persistent clustering of uninsured populations that decreased over time. GARCH modeling identified regional volatility patterns. Discussion: Spatial models complement non-spatial ones by identifying clusters of uninsured populations, providing valuable context for interpreting predictors' effects. Even with error-prone data, the utilization of diagnostics in geostatistical modeling can detect overdispersion, spatial dependencies, and temporal heteroskedasticity and allow for making informed decisions about the analysis. These methods address linear, spatial, and temporal errors in geostatistical modeling, improving model stability.

N. B. Stanley (✉)
Office of Community Outreach and Engagement, Moffitt Cancer Center, Tampa, FL, USA
e-mail: Nathanael.Stanley@moffitt.org

N. R. Choudhari
School of Geosciences, University of South Florida, Tampa, FL, USA

B. G. Jacob
College of Public Health, University of South Florida, Tampa, FL, USA

B. Jacob et al. (eds.), *Tech Enabled Global Health Security*, Advanced Sciences and Technologies for Security Applications, https://doi.org/10.1007/978-3-031-86997-6_8

Keywords Generalized linear model · Geostatistics · Geotemporal · Heteroskedasticity · Regression · Secondary data · Time-series · Uninsured · Variable selection

1 Introduction

Geostatistical analyses are integral to examining spatial and temporal patterns in count data, particularly in public health, epidemiology, and environmental studies. These analyses rely on statistical modeling methods to quantify relationships between predictors (independent variables) and outcomes (dependent variables). Predictors are used to explain or predict the variability in the outcome variable [28]. The validity of these models depends on satisfying key assumptions, including equidispersion, which is a probability distribution typically associated with count data, where the mean and variance of the distribution are equal independence of observations, and the absence of spatial or temporal autocorrelation in residuals [24]. Overdispersion, where the variance exceeds the mean, violates the equidispersion assumption of Poisson regression [24] and necessitates alternative approaches. Similarly, zero spatial autocorrelation, where residuals are random across geographic space and temporal heteroskedasticity i.e., unequal variance changes over time, can bias estimates and reduce model reliability [27]. Early models, including the Poisson regression, assumed equidispersion—that the mean and variance of the response variable were equal. While this assumption simplifies computation, real-world count data often exhibit overdispersion due to unobserved heterogeneity, omission of key predictors, and outliers, necessitating alternative approaches like the Negative Binomial regression [10]. Payne et al. [41] demonstrated through simulation studies that Negative Binomial regression and Negative Binomial regression- Generalized Linear Mixed Models consistently outperformed other methods for handling overdispersion regardless of its source, including population heterogeneity, omitted variables, and outliers. Scale-adjusted Poisson models, while computationally simpler, often produce biased standard-error estimates, with Double-Stirring Poisson (DS-Poisson) underestimating and Power-Stabilized Poisson (PS-Poisson) overestimating standard errors. The PS-Poisson modifies the mean–variance relationship to handle non-equidispersion using a power parameter, while the DS-Poisson introduces additional parameters to flexibly model overdispersion or underdispersion in count data [41]. These findings support the use of flexible model-based approaches, such as Negative Binomial regression and Negative Binomial regression-Generalized Linear Mixed Models, in geostatistical analyses where overdispersion is likely to occur.

As highlighted by Hanks et al. [25], spatial confounding is extensively studied in discrete spatial data (e.g., areal data) and continuous geostatistical settings. Restricted Spatial Regression (RSR) has been proposed as a method to address spatial confounding by constraining random effects to be orthogonal to fixed effects, offering computational benefits. However, the study by Hanks et al. underscores potential risks with RSR, such as overly narrow Bayesian credible intervals under model misspecification. This highlights the importance of careful consideration and diagnostics when employing Spatial Generalized Linear Mixed Models (SGLMMs) in geostatistical modeling.

Advances in computational power enabled the development of flexible modeling frameworks capable of handling non-Gaussian distributions, overdispersion, and unequal variance. Negative Binomial regression, for instance, introduces a dispersion parameter with a non-homogeneous gamma-distributed mean to address overdispersion [10]. Temporal models such as Generalized Autoregressive Conditional Heteroscedasticity (GARCH) accommodate conditional variance in longitudinal data, while spatial diagnostics like Moran's I detect clustering or dispersion in residuals [5, 17]. GARCH models are used when the variance of the error term is not constant i.e., heteroskedastic. Additionally, advancements in Generalized Linear Mixed Models (GLMMs) extended count-based modeling to account for random effects, offering a flexible framework for analyzing correlated or hierarchical count data [53]. In statistics, a generalized linear mixed model (GLMM) is an extension to the generalized linear model (GLM) in which the linear predictor contains random effects in addition to the usual fixed effects [8]. These tools allow researchers to rigorously evaluate assumptions and improve model validity.

By observing the data linearly, spatially, and temporally, erroneous predictors may be quantified in varying dimensions and how they interact within sentinel site datasets. The methodology utilizes diagnostics such as Akaike Information Criterion (AIC), Bayesian Information Criterion (BIC), deviance, and spatial metrics like Moran's I to inform model selection and refinement. Akaike's information criterion (AIC) and Bayesian information criterion (BIC) are criteria for model selection among a finite set of models. Models with lower BIC are generally preferred [3]. In statistics, Moran's I is a measure of spatial autocorrelation developed by Patrick Alfred Pierce Moran [3]. Spatial autocorrelation is characterized by a correlation in a signal among nearby locations in space [40]. Spatial autocorrelation is more complex than one dimensional autocorrelation because spatial correlation is multi-dimensional (i.e. 2 or 3 dimensions of space) and multi-directional [22]. Hot and cold spot analysis is also included to reveal patterns of high and low intensities in spatial data, providing additional insights into geographic heterogeneity, and GARCH modeling gives insight into the potential volatility of the observations over time.

The primary aim of this chapter is to showcase a framework for diagnosing errors in geostatistical modeling—focusing on providing an example methodology for diagnosing spatial and temporal errors in geostatistical modeling using less-than-ideal datasets. To illustrate these methods, we utilize 5 datasets from the American Community Survey, each representing a 5-year estimate for years 2014–2022 (ACS 2018 5yr, ACS 2019 5yr, ACS 2020 5yr, ACS 2021 5yr, ACS 2022 5yr). The goal of this chapter is not to identify specific spatial and temporal trends in the uninsured population, although we do reveal space and time trends. The goal is to demonstrate methods for testing critical model assumptions, addressing overdispersion, and making informed analysis decisions based on diagnostic tests. This approach emphasizes the importance of leveraging statistical and spatial tools to refine models, ensuring robust interpretations even when working with data that exhibit limitations in quality or granularity.

2 Background and Context of Statistical Models

2.1 Correlation Assessment

A correlation assessment is a standard preliminary step when preparing for statistical modeling. The purpose of this assessment is to evaluate the pairwise relationships among predictor variables, typically quantified using Pearson or Spearman correlation coefficients. This evaluation identifies the presence and degree of multicollinearity, a condition in which predictor variables are highly correlated with each other [47]. Multicollinearity can lead to inflated standard errors for regression coefficients and potential instability in parameter estimates [16].

In stepwise regression, correlation analysis provides a systematic way to identify and possibly exclude highly correlated variables, which can complicate the iterative variable selection process. For Poisson and negative binomial regression, which are used for modeling count data, addressing multicollinearity is essential to avoid biased coefficient estimates or convergence issues during model fitting [16]. These methods assume independence among predictors, and significant correlations may interfere with the interpretability and accuracy of the resulting models.

2.2 Stepwise Backward Regression for Predictor Selection

Correlation assessment was followed by stepwise backward regression to reduce multicollinearity among predictors and generate initial Variance Inflation Factor (VIF) values. In statistics, the VIF is the ratio of the variance of a parameter estimate when fitting a full model that includes other parameters to the variance of the parameter estimate if the model is fit with only the parameter on its own [21]. This iterative procedure involves starting with a full model that includes all potential predictors and sequentially removing variables based on statistical criteria, such as p-values or Bayesian Information Criterion (BIC) [14]. By systematically eliminating non-significant predictors, stepwise regression simplifies the model and enhances its interpretability.

One primary benefit of stepwise backward regression is also its ability to address multicollinearity, a condition where two or more predictors are highly correlated [15]. Multicollinearity inflates the standard errors of regression coefficients, making it challenging to discern the independent contribution of each predictor. By removing redundant variables, stepwise regression minimizes multicollinearity, as evidenced by lower VIF values in the reduced model. VIF quantifies the degree of multicollinearity, with values above 10 typically indicating significant collinearity issues [15]. The process of variable elimination helps ensure that only the most independent and informative predictors remain.

Another advantage of this approach is model simplification. Including all available predictors can lead to overfitting, where the model captures random noise rather than true underlying patterns. Overfitting reduces the model's generalizability to new datasets [26]. By retaining only the most significant predictors, stepwise backward regression creates a more parsimonious model that balances complexity with explanatory power [45]. This step also improves computational efficiency, as models with fewer predictors require less time and resources to estimate, particularly when applied to large or complex datasets.

However, stepwise backward regression is not without risks. The reliance on BIC or other statistical criteria for variable elimination can inadvertently remove predictors that are important but appear non-significant in the context of the initial model. This is particularly problematic when there are complex interactions or non-linear relationships that the model does not account for. Additionally, the iterative nature of stepwise regression increases the risk of over-reliance on sample-specific patterns, potentially leading to model instability and reduced generalizability. BIC, while effective at penalizing model complexity, may prioritize overly simple models that fail to capture all relevant dynamics in the data. Despite these limitations, stepwise backward regression provides a useful starting point for identifying key predictors and addressing multicollinearity [20]. The initial VIF values generated during this process offer valuable diagnostic insights, guiding further refinement of the model. By carefully considering the results of stepwise regression and complementing it with robust diagnostic tests, researchers can build a solid foundation for subsequent analyses, such as Poisson and Negative Binomial regression.

2.3 *Poisson Regression to Test for Overdispersion*

Poisson regression is the foundational model for count data, offering simplicity and interpretability. However, its assumption of equidispersion often fails in practice, leading to biased estimates and invalid inference when overdispersion is present. The model assumes independence of observations, a linear relationship between predictors and the logarithm of the response variable, and absence of spatial and temporal autocorrelation in residuals [43]. These assumptions are critical for the model's validity and are often assessed using overdispersion tests and spatial diagnostics. The Poisson regression model assumes that the mean and variance of the response variable are equal, which is rarely the case in real-world data due to overdispersion [24]. To evaluate the performance of the Poisson model, several diagnostic tests are used. Akaike Information Criterion (AIC) is employed to assess the relative quality of the model by balancing goodness of fit with model complexity, where lower AIC values indicate better models. Bayesian Information Criterion (BIC) serves a similar purpose but applies a stricter penalty for additional parameters, making it particularly useful for model comparison [2]. Deviance, a measure of the discrepancy between the observed and predicted data, is calculated as twice the difference between the log-likelihoods of the saturated and fitted models [33]. Lower deviance values suggest better model performance. McFadden's R-squared, a pseudo-R-squared measure, quantifies the proportion of variance explained by the model, offering insight into its explanatory power [13]. These diagnostics are critical for evaluating whether the Poisson regression adequately captures the underlying structure of the data.

2.4 Negative Binomial Regression to Account for Overdispersion and Test Model Fitness

The Negative Binomial regression extends the Poisson model by introducing a non-homogeneous gamma-distributed mean to account for overdispersion [24]. This flexibility makes it particularly useful for real-world datasets with inherent variability. The model assumes independence of observations and no spatial or temporal auto-correlation in residuals but is better suited for overdispersed data due to its additional dispersion parameter. The Likelihood Ratio Test (LRT) is commonly used to compare Negative Binomial and Poisson models to determine whether the added complexity improves fit [32].

The Negative Binomial regression model addresses the limitation of overdispersion observed in Poisson regression by introducing a dispersion parameter [11]. This adjustment allows the model to account for variance that exceeds the mean, providing a more flexible approach for modeling count data. AIC and BIC are again applied to compare the Negative Binomial model to the Poisson model, with lower values indicating better performance. Deviance remains a key measure for assessing model fit, with reductions in deviance signaling improvements over the Poisson regression. McFadden's R-squared is used to evaluate the proportion of variance explained by the model. The Likelihood Ratio Test (LRT) plays a role in determining whether the inclusion of the dispersion parameter significantly improves model fit. By addressing overdispersion and utilizing these diagnostic tools, the Negative Binomial regression often provides a more reliable framework for analyzing count data in geostatistical contexts [14].

2.5 Bayesian Modeling

Bayesian models offer a probabilistic approach to statistical modeling by combining prior knowledge with observed data to make inferences about parameters. A key component of Bayesian modeling is the use of priors, which represent the researcher's belief about the parameters before observing the data. Priors can be informative, incorporating specific knowledge, or non-informative, reflecting minimal assumptions about parameter values. The combination of the prior and the likelihood (probability of the observed data given the parameters) produces the posterior distribution, which represents updated beliefs about the parameters after accounting for the data.

Posterior inference involves summarizing the posterior distribution to make decisions or predictions. Common summaries include the posterior mean, which represents the average value of the parameter, and credible intervals, which indicate the range of values that contain the parameter with a specified probability. Bayesian models are particularly useful in scenarios involving complex hierarchical data, uncertainty quantification, or the need to incorporate external knowledge into the analysis. By employing computational methods such as Markov Chain Monte Carlo (MCMC), Bayesian models provide a flexible framework for analyzing count data, spatial dependencies, and temporal variability.

2.6 Spatial Autocorrelation and Moran's I

Spatial autocorrelation measures the degree to which values of a variable cluster spatially. Moran's I is a widely used statistic for detecting clustering or dispersion in spatially referenced data. Positive values indicate clustering, while negative values suggest dispersion. Non-zero spatial autocorrelation reveals that the assumption of residual independence is violated, necessitating adjustments to the model or interpretation. Moran's I provides essential insights into the spatial structure of the data, helping researchers evaluate whether assumptions of independence are upheld. Spatial autocorrelation examines the degree to which residuals from a model are correlated across geographic space, violating the assumption of independence. Moran's I is the primary diagnostic tool used to measure spatial autocorrelation, with values close to $+1$ indicating strong clustering, values near -1 suggesting dispersion, and values around 0 implying randomness. Moran's I is calculated by analyzing the spatial arrangement of residuals, providing insights into the presence of spatial dependencies that might undermine the validity of a model. Detecting significant spatial autocorrelation suggests that adjustments are necessary, either by incorporating spatial effects into the model or reevaluating its assumptions. To maintain spatial integrity, the data used in the analysis must be projected to account for the curvature of the earth and potential warping of space as a result of superimposing a spherical object on a 2-dimensional surface [5].

2.7 Getis-Ord Gi, Hot/Cold Spot Analysis

The Getis-Ord Gi* statistic and Hot Spot Analysis also provide spatial insights that complement regression and spatial analyses, especially when handling count data models like Poisson and Negative Binomial regressions, which assume independence among observations. If spatial autocorrelation—where similar values cluster together—is present, these models may yield biased results. By identifying statistically significant clusters, Hot Spot Analysis flags areas where spatial dependencies may need to be considered. Further spatial autocorrelation analysis, such as Moran's I or Local Moran's I, complements Hot Spot Analysis by quantifying the degree of similarity between neighboring areas across the study region. These layered spatial analyses provide a robust understanding of clustering patterns, with Hot Spot Analysis highlighting concentrated areas of uninsured populations and spatial autocorrelation and regression models exploring the relationships and drivers of these patterns.

2.8 Temporal Variability and GARCH Models

Temporal analyses often reveal heteroskedasticity, or unequal variance over time. The Cochran-Armitage Trend Test is a statistical method commonly used to evaluate trends in proportions over ordered groups. In this analysis, it was applied to assess whether there were statistically significant trends—either increases or decreases—in uninsured rates across demographic groups defined by race, age, income, and educational attainment over 5 ACS intervals from ACS 2018 5yr (2014–2018) to ACS 2022 5yr (2018–2022). By examining each demographic category across ordered survey intervals, the test identified significant changes in uninsured proportions.

This test's utility lies in its sensitivity to categorical data trends across ordered intervals, using a chi-square approximation to evaluate whether there is a linear trend in proportions over time. This approach is widely used in public health and social science research to analyze temporal trends in health outcomes, such as insurance coverage, where categorical data vary across time intervals [1].

Generalized Autoregressive Conditional Heteroskedasticity (GARCH) models, developed by Engle [17], are designed to model such volatility. These models are particularly useful in datasets where temporal patterns exhibit persistence or irregular fluctuations, offering a more nuanced understanding of temporal dynamics. By capturing conditional heteroskedasticity, GARCH models address non-constant variance and improve the reliability of temporal analyses. The GARCH model is used to address temporal variability, particularly heteroskedasticity, which refers to unequal variance across time. This model captures temporal autocorrelation and conditional variance, providing a nuanced understanding of trends and volatility in count data. AIC and BIC are utilized to evaluate the fit of the GARCH model, with lower values indicating better performance. By modeling temporal dependencies, the GARCH framework identifies periods of volatility, offering insights into temporal patterns that could influence the interpretation of results. The ability to model conditional variance enhances the robustness of analyses, particularly in longitudinal datasets with irregular fluctuations.

Using GARCH modeling to study temporal volatility in the rates of uninsured individuals provides insights into the stability of these rates over time. Health insurance coverage levels often reflect underlying economic, policy, and demographic shifts, which can introduce irregularities and variability in aggregated data over time. By applying a GARCH model, these temporal variances are directly modeled, providing a clearer picture of fluctuations in the uninsured population. This approach captures not only average trends but also the extent and persistence of volatility, helping to identify ZCTAs with particularly unstable coverage rates over time.

One limitation of this approach is that the GARCH model is applied to estimated counts of uninsured individuals within each ZCTA, lacking a spatial dimension that would allow for cross-regional interaction effects. However, this limitation is mitigated by the inclusion of spatial analyses in preceding steps of the overall study. Previous analyses focused on spatial clustering, helping to contextualize the volatility findings of the GARCH model. While GARCH alone does not account for spatial dependencies, combining it with spatial analysis provides a more comprehensive view, linking both temporal and geographic patterns in uninsured rates across the population.

2.9 Importance of Diagnostic Tests

If a researcher fails to assess for overdispersion or ignores diagnostic tests while analyzing count-based, geographically and temporally specific data, or uses non-count-based modeling, several critical issues could arise. Ignoring overdispersion, which occurs when the variance of the count data exceeds its mean, could violate the assumptions of models like Poisson regression [41]. This can result in underestimated standard errors and inflated significance levels, leading to Type I errors where predictors are falsely identified as statistically significant. Furthermore, without testing for assumptions such as spatial independence or temporal homogeneity, there is a risk of misinterpreting the model's output. For instance, failing to address spatial autocorrelation may lead to the incorrect conclusion that geographic patterns are random when they are not, while neglecting temporal heteroskedasticity could bias estimates in longitudinal analyses.

Using non-count-based models, such as ordinary least squares (OLS), for count data can also lead to model misspecification [50]. Count data typically follow discrete probability distributions, such as Poisson or Negative Binomial, whereas OLS assumes a continuous Gaussian distribution. This mismatch can result in poorly fitted models, residuals with non-normality or heteroskedasticity, and biased coefficients, particularly when the dataset includes extreme values. Additionally, count-based data that are geographically and temporally specific often exhibit dependencies. Ignoring spatial or temporal diagnostics, such as Moran's I for spatial autocorrelation or GARCH models for temporal variance, may result in overlooking important geographic clusters or temporal patterns. This could lead to misattributing variability to predictors when it is due to inherent dependencies in the data structure.

These issues collectively undermine the validity and interpretability of the analysis. Misinterpreted relationships and model misspecifications can obscure key insights about the data, potentially leading to incorrect conclusions about the relationships between variables and the dynamics underlying geographic and temporal trends. This highlights the importance of using diagnostic tests and appropriate models tailored to the characteristics of the data to ensure accurate and reliable findings.

2.10 Study Area and Data Description

The study area encompasses ZCTA tabulation areas (ZCTAs) within the 23-county catchment area of an NCI-designated comprehensive cancer center in Florida (Fig. 1). The analysis uses 5 datasets of ZCTA level ACS 5-year estimates for uninsured populations, census table S2701. These data, representing count estimates, were selected for their relevance to understanding geographic and temporal trends in healthcare access across a large service area.

Key variables from ACS table S2701 include total uninsured population (S2701 C04 001E), stratified by demographic categories that reflect nuances and disparities in uninsured populations, including age [52], race and ethnicity [35], gender [9], disability [30, 44], nativity [4], income [19], and educational attainment [39]. These variables provide insights into subgroup-specific disparities, and their geographic and temporal characteristics support the evaluation of both spatial and temporal patterns in uninsured populations.

3 Research Design and Methodology

The analysis to diagnose heteroskedasticity in uninsured populations over time and across geographic space involved examining ZCTA trends using statistical and spatial models. Each of the models described were conducted with on all five of the ACS 5yr datasets independently, except for the GARCH test which required all datasets to be merged into one.

First, multicollinear variables were removed via correlation assessment, followed by stepwise backward regression. Poisson regression served as the baseline model for estimating uninsured counts, with negative binomial regression addressing overdispersion by introducing a dispersion parameter. Model fitness was compared using a likelihood ratio test, Pseudo R^2, and adjusted Pseudo R^2, while Bayesian regression leveraged prior knowledge to identify explanatory variables. Spatial autocorrelation tests, Moran's I, assessed geographic clustering of uninsured rates, and hot (high) and cold (low) spots of uninsured populations using Getis-Ord Gi* and z-scores highlighted regions with significantly higher or lower rates. Mapping these clusters revealed consistent geographic health disparities. Additionally, GARCH modeling

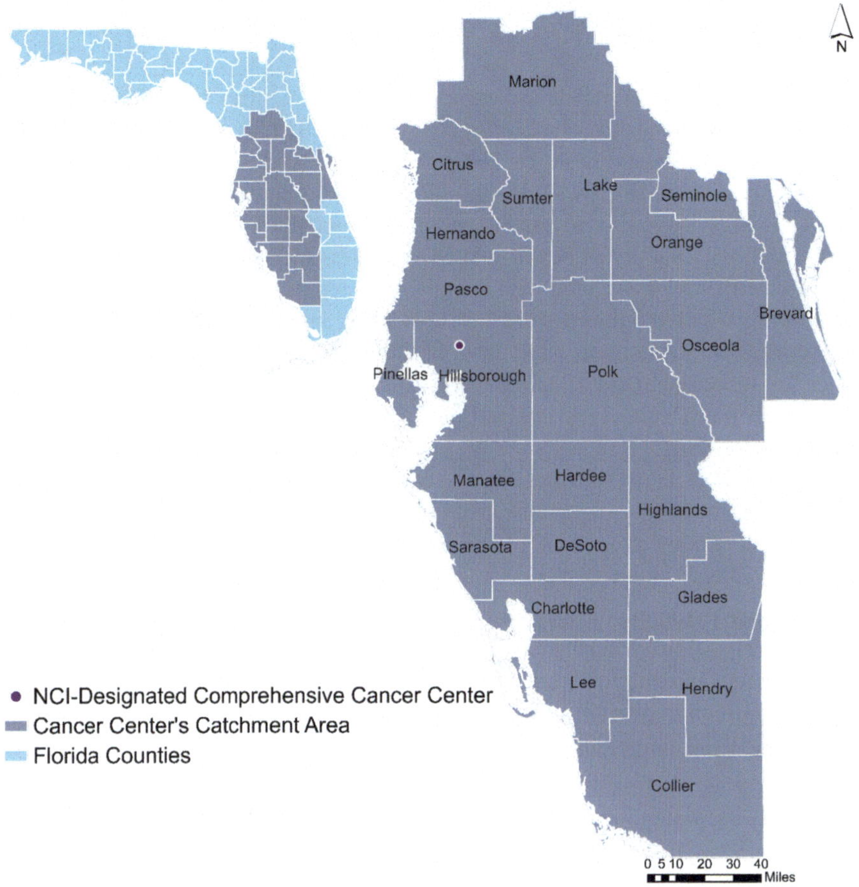

Fig. 1 Study Site

captured year-over-year temporal volatility, identifying areas with significant fluctuations in uninsured rates, providing a comprehensive understanding of both spatial and temporal patterns.

Together, these methods provide a methodological framework to diagnose heteroskedasticity and overdispersion in less than ideal datasets: regression models capture central trends while addressing count-specific issues like overdispersion, with deviance residuals and dispersion tests assessing model fit. Spatial analyses reveal geographic clustering through Moran's I and Getis-Ord Gi*, while the GARCH model highlights temporal volatility. This combined approach supports a comprehensive assessment of categorical, spatial and temporal patterns in uninsured rates across ZCTAs. These analyses utilize R and ArcPro 2.9.

3.1 Data Preparation

Data for the uninsured population comes from the American Community Survey (ACS) 5-year estimates for year groups 2018 5-year, 2019 5-year, 2020 5-year, 2021 5-year, and 2022 5-year (Table 1). These year groups were selected based on being the most up to date data available that allows for consecutive 5-year groups to be analyzed. ACS is a highly accessible source for population-level data and includes information on a variety of health-related variables. Although ACS count data are estimates with error associated with them, we wanted to utilize a data source widely accessible—which is important for low resource teams that need to analyze health metrics at scales lower than county level [49]. We utilize ZCTAs as the spatial scale, instead of more stable areas such as census tracts [34], due to ZCTAs having larger populations and due to their application to regional outreach strategies for large health systems [46].

The ACS table S2701 provides detailed characteristics of the uninsured population and serves as the primary data table for this analysis. The key variable of interest, the total uninsured population (S2701 C04 001E), is supplemented by various covariates within the uninsured, including:

- **Race and Ethnicity**: African American (S2701 C04 017E), American Indian/ Alaska Native (S2701 C04 018E), Asian (S2701 C04 019E), Native Hawaiian/ Pacific Islander (S2701 C04 020E), Non-Hispanic White (S2701 C04 024E), Hispanic (S2701 C04 023E), and those of two or more races (S2701 C04 022E).
- **Gender**: Uninsured Male (S2701 C04 014E) and Uninsured Female (S2701 C04 015E).
- **Poverty Level**: Uninsured individuals living 100% below the poverty line (S2701 C04 061E).
- **Age Groups**: Uninsured individuals aged 19–64 (S2701 C04 012E) and those 65 and older (S2701 C04 013E).
- **Employment Status**: Uninsured Employed (S2701 C04 044E) and Uninsured Unemployed (S2701 C04 045E).
- **Nativity**: Uninsured Native-Born (S2701 C04 031E) and Foreign-Born (S2701 C04 032E).
- **Disability Status**: Uninsured individuals with a disability (S2701 C04 035E) and without a disability (S2701 C04 036E).
- **Educational Attainment**: Ranges from those with less than a high school education (S2701 C04 038E) to high school graduates (S2701 C04 039E), those with some college or an associate degree (S2701 C04 040E), and those with a bachelor's degree or higher (S2701 C04 041E).

Data files were accessed via data.census.gov in excel format, cleaned to remove duplicate headings and to ensure the variables were in a format that matched the tiger line GEOIDs from the ZCTA boundaries from the U.S. Census. We utilized the 2022 Tigerline ZCTA boundary file to join the excel files for all year groups. There were 983 Florida ZCTAs in 2018 5yr, 2019 5yr, and 2020 5yr, and they increased

Table 1 Demographics of the uninsured population, ACS 5-year estimates for 2018, 2019, 2020, 2021, 2022

	ACS 2018 5yr		ACS 2019 5yr		ACS 2020 5yr		ACS 2021 5yr		ACS 2022 5yr	
	Total population	Uninsured population	Total population	Uninsured population	Total population	Uninsured population	Total population	Uninsured population	Total population	Uninsured population
Total population*	9,576,399	1,232,036	9,790,586	1,187,624	9,987,703	1,195,896	10,060,555	1,204,489	10,244,912	1,201,921
% uninsured		12.87%		12.13%		11.97%		11.97%		11.73%
Non-Hispanic White*	5,993,520	579,077	6,057,228	551,508	6,104,270	556,398	6,041,619	554,885	6,063,697	541,853
% uninsured		9.66%		9.10%		9.11%		9.18%		8.94%
Black/African American*	1,137,050	167,252	1,168,847	161,838	1,185,070	160,568	1,181,747	162,868	1,189,333	161,173
% uninsured		14.71%		13.85%		13.55%		13.78%		13.55%
Hispanic*	1,992,453	428,298	2,096,929	418,173	2,181,813	418,310	2,262,192	418,781	2,353,396	425,407
% uninsured		21.50%		19.94%		19.17%		18.51%		18.08%
2 or more races^	262,866	31,332	276,240	31,785	546,163	76,628	810,709	119,380	1,108,081	168,007
% uninsured		11.92%		11.51%		14.03%		14.73%		15.16%
Asian*	275,460	36,379	284,682	34,852	297,390	35,625	300,605	33,433	309,505	32,241
% uninsured		13.21%		12.24%		11.98%		11.12%		10.42%
American Indian/ American Indian/Alaska Native^	28,288	1,268	28,332	1,174	26,077	1,392	26,092	1,257	31,066	1,269
% uninsured		21.69%		19.54%		21.44%		22.64%		28.89%

(continued)

Table 1 (continued)

	ACS 2018 5yr		ACS 2019 5yr		ACS 2020 5yr		ACS 2021 5yr		ACS 2022 5yr	
	Total population	Uninsured population	Total population	Uninsured population	Total population	Uninsured population	Total population	Uninsured population	Total population	Uninsured population
Native Hawaiian/ Pacific Islander	6,257	1,268	6,306	1,174	6,362	1,392	6,664	1,257	6,720	1,269
% uninsured		20.27%		18.62%		21.88%		18.86%		18.88%
Male*	4,651,057	660,519	4,755,002	638,610	4,850,271	638,863	4,910,898	646,368	5,014,606	652,522
% uninsured		14.20%		13.43%		13.17%		13.16%		13.01%
Female*	4,925,342	571,517	5,035,584	549,014	5,137,432	557,033	5,149,657	558,121	5,230,306	549,399
% uninsured		11.60%		10.90%		10.84%		10.84%		10.50%
Age 19–64*	5,467,782	1,055,805	5,579,252	1,021,817	5,674,746	1,028,724	5,724,401	1,033,141	5,832,245	1,031,424
% uninsured		19.31%		18.31%		18.13%		18.05%		17.68%
Age 65 + *	2,081,528	19,274	2,160,974	19,661	2,240,886	21,124	2,239,408	19,299	2,312,195	20,069
% uninsured		0.93%		0.91%		0.94%		0.86%		0.87%
Less than HS*	746,124	189,713	747,888	184,310	744,208	179,271	711,683	173,778	705,852	172,883
% uninsured		25.43%		24.64%		24.09%		24.42%		24.49%
HS Grad/ GED*	1,999,562	334,157	2,036,542	329,788	2,057,248	333,164	2,049,141	336,524	2,063,691	333,447
% uninsured		16.71		16.19		16.19		16.42		16.16

(continued)

Table 1 (continued)

	ACS 2018 5yr		ACS 2019 5yr		ACS 2020 5yr		ACS 2021 5yr		ACS 2022 5yr	
	Total population	Uninsured population	Total population	Uninsured population	Total population	Uninsured population	Total population	Uninsured population	Total population	Uninsured population
Associate/ some college*	2,069,630	243,481	2,112,766	233,166	2,171,096	241,340	2,184,026	240,021	2,202,350	237,298
% uninsured		11.76		11.04		11.12		10.99		10.77
Bachelor or higher*	1,948,094	118,463	2,050,447	120,353	2,149,962	126,442	2,235,255	133,141	2,343,291	134,965
% uninsured		6.08%		5.87%		5.88%		5.96%		5.76%
Employed*	3,878,027	677,914	4,007,460	666,673	4,106,278	676,859	4,166,964	683,698	4,268,996	684,123
% uninsured		17.48%		16.64%		16.48%		16.41%		16.03%
Unemployed*	244,936	101,894	221,740	89,655	223,811	88,614	219,696	84,176	214,374	81,752
% uninsured		41.60%		40.43%		39.59%		38.31%		38.14%
Native born*	8,211,744	881,400	8,369,148	849,160	8,521,168	863,106	8,555,488	871,977	8,684,826	869,197
% uninsured		10.73%		10.15%		10.13%		10.19%		10.01%
Foreign born*	1,364,655	350,636	1,421,438	338,464	1,466,535	332,790	1,505,067	332,512	1,560,086	332,724
% uninsured		25.69%		23.81%		22.69%		22.09%		21.33%
100% Below Poverty*	1,363,805	313,029	1,310,819	284,620	1,269,418	268,218	1,255,675	261,227	1,251,268	255,242
% uninsured		22.95%		21.71%		21.13%		20.80%		20.40%
With a disability*	1,357,957	95,160	1,398,280	93,753	1,445,103	101,165	1,427,569	99,961	1,461,031	98,281

(continued)

Table 1 (continued)

	ACS 2018 5yr		ACS 2019 5yr		ACS 2020 5yr		ACS 2021 5yr		ACS 2022 5yr	
	Total population	Uninsured population	Total population	Uninsured population	Total population	Uninsured population	Total population	Uninsured population	Total population	Uninsured population
% uninsured		7.01%		6.70%		7.00%		7.00%		6.73%
Without a disability*	8,218,442	1,136,876	8,392,306	1,093,871	8,542,600	1,094,731	8,632,986	1,104,528	8,783,881	1,103,640
% uninsured		13.83%		13.03%		12.81%		12.79%		12.56%

^ = significant *increase* in uninsured population at 0.05 with Cochran-Armitage Trend Test, provides a statistical measure to detect trends in proportions across ordered groups (years), assessing whether there is a significant increase or decrease over time.
* = significant *decrease* in uninsured population at 0.05 with Cochran-Armitage Trend Test.

to 1,013 in 2021 5yr and 2022 5yr datasets. The ZCTA layers were projected using USA Contiguous Albers Equal Area Conic projection in ArcPro 2.9. ZCTAs that had their geometric center within the 23-county catchment area were selected using the Select by Location tool in Arc Pro 2.9, and the selection adapted to also include those that fell with the 23-county area. Since we used the 2022 tiger line file to create the catchment area ZCTA layer, all 5 years groups had 498 catchment area ZCTAs. Joining the excel uninsured population data to the Tigerline files reached 96—100% join success. There were twenty ZCTAs in 2018 5yr, 2019 5yr, and 2020 5yr that could not be joined due to not being present in the dataset, due to the increase in ZCTAs in 2021. Those 20 NULL ZCTAs were removed from the analysis files. The removed ZCTAs for years 2018 5yr, 2019 5yr, and 2020 5yr represent an average of 1,940 uninsured people, ranging from 0—24,467, where eight ZCTAs had a population of 0 and 5 of the remaining 12 non-zero ZCTAs had populations less than 70.

3.2 Non-Parametric Testing of Uninsured Populations Over Time

Specifically, the Cochran-Armitage Trend Test from the DescTools package in R assessed uninsured rate trends across demographic characteristics like race, gender, income, and education, revealing where significant changes occurred over time.

A conditional check ensured that each group had at least two years of data with positive counts; otherwise, a warning was issued, and the result for that group was marked as missing. When data was sufficient, the CochranArmitageTest function from the DescTools package calculated the chi-square trend statistic and associated p-value for each group. This procedure was repeated for all demographic variables, compiling results that included each group's trend statistic and p-value, highlighting where statistically significant changes in uninsured rates occurred.

3.3 Correlation Assessment

The dataset was prepared by renaming variables to improve clarity and facilitate interpretation. To assess multicollinearity, a subset of key variables was selected for analysis. A correlation matrix was computed using pairwise Pearson correlations to identify predictors with high correlations exceeding a threshold of $r > 0.9$. Variables with substantial correlations were iteratively removed, while ensuring that Total uninsured pop remained in the dataset. The final selection of variables excluded redundant predictors, reducing the risk of multicollinearity, and ensuring that subsequent modeling efforts would yield more reliable results.

We created correlation heatmaps to visualize the relationships among the retained predictors (Figs. 2, 3, 4, 5, 6). This visualization used pairwise Pearson correlations, rounded to two decimal places. The heatmaps provided an overview of the interdependence among the variables, highlighting potential areas of redundancy.

To further diagnose multicollinearity, a linear regression model was fitted with the total uninsured population as the dependent variable. Variance inflation factor (VIF) values were calculated for each predictor, to observe multicollinearity. When detected, we utilized variable reduction based on high values to address multicollinearity in the data. We elected not to use transformation to maintain the estimated count distributions existing in the data.

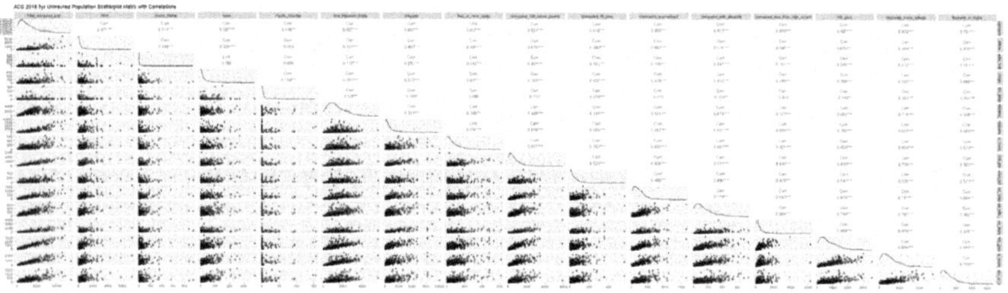

Fig. 2 ACS 2018 5yr correlation Scatterplot

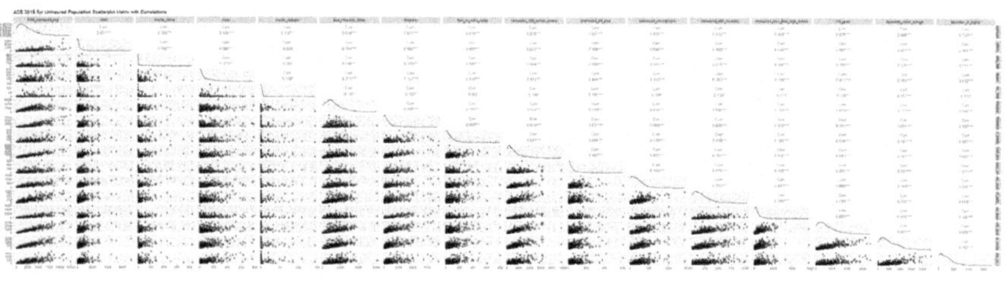

Fig. 3 ACS 2019 5yr correlation scatterplot

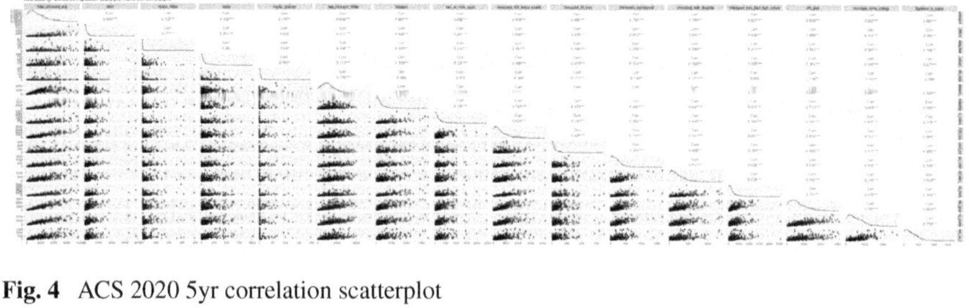

Fig. 4 ACS 2020 5yr correlation scatterplot

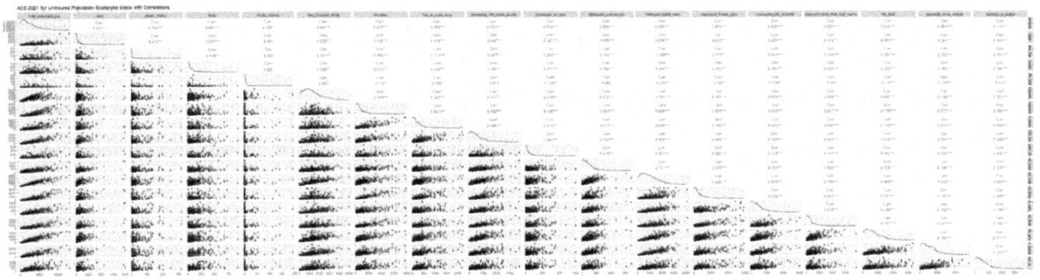

Fig. 5 ACS 2021 5yr correlation scatterplot

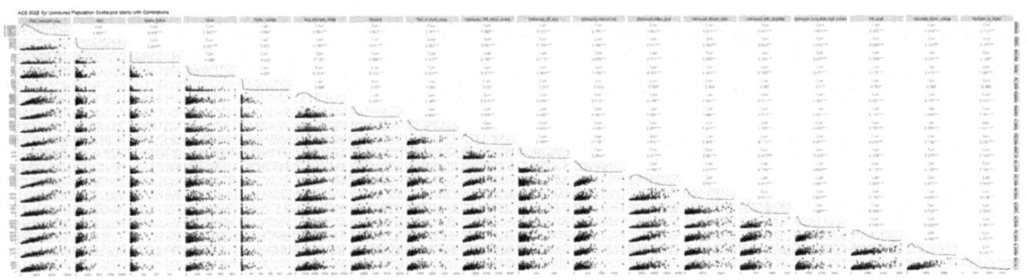

Fig. 6 ACS 2022 5yr correlation scatterplot

3.4 Stepwise Backward Regression

The analysis utilized stepwise backward regression to model the total uninsured population. Stepwise regression, performed using the backward elimination method, iteratively removed predictors that contributed the least to the model's performance. The selection of variables at each step was guided by the Bayesian Information Criterion (BIC), calculated as BIC $= k \cdot \log(n)$, where k is the number of predictors and n is the number of observations. The BIC criterion penalizes models with more predictors unless they significantly improve model fit. Starting with a model that included all candidate predictors, the analysis proceeded by sequentially removing the least significant predictors based on their contribution to reducing the residual sum of squares (RSS) and improving the BIC. This process continued until further removal of predictors led to an increase in the BIC, indicating that no further simplification was beneficial.

At each step of the regression, several metrics were computed to evaluate the model's performance. The residual sum of squares (RSS) quantified the unexplained variance in the dependent variable, with lower values indicating better model fit. The Akaike Information Criterion (AIC) and BIC were used to evaluate the trade-off between goodness-of-fit and model complexity. While AIC imposes a lighter penalty for additional predictors, BIC is stricter, favoring simpler models when their explanatory power is comparable [2]. Changes in these metrics informed the decision to retain or exclude predictors at each step.

After the backward elimination process, the final model was evaluated including residual diagnostics to ensure the residuals were normally distributed and independent, which are key assumptions of linear regression [45, 29]. The residual standard error (RSE) was calculated to measure the average deviation of observed values from those predicted by the model, with lower values indicating better fit. Model performance was further assessed using R^2 and adjusted R^2, which respectively quantify the proportion of variance explained by the predictors and adjust for the number of predictors to prevent overfitting. A likelihood test assessed whether the predictors collectively explained a considerable proportion of the variance in the dependent variable, with a significant p-value confirming the model's overall significance.

The significance of individual predictors in the final model was evaluated using t-tests. These tests assessed whether the coefficients of each predictor were significantly different from zero, providing p-values to indicate the strength of their contribution. Predictors with p-values below 0.05 were considered statistically significant, and their coefficients were interpreted to understand their effects on the total uninsured population.

Additional metrics were computed to summarize the model's performance. AIC and BIC were recalculated for the final model to confirm its fit and simplicity, while residual deviance, representing the sum of squared residuals, was used to measure unexplained variance. The R^2 and adjusted R^2 values from the final model provided insights into the overall explanatory power.

3.5 *Poisson Regression*

Building on the results of the previous stepwise backward regression, the analysis utilized Poisson regression to model the total uninsured population based on significant predictors identified in the prior stepwise backward regression. The Poisson regression model was implemented using a generalized linear modeling framework with a log-link function, which assumes the dependent variable follows a Poisson distribution with the mean equal to the variance.

The predictors' significance was evaluated using z-tests, which assessed whether the regression coefficients were significantly different from zero. The null hypothesis for each predictor stated that its coefficient was equal to zero, implying no association with the dependent variable. Predictors with p-values below 0.05 were considered statistically significant. The coefficients in the Poisson model represent the natural logarithm of the relative rate of change in the total uninsured population for a one-unit change in the predictor, enabling an interpretation of how each variable influences the outcome.

The overall model fit was assessed using deviance measures, which compare the fit of the model to a null model (intercept-only). Residual deviance quantifies the discrepancy between observed and predicted values, with lower values indicating better fit. The reduction in deviance from the null model's deviance demonstrates the predictors' collective explanatory power. The Poisson model's performance was further evaluated using AIC and BIC. These metrics balance model complexity and goodness-of-fit, with lower values indicating a more parsimonious and effective model.

To provide a measure of the model's explanatory power, McFadden's pseudo-R^2 and the adjusted pseudo-R^2 were calculated. These values indicate the proportion of variability in the dependent variable that is explained by the model relative to the null model, with the adjusted pseudo-R^2 accounting for the number of predictors and sample size to prevent overfitting. While pseudo-R^2 values are lower than traditional R^2 values in linear regression, they offer meaningful insights into the goodness-of-fit for count data models [7].

All analyses were conducted in R, leveraging functions from the stats package. The Poisson regression was implemented using the glm() function with the family = "poisson" argument to specify the log-link function. Residual deviance and degrees of freedom were extracted to evaluate model fit, while AIC() and BIC() functions calculated information criteria. The z-tests for predictor significance were provided by the summary() function for the fitted model. The stepwise selection of predictors in the earlier analysis confirmed that only significant (0.05) and non-redundant variables were included in the Poisson regression, making the model both robust and interpretable. This methodological approach facilitated better modeling of the relationships between the predictors and the uninsured population while addressing the statistical assumptions inherent in count data.

3.6 Negative Binomial Regression

The analysis employed a Negative Binomial regression model to predict the total uninsured population, addressing overdispersion identified in the prior Poisson regression analysis. Overdispersion, indicated by a dispersion parameter exceeding 1, arises when the variance of count data exceeds the mean, violating the assumptions of the Poisson model. The Negative Binomial model incorporates an additional dispersion parameter to manage this excess variability, making it a more appropriate choice for modeling overdispersed count data.

The model was fitted using a log-link function, which establishes a nonlinear relationship between the predictors and the expected count of the dependent variable. Coefficients were estimated using maximum likelihood estimation, which identifies parameter values that maximize the likelihood of observing the given data. Statistical significance of the predictors was assessed using z-tests, which evaluate whether each predictor's coefficient differs significantly from zero under the null hypothesis.

To assess model performance and fit, deviance measures compared the residual deviance of the negative binomial model to the null deviance of an intercept-only model. A significant reduction in deviance indicates that the predictors collectively improve model fit. Information criteria, specifically the AIC and BIC, were computed to evaluate the trade-off between model complexity and explanatory power. Lower values of AIC and BIC suggest a better-fitting and more parsimonious model.

The negative binomial model was also compared to the Poisson model using a likelihood ratio test. This test evaluates whether the inclusion of the dispersion parameter in the negative binomial model results in a statistically significant improvement in fit. The null hypothesis assumes no difference in fit between the two models. A significant p-value supports the use of the negative binomial model over the Poisson model for the given data.

Model performance was further evaluated using McFadden's pseudo-R^2 and adjusted pseudo-R^2, which provide a measure of the proportion of variability explained by the model relative to the null model.

The analysis was conducted in R, leveraging the glm.nb() function from the MASS package to fit the negative binomial regression model. The summary() function provided the z-tests for coefficient significance and other model diagnostics. The anova() function, in conjunction with a likelihood ratio test, compared the Poisson and negative binomial models. AIC and BIC values were calculated using AIC() and BIC(), and residual deviance was extracted to evaluate the amount of unexplained variance. This methodological approach ensured the model's robustness and suitability for handling overdispersed count data.

3.7 Bayesian Regression

The Bayesian regression model was informed by the prior negative binomial model to incorporate uncertainty quantification and prior knowledge. The negative binomial model informed the selection of predictors, retaining only those that were statistically significant. Bayesian regression was implemented using a normal prior distribution (N(0,10)) for both the coefficients and the intercept, and the model was estimated via Markov Chain Monte Carlo (MCMC) using four chains to ensure convergence.

Model diagnostics and performance measures included Leave-One-Out Information Criterion (LOO-IC) for model comparison, posterior predictive residual analysis, and Bayesian R^2, providing an assessment of model fit and explained variance. The convergence of the MCMC chains was verified using R^ (r-hat) ensuring the reliability of posterior estimates.

3.8 Spatial Autocorrelation

Spatial autocorrelation (Global Moran's I) on the total uninsured population was conducted in ArcPro 2.9 using the Spatial Autocorrelation function in the Spatial Statistics toolbox. ZCTAs within the 23-county catchment area served as the spatial scale, where the centroid of the ZCTA is what the latitude and longitude values are based. We used the default Euclidean distance and weighted the ZCTAs using the total uninsured population, and maintained a consistent neighbor threshold of 34,895.5716 m (approximately 21.68 miles). The output includes spatially referenced Moran's index, ZScore, PValue, and a report file that includes the test and descriptive statistics of the model. In this analysis, spatial autocorrelation observes the distribution of uninsured rates across ZCTAs, informed by their geographic coordinates (latitude and longitude). Positive Moran's I indicates clustering of similar uninsured rates, while negative values indicate dispersion. This index reveals spatial patterns that deviate from the independence assumption in classical statistics, meaning values for uninsured populations are not randomly distributed but rather show spatial organization across ZCTAs.

Including spatial autocorrelation in the analysis provides a layer beyond traditional regression by identifying spatial clusters of uninsured populations. This is important because, without accounting for autocorrelation, we might assume that uninsured rates are geographically random and independent across ZCTAs. Non-zero spatial autocorrelation suggests that uninsured rates exhibit spatial organization, aligning with geographic socioeconomic disparities. Moran's I thereby supports the prior model refinements (stepwise backward regression, AIC criteria, and VIF adjustments) by adding a spatial dimension, ensuring that the model's predictors not only optimize statistical fit but also reflect real-world spatial clustering.

3.9 Getis-Ord Gi, Hot and Cold Spot Analysis

The Hot Spot Analysis using the Getis-Ord Gi* statistic was applied to answer the "where" of the spatial clustering of uninsured populations for each of five ACS 5-year groups: ACS 2018, ACS 2019, ACS 2020, ACS 2021, and ACS 2022. This method leveraged the Optimized Hot Spot Analysis tool in ArcPro 2.9, which calculates the Getis-Ord Gi* statistic to determine the degree of spatial clustering by comparing local concentrations of high or low values against a random distribution. Areas where values were significantly higher or lower than expected under random conditions were identified as "hot spots" (high values) or "cold spots" (low values). The Optimized Hot Spot Analysis tool selected an appropriate scale of analysis based on the spatial characteristics of each ACS dataset, ensuring accurate identification of statistically significant clusters of uninsured individuals.

3.10 Temporal Analysis with GARCH

To assess the volatility in the uninsured population across ZCTAs, a Generalized Autoregressive Conditional Heteroskedasticity (GARCH) model was used. The analysis was performed in R using the rugarch, dplyr, tidyr, and zoo libraries. GARCH models are widely used in time series analysis to capture volatility patterns within data that exhibit time-dependent changes in variance [36–42]. These models are particularly useful in situations where the variance, or "volatility," is not constant over time, a condition known as heteroskedasticity [27]. GARCH models extend the Autoregressive Conditional Heteroskedasticity (ARCH) model by incorporating lagged terms of past variance, allowing them to account for both immediate and persistent changes in volatility [18]. By modeling these fluctuations, GARCH enables a more precise understanding of time-based variability in a dataset.

3.10.1 Data Preparation and Handling of Missing Values

The GARCH analysis began by ensuring each ZCTA had a continuous yearly record. Missing counts in each ZCTA's time series were identified and filled as NA values. Afterward, linear interpolation was used to estimate missing values for the total uninsured population, creating a complete dataset for each ZCTA. This approach is consistent with the method discussed by Lepot et al. [36], who examined linear interpolation as a practical technique for handling missing data in time series. In their study, linear interpolation was employed to bridge data gaps smoothly, avoiding abrupt shifts and minimizing bias. Such interpolation provides a coherent time series, ensuring continuity without excessive extrapolation, thereby maintaining the integrity of observed trends.

3.10.2 Model Specification

A GARCH(1,1) model was specified, assuming a normal distribution for residuals and including a constant mean term. The chosen structure, a simple GARCH model, effectively captures conditional heteroskedasticity, which helps account for time-varying volatility in uninsured population data.

3.10.3 Fitting the GARCH Model by ZCTA

The model was then applied individually to each ZCTA's time series. Only ZCTAs with at least five interpolated observations were included in the modeling to ensure sufficient data for convergence. Each GARCH model fit was checked for convergence, and errors were noted for ZCTAs where the model did not converge, ensuring that only reliable estimates were retained.

For each ZCTA, parameter estimates, specifically omega, alpha1, and beta1, were extracted alongside model diagnostics (AIC and BIC). These metrics were used to assess the goodness-of-fit and relative performance of the GARCH models across different ZCTAs. The output was imported to ArcPro 2.9 and joined to the catchment area ZCTA file to compare the location of beta1 values in relation to the hot and cold ZCTAs.

4 Results

4.1 Description of the Uninsured Population

From ACS 2018 (2014–2018) to ACS 2022 (2018–2022), the population within the cancer center's catchment area expanded from approximately 9.6 million to 10.2 million (Table 1). Despite this growth, the uninsured population remained relatively stable at around 1.2 million, with the uninsured rate consistently near 12–13% across these five-year intervals. While the overall uninsured rate showed little change, certain demographic groups exhibited shifts in coverage.

Among racial and ethnic groups, Black or African American individuals saw a decrease in their uninsured rate, from 14.7% in ACS 2018 5yr to 13.6% in ACS 2022 5yr, even as their population increased. This trend was confirmed by the Cochran-Armitage Trend Test ($Z = -23.24$, $p < 0.001$), indicating statistically significant gains in insurance coverage within this group. The Asian population also displayed improvements in insurance coverage, with uninsured rates dropping from 13.2% to 10.4% ($Z = -35.56$, $p < 0.001$). The Hispanic population, which grew substantially, similarly experienced a significant decline in uninsured rates, from 21.5% to 18.1% ($Z = -96.23$, $p < 0.001$). However, certain groups faced worsening trends in uninsured rates. Alaska Native individuals, although a smaller demographic group, saw an increase in uninsured rates from 21.7% to 28.9% ($Z = 23.26$, $p < 0.001$), and individuals identifying as two or more races showed an increase from 11.9% to 15.2% ($Z = 57.55$, $p < 0.001$), suggesting rising coverage vulnerability within these populations. Pacific Islanders, on the other hand, showed no statistically significant trend ($Z = -1.66$, $p = 0.098$), indicating stable uninsured rates across this period.

Gender disparities in uninsured rates persisted between 2014–2022, with males more likely to be uninsured than females. Both groups experienced declines, where males dropped to 13.0% in ACS 2022 5yr, compared to 10.5% for females, yet males maintained higher uninsured rates. The trend tests confirmed these declines as significant, with values of ($Z = -53.43$, $p < 0.001$) for males and ($Z = -51.39$, $p < 0.001$) for females.

Individuals below the poverty line saw uninsured rates decrease from 22.9 to 20.4% ($Z = -53.32$, $p < 0.001$), indicating modest gains in insurance access for economically disadvantaged groups. Educational attainment also strongly influenced uninsured rates: individuals with less than a high school diploma experienced a consistently high rate around 24–25%, although it showed a small but significant decrease over time ($Z = -13.31$, $p < 0.001$). Conversely, those with a high school diploma ($Z = -10.61$, $p < 0.001$), some college education ($Z = -29.56$, $p < 0.001$), or a bachelor's degree or higher ($Z = -10.82$, $p < 0.001$) experienced a trend of reduced uninsured rates, reinforcing the positive association between educational attainment and insurance coverage [39].

Unemployed individuals had higher uninsured rates than the employed, although the uninsured rate for the unemployed decreased from 41.6% to 38.1% ($Z = -28.06$,

$p < 0.001$). Foreign-born individuals exhibited higher uninsured rates than native-born individuals, though their rates declined from 25.7% to 21.3% ($Z = -93.95$, $p < 0.001$). Among age groups, adults aged 19–64 had the highest uninsured rates, but showed a slight yet significant decrease from 19.3% to 17.7% ($Z = -67.91$, $p < 0.001$), while those aged 65 and older maintained much lower uninsured rates.

While overall uninsured rates remained stable across the catchment area, the data reveal persistent disparities and significant trends within specific demographic groups. Vulnerable groups—including certain racial and ethnic minorities, lower-income individuals, those with less education, the unemployed, and the foreign-born—continued to experience disproportionately high uninsured rates, despite overall improvements in coverage.

4.2 Correlation Assessment

Each year, a correlation matrix (Figs. 2, 3, 4, 5, 6) and Variance Inflation Factor (VIF) diagnostics were applied to identify and address multicollinearity. Variables such as Hispanic, Non-Hispanic White, and uninsured native-born consistently exhibited high VIF values across multiple years, leading to their removal. Conversely, Alaska Native, Asian, Pacific Islander, and Uninsured 65 plus were frequently retained due to acceptable VIF levels (Table 2).

In 2018, a total of 15 variables were retained after removing those with high multicollinearity, such as Hispanic (VIF = 34.06) and Non-Hispanic White (VIF = 17.80). Retained variables included Alaska Native, Asian, and Bachelor or higher.

In 2019, the refined dataset also retained 15 variables, with Hispanic (VIF = 30.56) and Non-Hispanic White (VIF = 15.67) excluded. Retained variables were similar to 2018 and included AfAm, Uninsured with disability, and Two or more races.

By 2020, 15 variables were retained, while Hispanic (VIF = 25.13), Non-Hispanic White (VIF = 14.64), and Uninsured less than high school (VIF = 10.11) were removed. Key retained variables included Alaska Native, Uninsured 65 plus, and Bachelor or higher.

In 2021, severe multicollinearity led to the removal of Hispanic (VIF = 207.89) and native-born uninsured (VIF = 321.33). Fourteen variables were retained, including Pacific Islander, Uninsured unemployed, and HS grad.

In 2022, after diagnostic analyses, 13 variables were retained. Hispanic (VIF = 177.05) and Uninsured native born (VIF = 282.08) were removed, while key predictors such as Alaska Native, Associates or some college, and Uninsured with disability were preserved.

Across all years, Hispanic, Non-Hispanic White, and native-born uninsured consistently exhibited high VIF values, necessitating their exclusion. Variables such as Alaska Native, Asian, Pacific Islander, and Uninsured 65 plus were consistently retained, with the number of retained variables ranging from 13 to 15 per year.

Table 2 Variable performance in each model

Year group	Model name	Variables maintained	Number of variables maintained
ACS 2018 5yr	Initial Dataset	Total uninsured population, Black/African American, American Indian/Alaska Native, Asian, Native Hawaiian/Pacific Islander, Non-Hispanic White, Hispanic, Two or more races, Uninsured Male, Uninsured Female, Uninsured 100 below poverty, Uninsured 19–64, Uninsured 65 plus, Uninsured employed, Uninsured unemployed, Uninsured native born, Uninsured foreign born, Uninsured with disability, Uninsured without disability, Uninsured less than high school, HS grad, Associate some college, Bachelor or higher	23
	After Correlation Filter	Total uninsured population, Black/African American, American Indian/Alaska Native, Asian, Native Hawaiian/Pacific Islander, Non Hispanic White, Hispanic, Two or more races, Uninsured 100 below poverty, Uninsured 65 plus, Uninsured unemployed, Uninsured with disability, Uninsured less than high school, HS grad, Associate some college, Bachelor or higher	16
	After VIF Filter	Total uninsured population, American Indian/ Alaska Native, Asian, Native Hawaiian/Pacific Islander, Two or more races, Uninsured 65 plus, Uninsured unemployed, Uninsured with disability	8
	Stepwise Regression	Asian, Two or more races, Uninsured 65 plus, Uninsured unemployed, Uninsured with disability	5
	Poisson Regression	Asian, Two or more races, Uninsured 65 plus, Uninsured unemployed, Uninsured with disability	5
	Negative Binomial Regression	Asian, Two or more races, Uninsured 65 plus, Uninsured unemployed, Uninsured with disability	5
	Bayesian Regression	Asian, Two or more races, Uninsured 65 plus, Uninsured unemployed, Uninsured with disability	5

(continued)

Table 2 (continued)

Year group	Model name	Variables maintained	Number of variables maintained
ACS 2019 5yr	Initial Dataset	Total uninsured population, Black/African American, American Indian/Alaska Native, Asian, Native Hawaiian/Pacific Islander, Non Hispanic White, Hispanic, Two or more races, Uninsured Male, Uninsured Female, Uninsured 100 below poverty, Uninsured 19–64, Uninsured 65 plus, Uninsured employed, Uninsured unemployed, Uninsured native born, Uninsured foreign born, Uninsured with disability, Uninsured without disability, Uninsured less than high school, HS grad, Associate some college, Bachelor or higher	23
	After Correlation Filter	Total uninsured population, Black/African American, American Indian/Alaska Native, Asian, Native Hawaiian/Pacific Islander, Non Hispanic White, Hispanic, Two or more races, Uninsured 100 below poverty, Uninsured 65 plus, Uninsured unemployed, Uninsured with disability, Uninsured less than high school, HS grad, Associate some college, Bachelor or higher	16
	After VIF Filter	Total uninsured population, American Indian/Alaska Native, Asian, Native Hawaiian/Pacific Islander, Two or more races, Uninsured 65 plus, Uninsured unemployed, Uninsured with disability	8
	Stepwise Regression	Asian, Two or more races, Uninsured 65 plus, Uninsured unemployed, Uninsured with disability	5
	Poisson Regression	Asian, Two or more races, Uninsured 65 plus, Uninsured unemployed, Uninsured with disability	5
	Negative Binomial Regression	Asian, Two or more races, Uninsured 65 plus, Uninsured unemployed, Uninsured with disability	5
	Bayesian Regression	Asian, Two or more races, Uninsured 65 plus, Uninsured unemployed, Uninsured with disability	5

(continued)

Table 2 (continued)

Year group	Model name	Variables maintained	Number of variables maintained
ACS 2020 5yr	Initial Dataset	Total uninsured population, Black/African American, American Indian/Alaska Native, Asian, Native Hawaiian/Pacific Islander, Non Hispanic White, Hispanic, Two or more races, Uninsured Male, Uninsured Female, Uninsured 100 below poverty, Uninsured 19–64, Uninsured 65 plus, Uninsured employed, Uninsured unemployed, Uninsured native born, Uninsured foreign born, Uninsured with disability, Uninsured without disability, Uninsured less than high school, HS grad, Associate some college, Bachelor or higher	23
	After Correlation Filter	Total uninsured population, Black/African American, American Indian/Alaska Native, Asian, Native Hawaiian/Pacific Islander, Hispanic, Two or more races, Uninsured 100 below poverty, Uninsured 65 plus, Uninsured unemployed, Uninsured with disability, Uninsured less than high school, HS grad, Associate some college, Bachelor or higher	15
	After VIF Filter	Total uninsured population, American Indian/Alaska Native, Asian, Two or more races, Uninsured 65 plus, Uninsured unemployed, Uninsured with disability, Bachelor or higher	8
	Stepwise Regression	American Indian/Alaska Native, Two or more races, Uninsured unemployed, Uninsured with disability, Bachelor or higher	5
	Poisson Regression	American Indian/Alaska Native, Two or more races, Uninsured unemployed, Uninsured with disability, Bachelor or higher	5
	Negative Binomial Regression	American Indian/Alaska Native, Two or more races, Uninsured unemployed, Uninsured with disability, Bachelor or higher	5
	Bayesian Regression	American Indian/Alaska Native, Two or more races, Uninsured unemployed, Uninsured with disability, Bachelor or higher	5

(continued)

Table 2 (continued)

Year group	Model name	Variables maintained	Number of variables maintained
2021 ACS 5yr	Initial Data	Total uninsured population, Black/African American, American Indian/Alaska Native, Asian, Native Hawaiian/Pacific Islander, Non Hispanic White, Hispanic, Two or more races, Uninsured Male, Uninsured Female, Uninsured 100 below poverty, Uninsured 19–64, Uninsured 65 plus, Uninsured employed, Uninsured unemployed, Uninsured native born, Uninsured foreign born, Uninsured with disability, Uninsured without disability, Uninsured less than high school, HS grad, Associate some college, Bachelor or higher	23
	After Correlation Analysis	Total uninsured population, Black/African American, American Indian/Alaska Native, Asian, Native Hawaiian/Pacific Islander, Non Hispanic White, Hispanic, Two or more races, Uninsured 100 below poverty, Uninsured 65 plus, Uninsured unemployed, Uninsured native born, Uninsured foreign born, Uninsured with disability, Uninsured less than high school, HS grad, Associate some college, Bachelor or higher	16
	After VIF Analysis	Total uninsured population, American Indian/Alaska Native, Asian, Native Hawaiian/Pacific Islander, Uninsured 65 plus, Uninsured unemployed, Uninsured with disability	7
	Stepwise Regression	Asian, American Indian/Alaska Native, Uninsured 65 plus, Uninsured unemployed, Uninsured with disability	5
	Poisson Regression	Asian, American Indian/Alaska Native, Uninsured 65 plus, Uninsured unemployed, Uninsured with disability	5
	Negative Binomial Regression	Asian, Uninsured 65 plus, Uninsured unemployed, Uninsured with disability	4
	Bayesian Regression	Asian, Uninsured 65 plus, Uninsured unemployed, Uninsured with disability	4

(continued)

Table 2 (continued)

Year group	Model name	Variables maintained	Number of variables maintained
ACS 2022 5yr	Initial Data	Total uninsured population, Black/African American, American Indian/Alaska Native, Asian, Native Hawaiian/Pacific Islander, Non Hispanic White, Hispanic, Two or more races, Uninsured Male, Uninsured Female, Uninsured 100 below poverty, Uninsured 19–64, Uninsured 65 plus, Uninsured employed, Uninsured unemployed, Uninsured native born, Uninsured foreign born, Uninsured with disability, Uninsured without disability, Uninsured less than high school, HS grad, Associate some college, Bachelor or higher	23
	After Correlation Assessment	Total uninsured population, Black/African American, American Indian/Alaska Native, Asian, Native Hawaiian/Pacific Islander, Non Hispanic White, Hispanic, Two or more races, Uninsured 100 below poverty, Uninsured 65 plus, Uninsured unemployed, Uninsured native born, Uninsured foreign born, Uninsured with disability, Uninsured less than high school, HS grad, Associate some college, Bachelor or higher	18
	After VIF Analysis	Total uninsured population, American Indian/Alaska Native, Asian, Native Hawaiian/Pacific Islander, Two or more races, Uninsured 65 plus, Uninsured employed, Uninsured unemployed, Uninsured with disability, Uninsured without disability, HS grad, Associate some college, Bachelor or higher	13
	Stepwise Regression	American Indian/Alaska Native, Asian, Uninsured 65 plus, Uninsured unemployed, Uninsured with disability	5
	Poisson Regression	American Indian/Alaska Native, Asian, Uninsured 65 plus, Uninsured unemployed, Uninsured with disability	5
	Negative Binomial Regression	American Indian/Alaska Native, Asian, Uninsured 65 plus, Uninsured unemployed, Uninsured with disability	5
	Bayesian Regression	American Indian/Alaska Native, Asian, Uninsured 65 plus, Uninsured unemployed, Uninsured with disability	5

This iterative process ensured the datasets were statistically robust and free from multicollinearity issues, enabling reliable modeling efforts.

4.3 Stepwise Backward Regression

Across 2018–2022 5-year groups, stepwise backward regression refined models predicting the total uninsured population, guided by multicollinearity diagnostics and Bayesian Information Criterion (BIC). Each year began with a set of variables identified from prior analyses, iteratively excluding those with high multicollinearity (Variance Inflation Factor (VIF) > 5 or Pearson's correlation > 0.9) or low contributions to model fit. The final models consistently retained Alaska Native, Asian, Uninsured 65 plus, Uninsured unemployed, and Uninsured with disability due to their significant explanatory power (Table 2). Conversely, variables such as Hispanic, Non-Hispanic White, native-born uninsured, and Pacific Islander consistently exhibited high VIF values and were excluded.

In 2018, the analysis began with seven variables: Alaska Native, Asian, Pacific Islander, Two or more races, Uninsured 65 plus, Uninsured unemployed, and Uninsured with disability. The final model retained five predictors (Asian, Two or more races, Uninsured 65 plus, Uninsured unemployed, and Uninsured with disability) and excluded Pacific Islander and Alaska Native. This model explained 9.88% of the variance in the uninsured population based on McFadden's Pseudo R^2 (adjusted = 8.85%). The BIC value was 7,405.6 (Table 3).

In 2019, a similar set of variables was used, and the final model again retained Asian, Two or more races, Uninsured 65 plus, Uninsured unemployed, and Uninsured with disability, while excluding Alaska Native and Pacific Islander. The McFadden's Pseudo R^2 for this model was 9.57% (adjusted = 8.59%), and the BIC value was 7,747.6.

By 2020, the analysis introduced Bachelor or higher to the initial variables, while retaining Alaska Native, Asian, Pacific Islander, Two or more races, Uninsured 65 plus, Uninsured unemployed, and Uninsured with disability. The final model included Alaska Native, Two or more races, Uninsured unemployed, Uninsured with disability, and Bachelor or higher, removing Asian and Pacific Islander. This model explained 10.13% of the variance (adjusted = 9.15%), with a BIC value of 7,682.9.

In 2021, the initial set of variables was refined to include Alaska Native, Asian, Uninsured 65 plus, Uninsured unemployed, and Uninsured with disability. Two or more races and Pacific Islander were removed, leaving a final model with five significant predictors. The McFadden's Pseudo R^2 was 8.41% (adjusted = 7.43%), with a BIC value of 7,977.7.

In 2022, the analysis revisited predictors identified in earlier years, initially including Pacific Islander but excluding it early in the stepwise regression process due to minimal improvement in BIC. The final model retained Alaska Native, Asian, Uninsured 65 plus, Uninsured unemployed, and Uninsured with disability, explaining 8.73% of the variance (adjusted = 7.75%). The BIC value for this model was 7,876.7.

Table 3 Summary table of model fitness statistics

Year	Model	AIC	BIC	Residual Deviance (null deviance)	McFadden R2 (Pseudo R2)/Bayesian R2
ACS 2018 5yr	Stepwise Backward Regression	7,376.9	7,405.602	385,937,602	Pseudo-R2 (McFadden): 0.0988 Adjusted Pseudo-R2: 0.0885
	Poisson	256,533.3	256,557.9	252,360 (776,582.8)	Pseudo R^2 (McFadden): 0.6714 Adjusted Pseudo R^2: 0.6677
	Negative Binomial	7,516.85	7,545.55	472.3462 (499.9974)	Pseudo R^2 (McFadden): 0.0548 Adjusted Pseudo R^2: 0.0441
	Bayesian	8,196.04	8,195.52	3,348,284,074	Bayesian R^2 (mean): 0.503 Bayesian R^2 (95% CI): 0.3467, 0.6293
ACS 2019 5yr	Stepwise Backward Regression	7,718.538	7,747.562	399,766,872	Pseudo-R2 (McFadden): 0.0957 Adjusted Pseudo-R2: 0.0859
	Poisson	292,918.4	292,943.3	288,631 (831,744.9)	Pseudo R2 (McFadden): 0.6496 Adjusted Pseudo R2: 0.6458
	Negative Binomial	7,877.293	7,906.318	503.504 (533.6307)	Pseudo R2 (McFadden): 0.0475 Adjusted Pseudo R2: 0.0372
	Bayesian	8,501.99	8,501.77	2,973,746,618	Bayesian R2 (mean): 0.5245 Bayesian R2 (95% CI): 0.3901, 0.6405
ACS 2020 5yr	Stepwise Backward Regression	7,653.852	7,682.861	359,721,560	Pseudo-R2 (McFadden): 0.1013 Adjusted Pseudo-R2: 0.0915
	Poisson	304,934	304,958.9	300,658.9	Pseudo R2 (McFadden): 0.635 Adjusted Pseudo R2: 0.631
	Negative Binomial	7,891.063	7,920.072	505.3968 (534.096)	Pseudo R2 (McFadden): 0.0449 Adjusted Pseudo R2: 0.03450.45
	Bayesian	8,695.55	8,695.51	3,391,535,991	Bayesian R2 (mean): 0.3607 Bayesian R2 (95% CI): 0.2967, 0.4223
ACS 2021 5yr	Stepwise Backward Regression	7948.562	7977.705	499,829,961	Pseudo-R2 (McFadden): 0.0841 Adjusted Pseudo-R2: 0.0743
	Poisson	405,088.1	405,113.1	400,755.6 (861,139)	Pseudo R2 (McFadden): 0.532 Adjusted Pseudo R2: 0.527
	Negative Binomial	9,205.50	36,737,297	36,737,173	95% Credible Interval: 0.41 - 0.51
	Bayesian	8,872.52	8,872.48	3,544,925,592	Bayesian R2 (mean): 0.3382 Bayesian R2 (95% CI): 0.2755, 0.3986

(continued)

Table 3 (continued)

Year	Model	AIC	BIC	Residual Deviance (null deviance)	McFadden R2 (Pseudo R2)/Bayesian R2
ACS 2022 5yr	Stepwise Backward Regression	7,847.551	7,876.65	445,901,536	Pseudo-R2 (McFadden): 0.0873 Adjusted Pseudo-R2: 0.0775
	Poisson	380,901.9	380,926.9	376,571.9 (818,554.1)	Pseudo R2 (McFadden): 0.5371 Adjusted Pseudo R2: 0.5322
	Negative Binomial	8,018.759	8,047.858	512.3727 (540.1462)	Pseudo R2 (McFadden): 0.0409 Adjusted Pseudo R2: 0.0306
	Bayesian	8,813.4	8,813.28	3,461,658,240	Bayesian R2 (mean): 0.333 Bayesian R2 (95% CI): 0.2702, 0.3968

Across all five years, the models consistently retained five significant predictors (Alaska Native, Asian, Uninsured 65 plus, Uninsured unemployed, and Uninsured with disability), ensuring robust and parsimonious models. The excluded variables, notably Hispanic, Non-Hispanic White, native-born uninsured, and Pacific Islander, exhibited high VIF values and were systematically removed. The iterative refinement process balanced model complexity and predictive strength, achieving McFadden's Pseudo R^2 values ranging from 8.41% to 10.13%.

4.4 Poisson Regression

Across the years 2018 to 2022, Poisson regression models were developed to predict the total uninsured population, leveraging insights from prior stepwise backward regression analyses. These models consistently refined predictors to ensure robust fit by removing variables with high multicollinearity or weak contributions, as determined by variance inflation factor (VIF > 5), z-tests, and changes in information criteria. The analysis focused on retaining significant predictors while evaluating model performance using residual deviance, Akaike Information Criterion (AIC), Bayesian Information Criterion (BIC), and Pseudo-R^2 metrics.

In 2018, the Poisson regression model identified Asian, Two or more races, Uninsured 65 plus, Uninsured unemployed, and Uninsured with disability as the significant predictors, which were retained in the final model. The removed variables included Pacific Islander and Alaska Native due to their minimal contributions. The model achieved a residual deviance of 252,360 compared to a null deviance of 776,583, demonstrating a substantial reduction in unexplained variability. The

AIC for this model was 256,533.3, while McFadden's pseudo-R^2 was 0.6714, indicating that the model explained approximately 67% of the variability in the uninsured population. The adjusted pseudo-R^2 of 0.6677 further underscored the model's robustness.

In 2019, the same predictors from 2018—Asian, Two or more races, Uninsured 65 plus, Uninsured unemployed, and Uninsured with disability—were retained, with Alaska Native and Pacific Islander again excluded. This model achieved a residual deviance of 288,631, reduced from a null deviance of 831,745, highlighting strong model performance. The AIC was 292,918, while McFadden's pseudo-R^2 and adjusted pseudo-R^2 were 0.6496 and 0.6458, respectively, reflecting a slightly lower but still substantial proportion of explained variability compared to 2018.

In 2020, the predictors retained were Alaska Native, Two or more races, Uninsured unemployed, Uninsured with disability, and Bachelor or higher, with Asian and Pacific Islander removed. The residual deviance decreased to 300,659 from a null deviance of 831,080, confirming the model's effectiveness. The AIC was calculated as 304,934, while McFadden's pseudo-R^2 and adjusted pseudo-R^2 values were 0.635 and 0.631, respectively, showing that the model explained approximately 63% of the variability in the uninsured population.

In 2021, the retained predictors were Alaska Native, Asian, Uninsured 65 plus, Uninsured unemployed, and Uninsured with disability, with Two or more races and Pacific Islander excluded. The residual deviance for this model was 400,756, reduced from a null deviance of 861,139, with an AIC of 405,088.1. The pseudo-R^2 and adjusted pseudo-R^2 values were 0.532 and 0.527, respectively, indicating that the model explained approximately 53% of the variability in the uninsured population.

By 2022, the predictors remained consistent with 2021, retaining Alaska Native, Asian, Uninsured 65 plus, Uninsured unemployed, and Uninsured with disability, with no additional variables excluded. The residual deviance was 376,572, reduced from a null deviance of 818,554, and the AIC was 380,902. McFadden's pseudo-R^2 was 0.5371, while the adjusted pseudo-R^2 was 0.5322, showing a similar proportion of explained variability as in 2021.

Across all years, variables including Alaska Native, Asian, Uninsured 65 plus, Uninsured unemployed, and Uninsured with disability were consistently retained, highlighting their strong association with the uninsured population. In contrast, variables like Pacific Islander, Hispanic, Non-Hispanic White, and native-born uninsured were removed due to high multicollinearity or weak statistical significance. The models consistently maintained five significant predictors annually, achieving pseudo-R^2 values ranging from 0.532 to 0.671, with residual deviance reductions indicating substantial improvements over null models. This iterative refinement process ensured that each model was statistically valid, parsimonious, and predictive, offering robust insights into the factors driving variations in the uninsured population.

4.5 Negative Binomial Regression

Between the 2018 to 2022 5yr groups, Negative Binomial regression models built on prior stepwise backward regression and Poisson regression analyses. These models addressed overdispersion, identified by variance exceeding the mean in Poisson models, and incorporated an additional dispersion parameter to capture the variability in count data more effectively. Each year's analysis retained significant predictors while removing variables that contributed minimally or failed to achieve statistical significance.

In 2018, the predictors included in the Negative Binomial model were Asian, Two or more races, Uninsured 65 plus, Uninsured unemployed, and Uninsured with disability. The model demonstrated strong associations for all predictors except Asian ($p = 0.060$), with significant positive contributions from Uninsured 65 plus ($\beta = 0.0023$, $p < 0.0001$), Uninsured unemployed ($\beta = 0.0018$, $p < 2 \times 10^{-16}$), and Two or more races ($\beta = 0.0011$, $p = 0.0062$). The residual deviance of 472.35, reduced from a null deviance of 499.99, supported the model's improved fit, while an AIC of 7516.8 and BIC of 7545.5 indicated model efficiency. McFadden's pseudo-R^2 was 0.0548, confirming a modest level of explanatory power. This year's analysis validated the superiority of the Negative Binomial model over the Poisson model through a likelihood ratio test ($p = 0$).

In 2019, the Negative Binomial regression retained Asian, Two or more races, Uninsured 65 plus, Uninsured unemployed, and Uninsured with disability as predictors, all of which were statistically significant. The model achieved a residual deviance of 503.5 (null deviance = 533.6) and an AIC of 7,877.3 with a BIC of 7,906.3, reflecting a good balance between fit and parsimony. The pseudo-R^2 and adjusted pseudo-R^2 values were 0.0475 and 0.0372, respectively. The likelihood ratio test confirmed that the Negative Binomial model outperformed the Poisson model ($p < 0.0001$). Predictors such as Uninsured unemployed ($\beta = 0.0021$) and Uninsured with disability ($\beta = 0.0016$) continued to contribute significantly.

In 2020, Alaska Native, Two or more races, Uninsured unemployed, Uninsured with disability, and Bachelor or higher were included as predictors. Except for Alaska Native, all predictors were statistically significant. For example, Uninsured unemployed ($\beta = 0.0015477$, $p < 0.0001$) and Uninsured with disability ($\beta = 0.0013679$, $p < 2 \times 10^{-9}$) showed strong positive effects. The model achieved a residual deviance of 505.4 (null deviance = 534.1), with an AIC of 7,891.1 and BIC of 7,920.1. The pseudo-R^2 values (0.0449 adjusted to 0.0345) reflected a modest improvement in explanatory power. A likelihood ratio test again validated the superiority of the Negative Binomial model over the Poisson alternative ($p < 0.001$).

In 2021, the model included Alaska Native, Asian, Uninsured 65 plus, Uninsured unemployed, and Uninsured with disability, though Alaska Native did not achieve statistical significance ($p = 0.190$). Significant predictors included Asian ($\beta = 0.0009975$, $p = 0.003$), Uninsured 65 plus ($\beta = 0.0014657$, $p = 0.021$), and Uninsured with disability ($\beta = 0.0018559$, $p < 0.001$). The residual deviance of 520.53 (null deviance = 962.74) demonstrated a strong improvement in model fit, supported by an AIC of 8,083.1 and BIC of 8,112.2. The pseudo-R^2 and adjusted pseudo-R^2 values were modest at 0.039 and 0.029, respectively, with a likelihood ratio test confirming the Negative Binomial model's advantage over the Poisson model.

In 2022, the predictors Alaska Native, Asian, Uninsured 65 plus, Uninsured unemployed, and Uninsured with disability were retained, with all but Alaska Native achieving statistical significance. Significant coefficients included Uninsured with disability ($\beta = 0.0019321$, $p < 0.001$), Uninsured unemployed ($\beta = 0.0020340$, $p < 0.001$), and Uninsured 65 plus ($\beta = 0.0013405$, $p = 0.026$). The model achieved a residual deviance of 512.373 (null deviance = 540.146), with an AIC of 8,018.8 and BIC of 8,047.9. The pseudo-R^2 was 0.0409, adjusted to 0.0306, and the likelihood ratio test confirmed the model's suitability for addressing overdispersion ($p < 0.001$).

Across the five years, the models consistently retained Asian, Uninsured 65 plus, Uninsured unemployed, and Uninsured with disability as significant predictors, while Alaska Native and Bachelor or higher were included in some years but not consistently significant. Deviance reductions and favorable AIC/BIC values underscored the robustness of these models, while pseudo-R^2 values highlighted the challenges of modeling complex demographic data. The Negative Binomial regression models effectively addressed overdispersion, offering improved fit and reliable insights into the uninsured population.

4.6 *Summary of Variable Interaction during Regressions*

From ACS 2018 5-year to ACS 2022 5-year analyses, unemployment and disability status emerged as consistent predictors of the uninsured population. Variables such as "Uninsured unemployed" and "Uninsured with disability" were reliably retained across various methods (stepwise, Poisson, Bayesian regressions), underscoring their strong associations. Conversely, variables like "HS grad," "Associate or some college," "Hispanic," "Uninsured Male," and "Non-Hispanic White" were frequently excluded due to multicollinearity or low statistical significance, indicating weaker or indirect associations (Fig. 7).

Race and ethnicity variables displayed varied patterns. Groups like Asians and individuals identifying as two or more races consistently appeared in final models (2018, 2019, 2020, 2022), indicating significant associations with uninsured rates. In contrast, Black/African American, Hispanic, and Non-Hispanic White variables were often excluded in later stages (2018–2022) due to multicollinearity with socioeconomic factors. Alaska Native and Pacific Islander variables showed inconsistent inclusion, reflecting context-dependent significance. These findings highlight the complexity of race and ethnicity's role in understanding uninsured rates. While certain groups, such as Asian and two or more races, consistently stand out as significant predictors from 2018 to 2022, others appear less stable, potentially due to their associations with other intersecting factors like employment, education, or income.

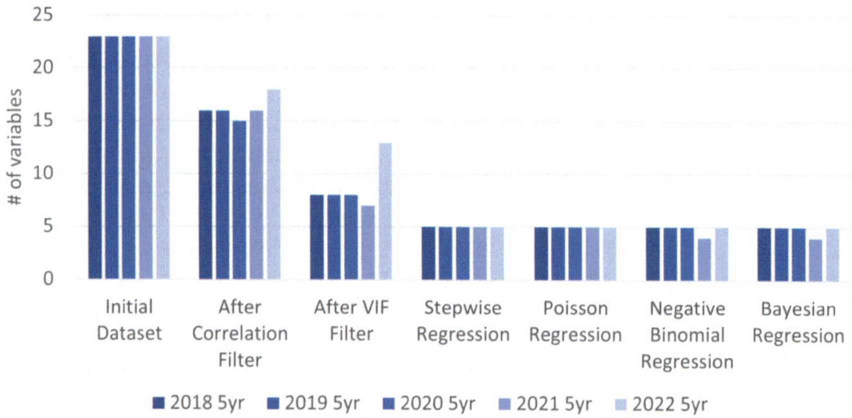

Fig. 7 Variable performance in each model

For education variables, "Bachelor or higher" appeared consistently in final models (2020–2022), highlighting its protective role against being uninsured. Meanwhile, lower education variables ("HS grad," "Associate or some college") were removed early (2018–2022), due to weaker direct associations or multicollinearity with other predictors. Their removal suggests weaker direct associations with the uninsured population or high multicollinearity with other socioeconomic factors, such as income or employment status.

Poverty and nativity variables showed limited retention. While the poverty indicator was included in early stages annually, it rarely appeared in final models after 2019, suggesting its effects were mediated by stronger predictors like employment or education. Similarly, nativity variables ("Uninsured native born," "Uninsured foreign born") were inconsistently retained (2020–2022), reflecting weaker associations with uninsured rates.

Disability-related variables, particularly "Uninsured with disability," consistently appeared in the final models across all years. From 2018 through 2022, this variable was retained in stepwise regression, Poisson regression, and Bayesian regression. Its consistent inclusion highlights the strong and stable association between disability status and being uninsured.

Overall, the analyses highlight unemployment and disability status as the most robust predictors of uninsured populations, consistently appearing across models and years. In contrast, variables related to educational attainment, race/ethnicity, and gender played less consistent roles, often being excluded due to weaker associations or multicollinearity. These findings underscore the importance of targeting populations with high unemployment and disability rates in future research and policy initiatives aimed at reducing uninsured rates.

4.7 Spatial Autocorrelation

Spatial autocorrelation was detected in all five ACS 5-year groups analyzed (Table 3). Spatial autocorrelation at the ZCTA level was applied to each ACS year group, using the default inverse distance method and using Euclidean distance with a threshold of 34,895.57 m, applying row standardization.

Across the ACS 5-year analyses from 2018 to 2022, there is a consistent, though gradually decreasing, spatial autocorrelation in the uninsured population, as indicated by the Moran's Index values. In 2018, the Moran's Index is 0.187 with a z-score of 12.81 ($p < 0.001$), showing an elevated level of spatial clustering. This value slightly declines each subsequent year, with the Moran's Index for 2019 at 0.177 and a z-score of 12.12 ($p < 0.001$). By 2020 and 2021, the Moran's Index stabilizes at 0.170 with z-scores of 11.65 and 11.71, respectively ($p < 0.001$). In 2022, the Moran's Index decreases further to 0.159, with a z-score of 10.95 ($p < 0.001$), indicating a subtle but consistent reduction in spatial clustering over the five-year period. This trend suggests a gradual dispersion in the uninsured population's spatial distribution across these years.

4.8 Hot Spot/Cold Spot Getis-Ord Gi

The hot spot analysis on uninsured populations, based on five-year ACS aggregates, shows the following trends from ACS 2018 to ACS 2022 (Figs. 8, 9, 10, 11, 12). The mean uninsured population decreases from 2,577.5 in ACS 2018 to approximately 2,413 in ACS 2022, with median values remaining stable around 1,912.5 to 1,977.5. Standard deviation stays consistent, around 2,200 to 2,300, indicating little change in variability across ZCTAs.

The mean Gi Bin increases gradually from 0.167 in ACS 2018 to 0.373 in ACS 2022, suggesting a slight rise in clustering intensity. Mean GiZScore starts at 2.68 in ACS 2018, decreasing to 2.47 by ACS 2022, indicating a slight reduction in clustering significance over time. Mean GiPValues remain low, between 0.19 and 0.25, with increasing standard deviation in later years, suggesting some variability in statistical significance across ZCTAs.

The hot spot/cold spot analysis shows stable patterns in uninsured population distribution across ZCTAs, with a consistent but slightly weakening intensity of clustering over the five-year aggregates. The minor upward trend in Gi Bin values and stable GiZScores suggest persistent clustering of uninsured populations, although with some indication of a potential decline in intensity in the later years. These results could imply a gradual geographic diffusion or redistribution of uninsured populations across the catchment area.

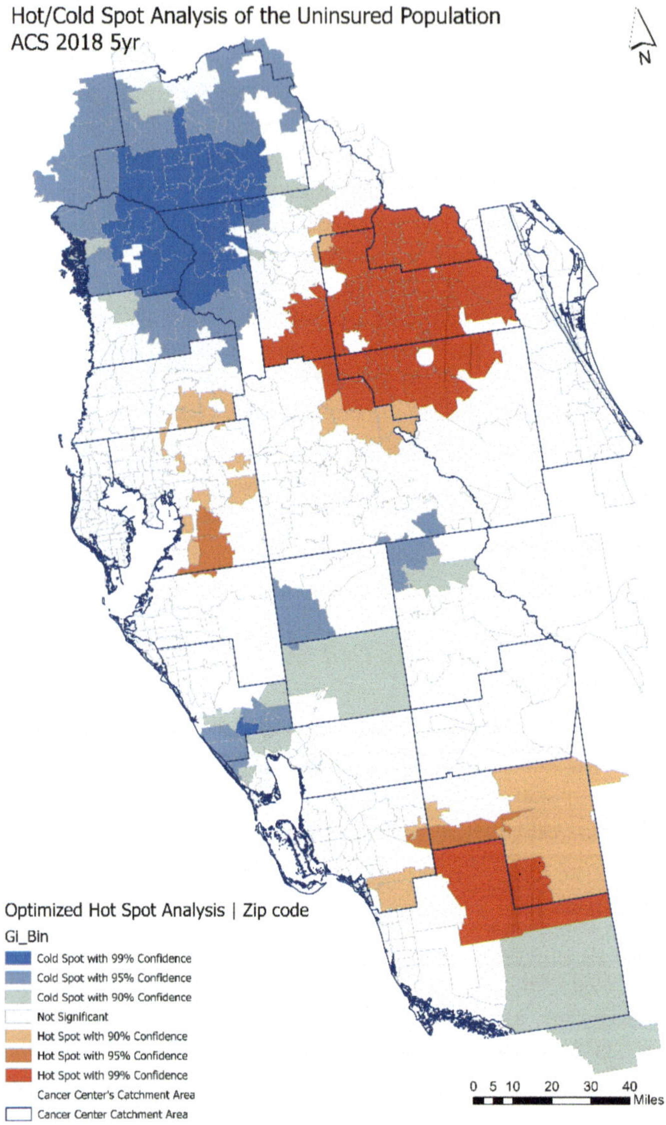

Fig. 8 ACS 2018 5yr hot spot analysis

4.9 GARCH Modeling

The GARCH analysis of uninsured populations across ZCTAs within a 23-county area, over five aggregate 5-year ACS periods, reveals insights into the volatility patterns of the uninsured population (Table 4). The table is ordered from smallest to largest AIC value, and colored based on beta1 value, where low values are green

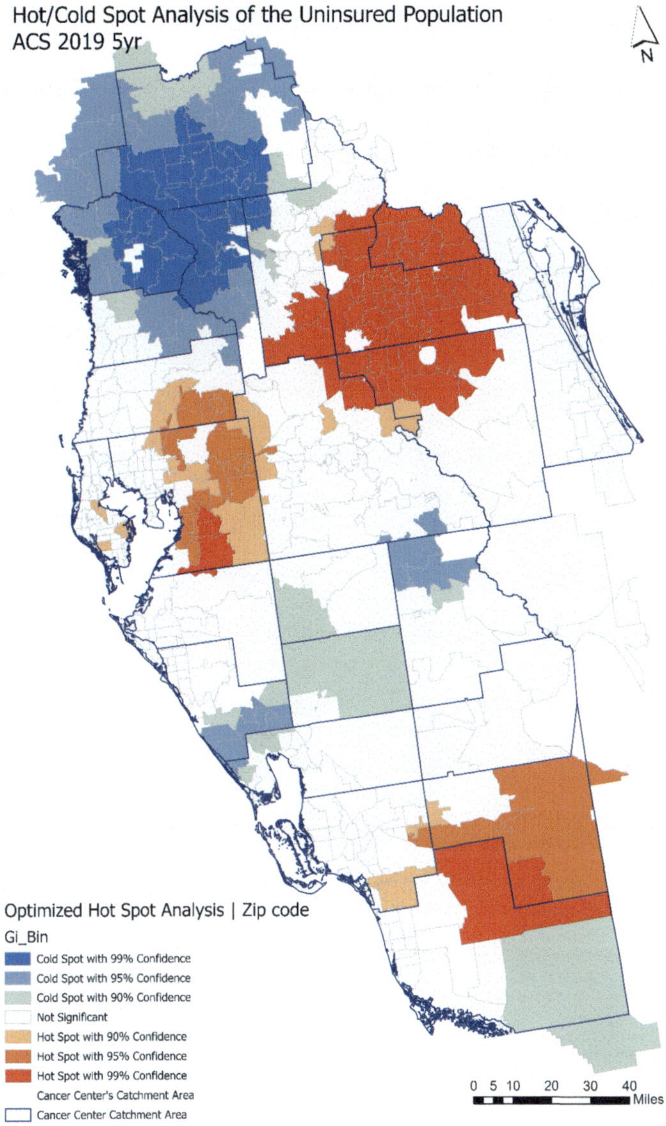

Fig. 9 ACS 2019 5yr hot spot analysis

and high values are red. ZCTAs with a high β_1 (close to 1) suggest that volatility, once it occurs, is likely to persist over time. ZCTAs with a low β_1 value indicate quicker returns to baseline variance, reflecting regions with more resilient or stable populations. From an initial sample of 498 ZCTAs, 470 (94.34%) had sufficient and complete data to be included in the final model, which led to the exclusion of twenty-eight ZCTAs with fewer than 100 data points or excessive missing values.

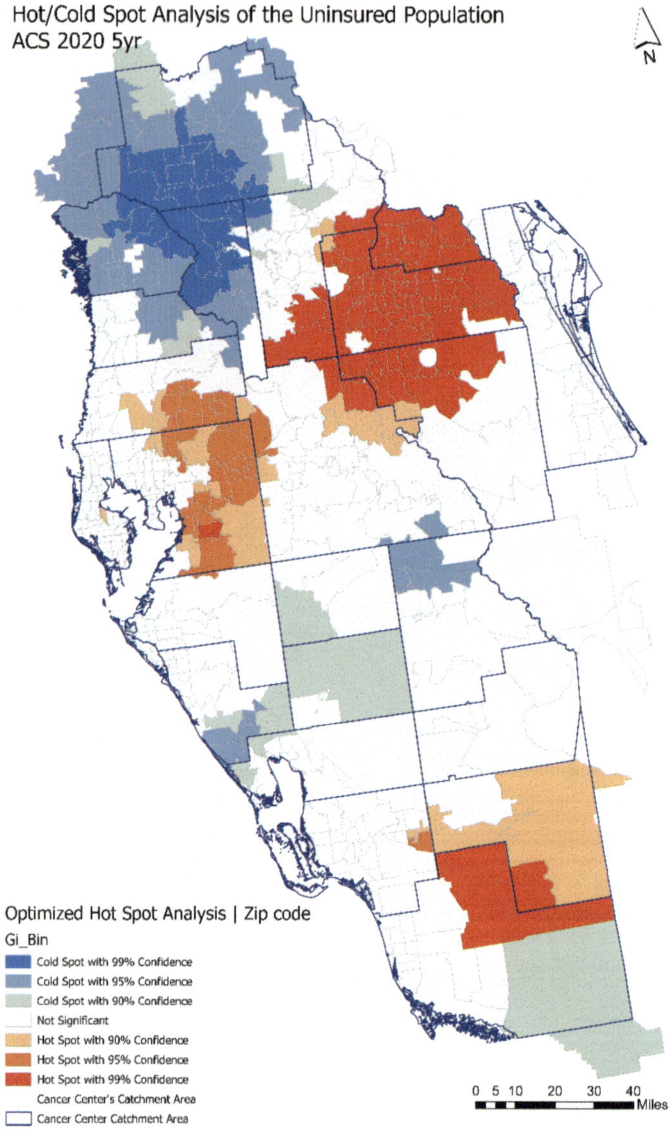

Fig. 10 ACS 2020 5yr hot spot analysis

This filtering ensured a more robust and reliable model fitting, minimizing biases from incomplete records. Figure 13 shows, at the ZCTA level, where volatility occurs within the 23-county area, with ZCTAs in Hernando County, FL experiencing the least amount of volatility. Table 5 lists the ZCTA level GARCH results sorted by AIC from smallest to largest.

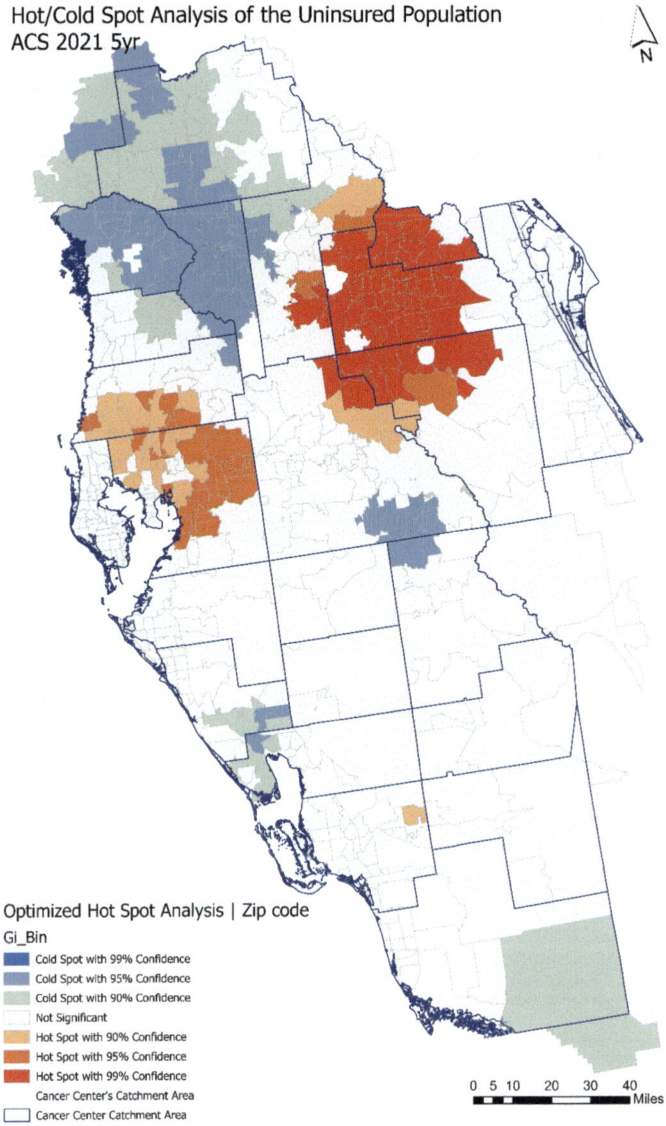

Fig. 11 ACS 2021 5yr hot spot analysis

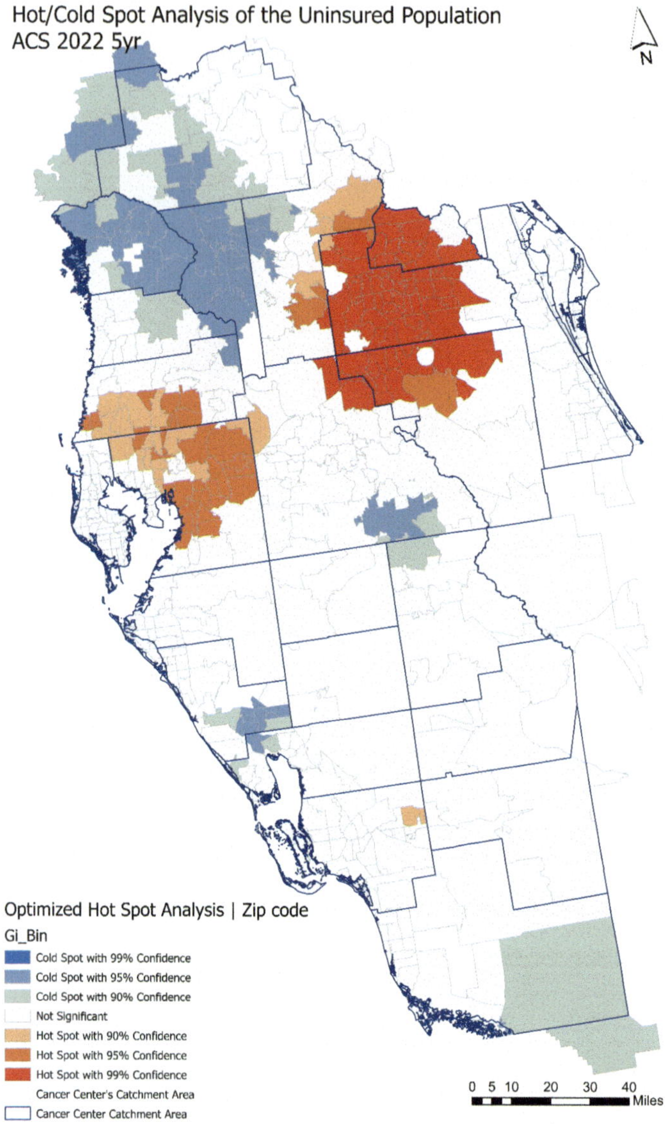

Fig. 12 ACS 2022 5yr hot spot analysis

Table 4 Spatial autocorrelation of the uninsured population

Year	Variable	Moran's Index	Expected Index	Variance	z-score	p-value	Neighbor threshold
ACS 2018 5 yr	Uninsured population	0.187190	−0.002096	0.000218	12.81471	< 0.001	34,895.5716 Meters
ACS 2019 5 yr	Uninsured population	0.177027	−0.002096	0.000218	12.12376	< 0.001	34,895.5716 Meters
ACS 2020 5 yr	Uninsured population	0.170118	−0.002096	0.000218	11.65269	< 0.001	34,895.5716 Meters
ACS 2021 5 yr	Uninsured population	0.169876	−0.002012	0.000216	11.70624	< 0.001	34,895.5716 Meters
ACS 2022 5 yr	Uninsured population	0.158900	−0.002012	0.000216	10.95478	< 0.001	34,895.5716 Meters

5 Discussion

The methodological framework employed across the analyses successfully integrated diverse statistical techniques to diagnose and address multicollinearity, heteroskedasticity, and overdispersion while identifying key predictors of the uninsured population. Variables such as *Uninsured unemployed and Uninsured with disability* consistently emerged as strong and stable predictors across models, underscoring their critical role in understanding disparities in health insurance coverage. Conversely, variables like *Hispanic, Non-Hispanic White*, and *native-born* were frequently excluded due to high multicollinearity or limited direct associations with uninsured rates. Spatial and temporal analyses provided additional layers of understanding, highlighting geographic and volatility patterns that contribute to the uninsured population's dynamics.

The spatial models in this analysis complement and enhance the non-spatial models by providing context about the geographic distribution and clustering of uninsured populations, which helps refine the interpretation and application of non-spatial statistical findings. While non-spatial models such as stepwise regression, Poisson regression, and Negative Binomial regression focus on identifying key predictors of uninsured rates and addressing issues like multicollinearity, overdispersion, and model fit, spatial models add an essential geographic layer of analysis that informs the broader methodological framework in several ways.

Firstly, the detection of spatial autocorrelation through Moran's Index and the Getis-Ord Gi hot spot analysis identifies regions where uninsured populations cluster. These spatial insights reveal areas of high or low insurance coverage that might be driven by underlying geographic factors, such as regional socioeconomic disparities,

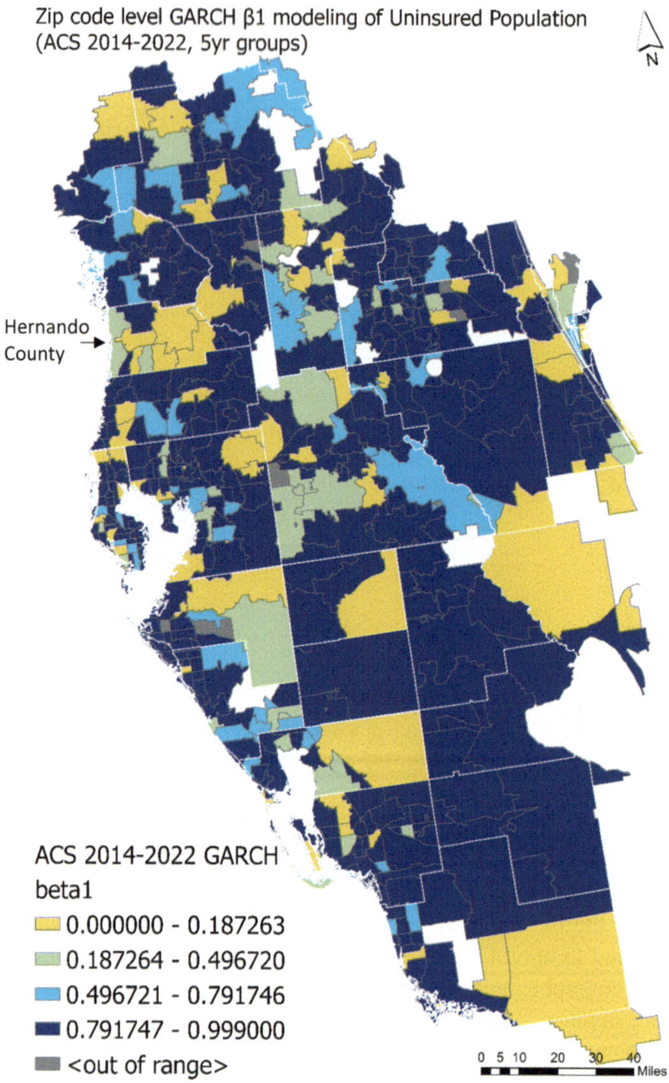

Fig. 13 ACS 2014–2022 GARCH β1 modeling: Persistence of volatility in the model

access to healthcare infrastructure, or demographic patterns. For instance, persistent spatial clusters identified in specific ZCTAs might indicate systemic barriers to insurance coverage, such as economic instability or policy gaps, which are not explicitly captured in non-spatial models. Incorporating these insights into non-spatial models allows analysts to contextualize the effects of predictors like unemployment or disability within their spatial environment, enhancing the explanatory power of the models.

Table 5 GARCH model summary: Uninsured population

ZCTA	Omega	Alpha1	Beta1	AIC	BIC
33621	0.000000642000	0.000000001150	0.895904900000	5.101784	4.789335
32925	3.581133110000	0.000000005590	0.998999985000	7.429352	7.116902
34762	3.138422946000	0.000000000963	0.998998088000	7.911920	7.599470
33503	0.000000733000	0.398991990000	0.000000000384	8.561843	8.249393
33924	0.000000853000	0.111601105000	0.149373088000	8.633487	8.321038
33867	0.000001410000	0.000000001460	0.260122970000	9.477636	9.165187
34137	67.350591670000	0.998999732000	0.000000265000	9.502363	9.189913
33827	31.024310130000	0.000000051700	0.998999948000	9.525696	9.213247
33744	7.369733411000	0.000000000291	0.998999905000	9.605483	9.293033
34681	0.000000734000	0.418508372000	0.001772317000	9.632346	9.319896
34139	36.402228550000	0.000000098600	0.026129148000	9.652258	9.339808
32664	0.000000274000	0.000000004060	0.187262794000	9.675211	9.362761
33715	12.956460780000	0.000000000000	0.998999938000	9.767492	9.455042
34141	0.000001830000	0.469993934000	0.003705722000	9.850381	9.537932
33854	105.653299300000	0.000000000000	0.000328981000	9.868179	9.555729
34737	0.000000308000	0.136438063000	0.000000000738	9.983753	9.671303
33960	27.545662680000	0.000000000297	0.998999999000	9.993629	9.681180
33865	0.337545483000	0.000000021300	0.955032482000	10.019100	9.706650
32133	51.619369710000	0.000000006440	0.998999993000	10.209051	9.896601
34228	24.575090900000	0.000000015000	0.998999985000	10.235611	9.923162
34679	11.926924070000	0.000000000011	0.998999996000	10.248496	9.936046
34684	32.878258650000	0.000000000012	0.998999984000	10.268311	9.955861
33848	216.939361600000	0.998999821000	0.000000179000	10.378215	10.065766
34215	0.508812640000	0.000000004900	0.998999550000	10.394330	10.081881
34216	0.000008980000	0.000000004400	0.608320091000	10.575319	10.262870
34140	0.000005170000	0.261832326000	0.000000015000	10.627682	10.315232
33576	0.000000511000	0.000000000626	0.103898948000	10.820604	10.508154
32778	23.031932100000	0.000000000000	0.998999972000	10.879574	10.567124
33786	97.069837550000	0.000000000001	0.998999999000	10.941326	10.628877
33877	81.652241630000	0.000000040900	0.998999959000	11.022335	10.709885
34269	126.518793900000	0.000000007470	0.998999992000	11.072118	10.759668
33921	0.000033600000	0.120820517000	0.140806854000	11.079036	10.766586
32798	0.000003650000	0.048741065000	0.278064366000	11.094000	10.781550
33965	75.342500100000	0.000000000015	0.998999981000	11.113602	10.801152
34138	41.826793680000	0.000000001090	0.998999998000	11.156960	10.844510
33708	63.554259870000	0.000000000002	0.998999985000	11.157866	10.845417

(continued)

Table 5 (continued)

ZCTA	Omega	Alpha1	Beta1	AIC	BIC
34739	121.130908900000	0.000000015500	0.998999984000	11.169151	10.856701
34285	0.000000129000	0.000000003020	0.385070067000	11.194896	10.882446
34797	147.176504100000	0.000000000125	0.998999999000	11.220223	10.907774
33521	0.000006300000	0.085293139000	0.104716128000	11.250855	10.938405
33847	128.539488800000	0.000000039100	0.998999960000	11.321799	11.009349
34217	0.000032900000	0.000000000195	0.615273987000	11.453180	11.140730
33859	0.000002330000	0.102786278000	0.000000000090	11.467792	11.155343
33704	182.948115200000	0.000000005670	0.998999994000	11.504816	11.192366
32709	47.552644980000	0.000000000000	0.998999944000	11.506147	11.193698
34705	2.555555468000	0.069979465000	0.251925361000	11.537954	11.225505
33839	161.153637900000	0.000000000000	0.998999984000	11.571984	11.259534
34661	0.000022400000	0.288962699000	0.000000151000	11.588669	11.276219
33857	1.683800930000	0.000000000003	0.997224738000	11.638477	11.326027
34760	0.000002810000	0.000000003050	0.280655218000	11.647356	11.334906
33629	140.268917500000	0.000000000127	0.998999996000	11.652013	11.339563
34655	0.000002250000	0.114958925000	0.000000002950	11.658622	11.346172
33785	2.240051740000	0.000000000000	0.769033191000	11.676327	11.363877
34449	1.822793937000	0.000000002560	0.901416165000	11.732236	11.419786
33946	155.046633600000	0.000000000000	0.998999998000	11.732785	11.420336
33585	184.654768900000	0.000000000000	0.998999963000	11.743591	11.431141
34695	0.000015800000	0.168191694000	0.092568251000	11.746976	11.434526
34601	0.000000252000	0.090289473000	0.005947390000	11.772014	11.459565
33912	1.972737862000	0.000000007150	0.905943182000	11.787225	11.474775
33755	2.025968623000	0.000000000168	0.975504669000	11.807642	11.495193
32931	0.000029000000	0.299429651000	0.077955751000	11.862219	11.549770
34461	2.210850881000	0.000000005050	0.998999882000	11.904111	11.591661
32702	2.245123593000	0.000000000052	0.988655573000	11.920213	11.607763
32686	0.000007760000	0.131073357000	0.000000000052	11.946156	11.633706
33770	0.000134614000	0.176668746000	0.000000000000	11.960326	11.647877
32732	2.413703205000	0.000000006530	0.998999750000	11.993181	11.681031
33834	254.358774500000	0.000000000003	0.998999998000	12.048319	11.735870
34201	408.364443800000	0.000000004220	0.998999995000	12.089308	11.776858
33549	114.336395100000	0.000000000000	0.998999977000	12.093507	11.781057
33760	0.000243482000	0.000000000027	0.670480721000	12.096864	11.784414
33548	0.000011400000	0.002893913000	0.387334711000	12.150609	11.838159
32976	0.000000000538	0.083867374000	0.225489569000	12.175833	11.863383

(continued)

Table 5 (continued)

ZCTA	Omega	Alpha1	Beta1	AIC	BIC
34434	0.000010200000	0.117918106000	0.000000000000	12.249473	11.937024
32681	248.091102700000	0.000000093000	0.998999907000	12.258972	11.946523
34219	0.000007250000	0.198138731000	0.000000000093	12.305896	11.993446
34773	319.705283900000	0.000000000000	0.998999998000	12.328839	12.016390
33777	0.000008570000	0.099678306000	0.000000011400	12.346253	12.033803
33851	258.348331400000	0.000000040200	0.998999959000	12.347301	12.034852
33767	3.438749152000	0.000000000000	0.959039767000	12.348960	12.036510
34222	1507.173820000000	0.998999959000	0.000000040500	12.353393	12.040943
33843	152.481404200000	0.000000000000	0.998999973000	12.366554	12.054104
32951	0.000015700000	0.197630537000	0.000000001310	12.371625	12.059175
34756	3.583479732000	0.000000004610	0.869670212000	12.382645	12.070195
33849	3.625624767000	0.000000000487	0.998999815000	12.409042	12.096593
32949	0.000017500000	0.000000007500	0.399303071000	12.453203	12.140753
33955	7.033083811000	0.053755711000	0.380445023000	12.468157	12.155708
34420	0.004934075000	0.000000000001	0.599198173000	12.472165	12.159715
33592	408.072128900000	0.000000003040	0.998999997000	12.473263	12.160813
33956	0.000000000005	0.077555519000	0.176918609000	12.490666	12.178216
34202	234.278478100000	0.000000006380	0.998999993000	12.609402	12.296953
34481	6.004271569000	0.000000012600	0.597711819000	12.628159	12.315710
34610	399.637917500000	0.000000004050	0.998999995000	12.647721	12.335271
32955	0.000000155000	0.110494479000	0.016762112000	12.694503	12.382053
33620	0.000015700000	0.146426978000	0.000000000205	12.698958	12.386508
33541	487.395502200000	0.000000000000	0.998999988000	12.720783	12.408333
32102	9.530514682000	0.148417298000	0.128022428000	12.740667	12.428217
32750	5.244893738000	0.000000037100	0.998999927000	12.767893	12.455443
34450	635.177095300000	0.000000035400	0.998999964000	12.778399	12.465949
33957	10.619620060000	0.017582620000	0.376693618000	12.860240	12.547791
33772	544.654971100000	0.000000000000	0.998999992000	12.871395	12.558946
32803	586.100363400000	0.000000000003	0.998999985000	12.995372	12.682922
34237	729.381014100000	0.000000000001	0.998999999000	12.997655	12.685205
33850	9.175873869000	0.000000001010	0.562108791000	13.022984	12.710534
33606	304.201114600000	0.000000000000	0.998999983000	13.029562	12.717112
33709	0.884081994000	0.000000000000	0.548659287000	13.033476	12.721027
33920	6.862016396000	0.000000010500	0.957650394000	13.042896	12.730447
33922	805.190630300000	0.000000000001	0.998999989000	13.143276	12.830826
33801	0.000009790000	0.092769439000	0.000000001190	13.160118	12.847668

(continued)

Table 5 (continued)

ZCTA	Omega	Alpha1	Beta1	AIC	BIC
34268	489.235608100000	0.000000000404	0.998999999000	13.203063	12.890613
34607	0.000306425000	0.037138210000	0.295189135000	13.254166	12.941716
33514	829.023511400000	0.000000038400	0.998999961000	13.262756	12.950307
34480	0.000006160000	0.109256813000	0.144812769000	13.276637	12.964187
32195	8.858904829000	0.000000001510	0.998999949000	13.290131	12.977682
32735	912.466192000000	0.000000058100	0.998999941000	13.293605	12.981155
34429	9.867299052000	0.000000000130	0.696380543000	13.314051	13.001602
34753	331.320628000000	0.000000000048	0.998999956000	13.327638	13.015188
33947	9.301179234000	0.000000000528	0.998999984000	13.336164	13.023714
32736	9.416851436000	0.000000000063	0.914523375000	13.348233	13.035783
33513	0.000705402000	0.170538684000	0.000000000001	13.359340	13.046890
33776	9.482917657000	0.000000027500	0.998999919000	13.371520	13.059071
32667	9.627984196000	0.000000000005	0.968321135000	13.382518	13.070068
33563	0.000019700000	0.077329613000	0.000000000369	13.385946	13.073497
34614	9.692411249000	0.000000012800	0.998999922000	13.388616	13.076167
33966	0.000030700000	0.000000019900	0.186244015000	13.392408	13.079958
33707	0.000034000000	0.084199816000	0.243528668000	13.409310	13.096860
33773	10.994503120000	0.000000000006	0.822382831000	13.448849	13.136399
34731	10.274345340000	0.000000003400	0.998983628000	13.451940	13.139490
34771	645.172183600000	0.000000000000	0.998999991000	13.453048	13.140599
33875	1227.987323000000	0.000000007080	0.998999992000	13.459639	13.147189
33954	845.491533300000	0.000000065100	0.998999934000	13.469863	13.157413
33624	1054.488116000000	0.000000000574	0.998999999000	13.473092	13.160642
33527	10.682229770000	0.000000000061	0.917598913000	13.479304	13.166854
32907	798.496582300000	0.000000001370	0.998999998000	13.488822	13.176372
32814	18.482175540000	0.000000997000	0.507296562000	13.519203	13.206754
33944	878.706274500000	0.000000074200	0.998999925000	13.552074	13.239624
34108	17.470765290000	0.000000156000	0.564779811000	13.554713	13.242263
34639	17.794900390000	0.000000024400	0.531192038000	13.556718	13.244269
32950	12.869170130000	0.000000024100	0.998999970000	13.565551	13.253102
34685	11.847094650000	0.000000000001	0.946060430000	13.587058	13.274608
33853	12.310971500000	0.000000000031	0.883325429000	13.605185	13.292736
34251	19.268873870000	0.084023656000	0.295041066000	13.608938	13.296488
33950	1616.303524000000	0.000000005690	0.998999994000	13.614646	13.302196
33838	13.420727370000	0.000000000003	0.695995419000	13.620332	13.307882
34604	0.000033800000	0.110204794000	0.000000000031	13.621951	13.309501

(continued)

Table 5 (continued)

ZCTA	Omega	Alpha1	Beta1	AIC	BIC
32827	14.743244150000	0.000000000032	0.998999997000	13.668436	13.355986
32948	0.000031400000	0.133067585000	0.070018293000	13.676204	13.363755
33907	13.336359940000	0.000000000241	0.998999970000	13.679928	13.367478
34103	1317.176421000000	0.000000054100	0.998999946000	13.682001	13.369552
33703	13.061173100000	0.000000000013	0.906362928000	13.682215	13.369765
34242	13.104909420000	0.000000000051	0.942186126000	13.687901	13.375451
33764	1214.569352000000	0.000000007760	0.998999992000	13.694634	13.382184
33712	16.422172310000	0.000000003700	0.682276934000	13.698772	13.386323
34453	16.993060030000	0.000000000105	0.998999986000	13.753462	13.441012
32730	14.450558890000	0.000000099900	0.998999833000	13.759166	13.446717
32953	24.047970360000	0.000000000536	0.451307244000	13.801534	13.489084
33701	838.001862700000	0.000000002860	0.998999996000	13.807924	13.495474
34436	1678.247600000000	0.000000023800	0.998999976000	13.812634	13.500184
33908	1583.825956000000	0.000000016400	0.998999983000	13.827426	13.514976
34442	1089.344966000000	0.000000002330	0.998999997000	13.830361	13.517912
33993	15.241525450000	0.000000000137	0.955792865000	13.843124	13.530675
33930	2076.882204000000	0.000000008440	0.998999991000	13.882281	13.569832
34241	15.939220430000	0.000000023200	0.998996144000	13.891151	13.578701
34229	1999.770872000000	0.000000014800	0.998999984000	13.891153	13.578703
34734	1214.054144000000	0.000000000000	0.998999993000	13.894896	13.582446
34109	16.035476610000	0.000000018200	0.998999626000	13.897176	13.584726
33890	0.000109790000	0.287090999000	0.000000009150	13.919073	13.606623
34683	2067.170125000000	0.000000004380	0.998999989000	13.919845	13.607396
33762	17.387782310000	0.000000027600	0.998999962000	13.927921	13.615471
32903	18.257307430000	0.000000013900	0.998999986000	13.929786	13.617337
33573	1855.004607000000	0.000000000885	0.998999999000	13.932648	13.620199
34292	1584.644501000000	0.000000008230	0.998999991000	13.934290	13.621840
34293	19.372373400000	0.000000000605	0.747511381000	13.941327	13.628877
32773	16.928741450000	0.000000000011	0.930777955000	13.944455	13.632005
34223	16.834211360000	0.000000020300	0.995631267000	13.945815	13.633365
34637	19.701367000000	0.000000000365	0.998999984000	13.958462	13.646012
32909	1744.555515000000	0.000000000000	0.998999999000	14.008029	13.695579
33609	22.580663910000	0.000000027300	0.510670842000	14.010863	13.698414
33540	18.660085350000	0.000000027300	0.998999911000	14.048728	13.736279
32767	2320.060477000000	0.000000002190	0.998999997000	14.060066	13.747616
33538	19.072544450000	0.000000000001	0.957718863000	14.066528	13.754079

(continued)

Table 5 (continued)

ZCTA	Omega	Alpha1	Beta1	AIC	BIC
33953	19.318764460000	0.000000000033	0.910436175000	14.069361	13.756911
33917	23.426555580000	0.000000031700	0.998999967000	14.074971	13.762522
34224	22.548351610000	0.000000000005	0.633824138000	14.097497	13.785047
32920	2229.117191000000	0.000000008110	0.998999991000	14.098275	13.785825
32779	20.543789310000	0.000000000385	0.998999995000	14.103434	13.790985
33825	2053.405485000000	0.000000005400	0.998999980000	14.106877	13.794427
32801	20.201827460000	0.000000001620	0.998999988000	14.107163	13.794713
33471	20.292750170000	0.000000042800	0.998999903000	14.132575	13.820125
34236	3016.261267000000	0.000000005540	0.998999994000	14.133709	13.821259
33765	20.774416570000	0.000000000046	0.909479941000	14.135843	13.823393
33870	21.459472530000	0.000000016200	0.998999983000	14.179586	13.867136
34688	0.007932434000	0.258998207000	0.000000000004	14.192274	13.879824
33914	21.918915000000	0.000000001070	0.998999991000	14.203676	13.891227
33618	37.795078010000	0.052452296000	0.371574549000	14.203856	13.891406
33876	24.349072070000	0.000000059100	0.998999941000	14.207194	13.894744
33613	22.798012790000	0.000000007490	0.878819324000	14.216588	13.904138
32820	0.000218685000	0.077436708000	0.142068588000	14.222493	13.910043
34428	22.454950730000	0.000000000009	0.936968961000	14.223730	13.911280
33928	2774.449305000000	0.000000004860	0.998999995000	14.232850	13.920400
32789	23.479575540000	0.000000000816	0.998999981000	14.249867	13.937417
34479	2296.536638000000	0.000000020700	0.998999979000	14.263374	13.950925
32179	23.395332230000	0.000000040900	0.998999888000	14.274938	13.962489
34690	24.463361480000	0.000000000017	0.869079606000	14.280324	13.967874
32754	24.864888940000	0.000000000900	0.998999998000	14.291438	13.978989
32821	37.438151250000	0.000000000243	0.998999975000	14.294835	13.982385
32937	2394.149970000000	0.000000090200	0.998999910000	14.307283	13.994833
32832	2488.517096000000	0.000000010800	0.998999988000	14.321797	14.009348
33565	0.000602067000	0.115862422000	0.000000000008	14.326693	14.014243
33605	24.685146900000	0.000000005520	0.998999923000	14.328543	14.016093
34669	25.143687310000	0.000000043200	0.998999939000	14.346843	14.034394
34602	61.321933640000	0.188126048000	0.034879596000	14.357346	14.044897
34240	27.913597910000	0.000000000047	0.791746009000	14.358055	14.045605
34488	2848.065029000000	0.000000013300	0.998999986000	14.384101	14.071651
32162	26.191899190000	0.000000003740	0.998999960000	14.387750	14.075300
34291	32.956776040000	0.000000000663	0.691179947000	14.399992	14.087542
33594	2898.279530000000	0.000000000000	0.998999999000	14.426379	14.113929

(continued)

Table 5 (continued)

ZCTA	Omega	Alpha1	Beta1	AIC	BIC
33616	27.300028780000	0.000000009880	0.998999968000	14.428907	14.116457
33584	3171.134543000000	0.000000002460	0.998999997000	14.434069	14.121619
32726	4402.649029000000	0.000000009330	0.998999990000	14.461118	14.148669
34432	29.338292660000	0.000000013600	0.998999918000	14.501080	14.188630
34238	3182.964037000000	0.000000000000	0.998999994000	14.505367	14.192918
33971	30.361063790000	0.000000000024	0.898981277000	14.510676	14.198227
33913	29.894922390000	0.000000018200	0.996007434000	14.520135	14.207685
34476	29.909338360000	0.000000021900	0.996070025000	14.520597	14.208148
32707	3324.885395000000	0.000000035200	0.998999964000	14.521810	14.209360
33884	35.804509210000	0.000000000000	0.998999976000	14.524685	14.212235
33980	35.462600450000	0.000000000007	0.737828025000	14.531566	14.219116
32927	30.836996520000	0.000000000001	0.981204090000	14.550319	14.237869
33534	31.195965540000	0.000000000994	0.942830861000	14.557277	14.244827
34231	3042.114831000000	0.000000015100	0.998999985000	14.560063	14.247614
33572	3894.244243000000	0.000000013600	0.998999986000	14.581551	14.269101
34239	53.145639770000	0.301575533000	0.104465802000	14.624246	14.311797
32668	33.208280360000	0.000000003500	0.998999953000	14.625223	14.312773
32776	36.745974000000	0.000000018100	0.998999979000	14.625236	14.312786
33813	86.418406700000	0.000000000332	0.362908803000	14.625898	14.313449
33637	35.831957860000	0.000000000009	0.838925915000	14.648794	14.336345
33919	4703.720980000000	0.000000007850	0.998999991000	14.669359	14.356909
34275	40.719547010000	0.000000084600	0.998999915000	14.674911	14.362461
33774	35.579611320000	0.000000026900	0.998999683000	14.694229	14.381780
32940	56.294250660000	0.375130207000	0.025605568000	14.708740	14.396290
32833	4710.122802000000	0.000000005880	0.998999994000	14.712895	14.400445
33916	5229.493667000000	0.000000021400	0.998999978000	14.713282	14.400832
33602	36.338183810000	0.000000002890	0.998999886000	14.715318	14.402868
32817	38.721846660000	0.000000005040	0.998999994000	14.720104	14.407654
33702	37.299602700000	0.000000000000	0.933363768000	14.731943	14.419493
34698	5298.859159000000	0.000000012500	0.998999987000	14.745318	14.432869
34470	37.605401340000	0.000000015600	0.998999388000	14.749614	14.437165
34102	37.755319240000	0.000000000014	0.982764676000	14.753003	14.440553
32796	47.331340880000	0.000000000224	0.998999992000	14.755443	14.442993
34785	38.200557160000	0.000000000000	0.926891727000	14.758099	14.445649
32159	4234.169204000000	0.000000029900	0.998999969000	14.758237	14.445787
33597	38.011515540000	0.000000000062	0.986258477000	14.759832	14.447382

(continued)

Table 5 (continued)

ZCTA	Omega	Alpha1	Beta1	AIC	BIC
32822	3505.388370000000	0.000000015700	0.998999980000	14.764956	14.452506
33558	4464.392790000000	0.000000015000	0.998999984000	14.772242	14.459792
33611	38.721308820000	0.000000000022	0.980025735000	14.777965	14.465515
32811	5630.618733000000	0.000000014900	0.998999985000	14.793836	14.481386
34608	0.000108233000	0.142950929000	0.000000004290	14.799239	14.486789
33635	1.674739636000	0.246039910000	0.000000000000	14.803634	14.491185
33873	2894.927506000000	0.000000012800	0.998999987000	14.807650	14.495200
34235	41.427963420000	0.000000002500	0.998999979000	14.810455	14.498005
34210	5350.408963000000	0.000000007670	0.998999992000	14.811466	14.499017
33868	54.931801500000	0.174375987000	0.226293717000	14.824380	14.511930
34134	40.550002100000	0.000000028200	0.998999921000	14.824952	14.512503
34433	44.696412240000	0.000000043300	0.552872094000	14.827669	14.515219
34286	61.094282070000	0.000000000203	0.496720205000	14.844160	14.531710
33716	43.106146150000	0.000000027500	0.998999971000	14.845168	14.532719
33909	0.000343797000	0.104384306000	0.000000000154	14.846125	14.533675
33809	42.276691890000	0.000000000816	0.887345566000	14.847960	14.535510
32819	42.202888360000	0.000000055300	0.998999929000	14.859653	14.547203
33763	42.790058370000	0.000000000013	0.905796206000	14.859761	14.547311
33782	54.204847430000	0.000000000007	0.998999989000	14.877186	14.564736
34431	43.775041430000	0.000000000000	0.774578369000	14.877508	14.565058
33556	43.932618260000	0.000000026100	0.998999943000	14.904940	14.592490
33705	44.291122520000	0.000000000660	0.998999962000	14.912848	14.600399
32809	46.405408870000	0.000000000095	0.828985684000	14.919124	14.606674
34113	53.058966820000	0.000000035100	0.998999959000	14.920559	14.608110
32766	47.294544680000	0.000000024800	0.998999975000	14.928610	14.616161
34473	13.745772980000	0.213096992000	0.000000000000	14.928764	14.616315
32714	4919.009268000000	0.000000000000	0.998999993000	14.941801	14.629351
33542	45.706684140000	0.000000000485	0.998999982000	14.943961	14.631511
34448	46.203035470000	0.000000003460	0.998999901000	14.955470	14.643021
33897	0.009013615000	0.204821460000	0.000000000004	14.961998	14.649548
32113	61.157699760000	0.000000036700	0.998999963000	14.975458	14.663008
33815	7419.210508000000	0.000000016300	0.998999983000	14.977715	14.665265
33567	52.100432360000	0.000000021700	0.998999974000	14.978041	14.665591
33931	47.264830100000	0.000000010600	0.998999959000	14.978152	14.665702
34714	51.512359210000	0.000000000121	0.824687909000	14.994334	14.681885
33714	5769.644288000000	0.000000014900	0.998999985000	14.998005	14.685555

(continued)

Table 5 (continued)

ZCTA	Omega	Alpha1	Beta1	AIC	BIC
34613	93.078843410000	0.215353181000	0.000000012400	14.998332	14.685883
32757	263.687019700000	0.090756731000	0.072795351000	15.004757	14.692307
32696	0.072880608000	0.245020624000	0.000000000000	15.015307	14.702858
33781	5697.874866000000	0.000000014300	0.998999986000	15.015637	14.703188
33778	97.815461570000	0.025988851000	0.399663013000	15.027088	14.714639
34748	68.644369840000	0.078005385000	0.323938306000	15.041629	14.729180
33981	51.557444190000	0.000000001350	0.998999980000	15.053301	14.740851
33972	51.858497530000	0.000000000000	0.960470801000	15.068938	14.756488
33812	52.020054670000	0.000000034900	0.998999962000	15.073692	14.761243
33904	98.306521330000	0.074216190000	0.317635657000	15.106093	14.793644
34972	1.001707525000	0.241269438000	0.000000000000	15.111793	14.799343
34747	7414.507634000000	0.000000016600	0.998999983000	15.120165	14.807715
32784	83.483769740000	0.000000000051	0.485256050000	15.122528	14.810078
32134	61.117666570000	0.000000000005	0.720878835000	15.131444	14.818994
33706	81.456148380000	0.000000000002	0.998999995000	15.144767	14.832317
33759	55.996933790000	0.000000023800	0.998999960000	15.147703	14.835253
33607	2.193396006000	0.243804064000	0.000000000000	15.160581	14.848131
32712	72.396132710000	0.000000038500	0.998999961000	15.162188	14.849738
33805	57.148563190000	0.000000021000	0.998999971000	15.168022	14.855573
33559	8349.234677000000	0.000000007670	0.998999992000	15.175536	14.863086
32751	60.533507570000	0.000000000075	0.872439310000	15.184500	14.872050
33982	0.145682541000	0.266850642000	0.000000000000	15.185951	14.873501
34119	87.958163460000	0.000000059400	0.572886972000	15.186369	14.873919
33860	134.881380600000	0.050493715000	0.273271335000	15.203416	14.890967
33756	74.038981030000	0.000000001060	0.696111182000	15.214968	14.902518
33545	60.921517930000	0.000000001680	0.998999979000	15.231773	14.919323
33543	69.085235420000	0.000000043200	0.998999956000	15.240118	14.927669
33547	127.277417100000	0.000000022000	0.998999969000	15.252730	14.940280
33647	63.950191220000	0.000000000111	0.998999974000	15.280405	14.967955
33872	65.915816650000	0.000000024200	0.998999919000	15.310830	14.998381
32934	66.027344740000	0.000000006630	0.998999983000	15.312465	15.000016
34974	67.375014610000	0.000000001330	0.998999634000	15.332747	15.020298
33991	6874.143284000000	0.000000028200	0.998999971000	15.335165	15.022715
32746	67.915229030000	0.000000006070	0.998999989000	15.340418	15.027968
32804	68.246139460000	0.000000002330	0.998999968000	15.345503	15.033053
33973	68.664010290000	0.000000000811	0.998999811000	15.351700	15.039251

(continued)

Table 5 (continued)

ZCTA	Omega	Alpha1	Beta1	AIC	BIC
33525	68.778376440000	0.000000000000	0.984197189000	15.353173	15.040723
33710	0.001818768000	0.141658847000	0.000000000035	15.353173	15.040723
34638	82.083821520000	0.000000000304	0.654653813000	15.355527	15.043077
33634	70.862554870000	0.000000000003	0.887662931000	15.367219	15.054770
33983	73.691978080000	0.000000000001	0.706989530000	15.368132	15.055683
32952	73.713466850000	0.000000000008	0.637614304000	15.368614	15.056165
34772	70.212884970000	0.000000003240	0.996171831000	15.373989	15.061539
34691	72.106454090000	0.000000004060	0.998999993000	15.377626	15.065176
34105	72.336361040000	0.000000000021	0.905570237000	15.393065	15.080615
34452	71.910695940000	0.000000035800	0.998999894000	15.397885	15.085435
33841	74.823804450000	0.000000000067	0.871444726000	15.414241	15.101792
33974	74.689622620000	0.000000014000	0.998999971000	15.435583	15.123133
34788	0.002809294000	0.269960731000	0.000000000014	15.468565	15.156115
34289	92.778142940000	0.000000014000	0.451669185000	15.472737	15.160287
34472	7872.345108000000	0.000000009660	0.998999990000	15.483501	15.171052
33803	131.576160600000	0.000000014000	0.998999984000	15.483745	15.171296
33898	95.292933310000	0.000000325000	0.551577396000	15.485796	15.173347
33570	0.014252643000	0.200383683000	0.000000000006	15.493156	15.180707
33596	87.804751280000	0.000000000000	0.557102368000	15.522766	15.210317
33612	83.169952000000	0.000000000152	0.998999988000	15.527697	15.215247
33948	102.326326600000	0.000000039400	0.437503300000	15.531610	15.219160
33510	112.891030900000	0.000000126000	0.475314739000	15.554106	15.241657
34654	102.822537800000	0.552721199000	0.000000029400	15.566431	15.253981
34689	6.346645391000	0.196861019000	0.000000000000	15.577836	15.265386
33544	89.316844650000	0.000000044500	0.998999952000	15.614563	15.302113
33711	89.870806250000	0.000000000001	0.957795340000	15.618265	15.305815
34653	94.628103420000	0.000000000094	0.855828912000	15.628701	15.316251
34668	94.104453970000	0.000000002590	0.879566164000	15.635939	15.323489
33569	91.795861200000	0.000000024600	0.998999949000	15.642005	15.329555
32805	204.076572600000	0.061263305000	0.275155732000	15.649600	15.337150
33952	132.773152500000	0.000000005610	0.998999992000	15.684029	15.371579
33771	110.452645700000	0.000000000140	0.998999995000	15.722368	15.409918
33880	127.612177300000	0.000000011800	0.998999985000	15.728239	15.415789
34609	191.441633600000	0.000000664000	0.460170129000	15.757187	15.444737
33566	131.897439000000	0.000000008890	0.998999987000	15.778479	15.466029
34465	105.612066100000	0.000000071200	0.998999926000	15.782078	15.469629

(continued)

Table 5 (continued)

ZCTA	Omega	Alpha1	Beta1	AIC	BIC
32926	106.600051200000	0.000000004520	0.998999958000	15.791533	15.479083
34475	115.121562800000	0.000000093400	0.998999886000	15.792576	15.480126
33511	16,963.642870000000	0.000000014800	0.998999985000	15.796752	15.484302
33579	139.615524000000	0.000000000005	0.672713489000	15.818193	15.505743
32812	109.763171700000	0.000000012000	0.998999981000	15.820375	15.507925
34112	8.140546258000	0.195527734000	0.000000000000	15.825936	15.513486
33935	291.411586200000	0.000000022600	0.998999971000	15.833916	15.521467
33625	113.177386300000	0.000000004510	0.931277892000	15.839951	15.527501
33603	137.838032100000	0.000000000000	0.553992382000	15.844190	15.531741
33905	112.989823200000	0.000000000002	0.983894134000	15.849098	15.536649
33604	114.004368800000	0.000000000138	0.998999983000	15.858664	15.546214
33810	367.576628200000	0.104972172000	0.110653998000	15.864222	15.551772
32836	119.181132900000	0.000000000000	0.899190343000	15.877199	15.564749
32922	117.560521000000	0.000000133000	0.998999847000	15.889293	15.576844
34715	120.457666600000	0.000000009740	0.998999986000	15.896767	15.584317
33617	118.601266600000	0.000000004350	0.998999436000	15.898248	15.585798
34484	127.526913200000	0.000000035400	0.998999964000	15.912530	15.600080
32701	124.174646100000	0.000000000001	0.947790387000	15.940429	15.627979
32765	189.328129800000	0.000000000000	0.572712179000	15.944153	15.631703
33967	148.058466100000	0.000000000421	0.615783855000	15.952084	15.639634
34287	132.869822200000	0.000000000003	0.609699220000	15.961306	15.648856
33976	172.834991800000	0.000000005400	0.410476415000	15.966579	15.654129
33523	128.033003700000	0.000000000003	0.933931591000	15.969664	15.657214
34652	129.482735400000	0.000000015700	0.997972549000	15.986032	15.673583
34114	131.836943400000	0.000000000124	0.881107918000	15.988319	15.675869
34474	166.323125000000	0.000000012100	0.998999986000	15.995913	15.683463
34232	181.034496900000	0.000000015000	0.998999984000	15.996962	15.684512
32708	141.286534800000	0.000000005760	0.840572800000	16.018075	15.705625
34207	134.827951100000	0.000000003710	0.998999946000	16.026444	15.713994
32780	211.199390600000	0.200385174000	0.116224901000	16.067992	15.755542
32908	141.851265400000	0.000000026900	0.998999907000	16.077206	15.764756
32935	144.080999400000	0.000000000001	0.950608522000	16.090059	15.777610
34104	145.332080100000	0.000000023000	0.998999967000	16.101365	15.788915
34203	149.390473200000	0.000000022000	0.998999966000	16.128906	15.816456
34288	178.489705200000	0.000000000426	0.722630868000	16.135820	15.823371
34243	350.358460000000	0.000000065400	0.998999934000	16.149531	15.837081

(continued)

Table 5 (continued)

ZCTA	Omega	Alpha1	Beta1	AIC	BIC
34761	156.498823600000	0.000000007900	0.998999989000	16.175403	15.862954
34482	355.993626100000	0.037098301000	0.308105171000	16.183758	15.871309
32901	165.871611400000	0.000000000007	0.861122023000	16.186627	15.874177
34234	161.101034200000	0.000000033500	0.998999928000	16.204483	15.892033
33901	195.523429600000	0.000000017000	0.998999977000	16.216301	15.903851
33881	174.479910700000	0.000000000001	0.844652158000	16.227610	15.915160
33852	166.589167800000	0.000000003100	0.998999880000	16.237988	15.925538
34145	170.860301500000	0.000000009780	0.998999976000	16.263269	15.950820
34677	172.688433100000	0.000000000001	0.984653153000	16.273677	15.961227
32807	174.714973000000	0.000000002020	0.998999885000	16.285629	15.973179
34266	177.471235600000	0.000000008960	0.998999975000	16.301159	15.988709
33626	177.785549600000	0.000000000002	0.963734653000	16.301760	15.989310
33598	180.298008500000	0.000000000000	0.948728676000	16.314548	16.002098
33837	184.448659200000	0.000000053900	0.998999924000	16.339478	16.027029
32829	298.390304000000	0.250799356000	0.000000221000	16.349825	16.037375
34471	186.743617800000	0.000000032000	0.998999957000	16.351983	16.039533
33578	260.095014500000	0.000000308000	0.455010770000	16.353274	16.040824
34736	270.383119400000	0.000000000164	0.628998957000	16.353873	16.041424
34110	191.329373000000	0.000000000000	0.967778428000	16.375299	16.062850
32905	8398.813134000000	0.000000013800	0.998999983000	16.378088	16.065638
34744	244.358553600000	0.000000000042	0.673122189000	16.379217	16.066767
34606	193.040560700000	0.000000019400	0.998999974000	16.385298	16.072849
33903	197.403350100000	0.000000011500	0.998999988000	16.407419	16.094969
33440	203.921110400000	0.000000021000	0.998999976000	16.440051	16.127601
33830	245.979945100000	0.000000000437	0.455392396000	16.446996	16.134546
34446	219.303257400000	0.000000000152	0.730609895000	16.452633	16.140183
34208	226.165895200000	0.000000036300	0.998999958000	16.543528	16.231078
33761	230.444043800000	0.000000049100	0.998999942000	16.562418	16.249968
33896	251.085691900000	0.000000000006	0.695573374000	16.565647	16.253197
33713	347.868369600000	0.000000000026	0.573417844000	16.580673	16.268223
32617	275.146027400000	0.000000000000	0.780958102000	16.582979	16.270529
32904	240.426242500000	0.000000000660	0.998999989000	16.604733	16.292283
34787	278.459396700000	0.000000000011	0.688568786000	16.606385	16.293936
34667	241.472806200000	0.000000000002	0.962522599000	16.607559	16.295110
34233	241.852699800000	0.000000040000	0.998999919000	16.610784	16.298334
34212	303.974760900000	0.000000000104	0.524044102000	16.657168	16.344719

(continued)

Table 5 (continued)

ZCTA	Omega	Alpha1	Beta1	AIC	BIC
34209	257.402434100000	0.000000030200	0.998999962000	16.673047	16.360598
34759	260.479193300000	0.000000000809	0.998999989000	16.684837	16.372388
32810	368.029845700000	0.000000000092	0.658791682000	16.763442	16.450992
32837	289.334015500000	0.000000011400	0.998999982000	16.789813	16.477364
32703	294.643077200000	0.000000011500	0.990012275000	16.808023	16.495573
34769	295.007669200000	0.000000047600	0.998999946000	16.809270	16.496820
32806	303.562325300000	0.000000002910	0.998999932000	16.838025	16.525575
34786	313.480302800000	0.000000003600	0.998999902000	16.870189	16.557740
33619	383.139412200000	0.000000000000	0.685979759000	16.904711	16.592261
33990	467.279547500000	0.491287556000	0.013650631000	16.917805	16.605355
32835	332.982866400000	0.000000026200	0.998999973000	16.930409	16.617960
34711	778.949000800000	0.000000000000	0.434232261000	16.961461	16.649011
33936	345.633551600000	0.000000014900	0.998999982000	16.967600	16.655150
32792	348.740037600000	0.000000019300	0.998999842000	16.976802	16.664352
32720	369.228854500000	0.000000028500	0.990303498000	17.033685	16.721236
34746	372.332284800000	0.000000000003	0.956541690000	17.038117	16.725667
34120	374.224483500000	0.000000000000	0.962085262000	17.045873	16.733423
32771	391.034299100000	0.000000000002	0.936905232000	17.081332	16.768883
33823	388.822906300000	0.000000038500	0.998999952000	17.085414	16.772964
33610	440.916508200000	0.000000000000	0.973189652000	17.210354	16.897904
33844	440.577862900000	0.000000108000	0.998999889000	17.210420	16.897970
34221	458.974934400000	0.000000000041	0.864800458000	17.215534	16.903085
34758	448.867015300000	0.000000050800	0.998999947000	17.229096	16.916646
32824	478.108339600000	0.000000024000	0.998999974000	17.292166	16.979717
33615	485.612573400000	0.000000042700	0.998999930000	17.307856	16.995406
32825	899.719827700000	0.011841101000	0.411526802000	17.320439	17.007990
32808	515.025592500000	0.000000007400	0.998999987000	17.366611	17.054162
32828	542.449772400000	0.000000000001	0.942648184000	17.411503	17.099054
32839	598.166539700000	0.000000010200	0.998999988000	17.516067	17.203618
34741	611.508543100000	0.000000001550	0.924503899000	17.524789	17.212340
33614	643.095951600000	0.000000024100	0.998999975000	17.588604	17.276154
34135	1082.956114000000	0.000000044900	0.998999954000	18.109808	17.797359
34491	1323.187727000000	0.000000025800	0.998998982000	18.310286	17.997836
32818	1602.791844000000	0.000000000001	0.475452724000	18.394883	18.082434
34142	3013.646291000000	0.000000000000	0.878269181000	19.124306	18.811857
34116	3002.715546000000	0.000000000000	0.963648069000	19.127894	18.815444
34743	3637.545708000000	0.000000004600	0.998999940000	19.321498	19.009049

* Beta1: Persistence in volatility; values near 1 suggest that periods of high volatility are likely to continue

Secondly, the gradual reduction in spatial autocorrelation over time, as shown by declining Moran's Index values, suggests that the geographic distribution of uninsured populations is becoming more dispersed. This dispersion can inform non-spatial model refinements by indicating potential changes in the relationships between key predictors and the uninsured population over time. For example, a decrease in clustering may suggest that socioeconomic predictors (e.g., unemployment, education) are exerting more uniform effects across regions, which could affect variable significance and interaction terms in non-spatial regression models.

Thirdly, the GARCH model, which examine temporal volatility across ZCTAs, offers insight into regions with persistent or transient fluctuations in uninsured rates. This temporal-spatial information helps refine non-spatial models by identifying areas where predictor effects might vary over time due to external shocks, such as economic downturns or policy changes. Non-spatial models could integrate these temporal patterns by including interaction terms or time-varying covariates to capture the dynamic nature of the uninsured population.

Lastly, spatial models highlight the potential limitations of non-spatial models by uncovering geographic heterogeneity that could lead to residual spatial autocorrelation in non-spatial regression results. For example, if a non-spatial model underexplains variability in certain regions, spatial clustering detected by Moran's Index or hot spot analysis may indicate the need for additional geographic or policy-relevant variables. This iterative feedback loop between spatial and non-spatial analyses ensures that the models remain both methodologically robust and contextually relevant.

By integrating the geographic and temporal dimensions of uninsured rates, spatial models provide a critical lens for interpreting the results of non-spatial models. Together, these approaches create a comprehensive framework that captures both the structural predictors and the spatial dynamics of the uninsured population, leading to more informed and actionable insights for policymakers and stakeholders.

5.1 Considering Other Impacts on Population Dynamics

It is important to include the context of the data sources, such as how data were collected, in the interpretation of results that span time. For example, the 2 or more-race category saw significant increases in the uninsured population, where, within the 23 county area of central and southern Florida, there was a 436.1% increase in uninsured population of 2 or more races between the ACS 2018 5 yr (n = 31,332) to ACS 2022 5 yr (n = 168,007). The 2 or more race population grew from 262,866 (ACS 2018 5 yr) to 1,108,081 (ACS 2022 5 yr) (Table 2). A possible reason is that, compared to previous years, the 2020 Census included more detailed options and instructions within the race and ethnicity questions, allowing respondents to identify with more than one racial origin [37]. For instance, the instructions encouraged individuals to write in specific racial or ethnic identities, leading to a more nuanced understanding of racial identity, especially for those identifying with more than one race. So, the increase in the uninsured multiracial population could show a disparity in this group, but the estimate may also be a product of the changes made to accurately track this population's growth over time—leading to more people identifying with this category.

5.2 Considering Stepwise Backward Regression

The inclusion of stepwise backward regression in this analysis was employed to reduce multicollinearity, refining the set of predictors used in subsequent modeling. By iteratively removing non-significant predictors based on BIC, the process yielded a more parsimonious model, improving computational efficiency and interpretability. This initial step influenced the Poisson and Negative Binomial models by reducing redundancy among predictors, as reflected in the retention of variables at the end of the analysis due to their small VIF values.

While stepwise backward regression offers clear advantages, such as simplifying the model and mitigating overfitting, it also carries inherent risks. The reliance on BIC for variable elimination may lead to the removal of predictors that appear non-significant but are, in fact, important for capturing complex interactions or non-linear relationships in the data [20]. In the absence of stepwise regression, the inclusion of highly collinear predictors could have compromised the analysis, leading to inflated standard errors and unreliable coefficient estimates. The presence of multicollinearity would have made it challenging to discern the independent effects of each predictor, potentially obscuring key insights into the factors influencing the response variable. But, a model with excessive predictors might have overfit the data, capturing noise rather than meaningful patterns and reducing the generalizability of the results. These insights underscore the importance of complementing stepwise regression with additional diagnostic tests to validate the final model.

5.3 Limitations and Benefits

This analysis uses 5 datasets of ACS 5yr aggregate data with overlapping sample years (e.g., the 2018 5yr estimate covering 2014–2018, followed by the 2019 5yr estimate covering 2015–2019), which can obscure recent changes and complicate short-term analyses. GARCH modeling can identify volatility within those year groups, but further investigation in those years with high volatility will be important to assess. Another limitation of ACS data is due to the nature of the counts being estimates that do not always reflect the larger population and become less stable as the spatial scale increases. ACS data come with Margins of Error (MOE), which give an indication of estimates stability [48]. Statistical methods, including Poisson and negative binomial regression and diagnostic tools like AIC and BIC, can address these complexities and strengthen model stability, optimizing the dataset's analytical utility within its limitations [2].

To address the MOE, a future direction for analysis could involve observing the MOEs along with the beta1 GARCH values to see if there is an association between MOE size and beta1 values approaching 1—where values close to 1 indicate higher volatility [18]. Incorporating MOEs as weights in a regression model could allow for a weighted regression approach, emphasizing more reliable estimates with smaller MOEs while minimizing the influence of less precise data.

The use of ZCTAs introduces another element of uncertainty due to being polygons created in a standardized way to represent ZIP code-like areas in a consistent geographic framework using census blocks [34, 23]. They are meant to be more stable, maintaining over time, compared to mail zip codes, but they also exclude zip codes that represent P.O. boxes [51]. ZCTAs, even with their error, are still utilized by multiple health entities, including insurance companies [38], the Centers for Disease Control and Prevention (CDC) [31], and the CDC's PLACES project [12]. There are a lot of health data aggregated at the ZCTA level, so we should have methods of analyzing them, which includes understanding their risks, including diagnostic tests, and including those assessments when interpreting the results. Methods used in this analysis can also be used on census tract or block group level data.

In terms of benefits, the inclusion of spatial tests, including Moran's I and Getis-Ord Gi*, add geographic context by identifying clustering patterns and hot or cold spots of uninsured rates. Moran's I detects overall spatial autocorrelation, while Getis-Ord Gi* identifies specific high- or low-uninsured areas, enhancing the demographic and temporal insights with a geographic lens [5]. GARCH modeling adds a temporal layer by capturing cycle-over-cycle volatility in uninsured rates, revealing areas with significant fluctuations.

This multidimensional approach balances each method's strengths and limitations: the linear trend assumption in the Cochran-Armitage Test, overdispersion in Poisson regression, potential bias in stepwise selection, and moderate fit in Bayesian models with stepwise or Poisson frameworks. The refit Poisson model, with high R^2 scores across years, provides a consistently strong fit for uninsured populations. Together, these methods provide a comprehensive view of uninsured trends, addressing count data challenges, spatial patterns, and temporal volatility.

6 Conclusion

This methodological framework provides a comprehensive process for diagnosing and addressing linear, spatial and temporal errors in geostatistical modeling. The proposed methods highlight the importance of addressing overdispersion, spatial dependencies, and temporal heteroskedasticity in health data analysis, increasing the reliability of the results for public health decision-making.

Observing test assumptions and diagnosing them during the analytical process is crucial for understanding how data interact with statistical models. Violations of assumptions such as equidispersion, independence, or absence of autocorrelation can significantly affect the interpretation and reliability of model outcomes. Diagnosing these issues early in the analysis allows for adjustments that enhance model accuracy and ensures that the relationships between predictors and outcomes are accurately represented. This process not only mitigates bias and improves predictive power but also provides deeper insights into the underlying data structure, guiding meaningful interpretation and robust conclusions.

References

1. Agresti A (2012) Categorical data analysis (Vol 792). John Wiley & Sons
2. Aho K, Derryberry D, Peterson T (2014) Model selection for ecologists: the worldviews of AIC and BIC. Ecology 95(3):631–636. https://doi.org/10.1890/13-1452.1
3. Akaike H (1992) Information theory and an extension of the maximum likelihood principle. In Kotz S, Johnson NL (Eds), Break Stat, (29: 6, pp 610–624). Springer. https://doi.org/10.1007/978-1-4612-0919-5_38
4. Alberto CK, Pintor JK, Langellier B, Tabb LP, Martínez-Donate AP, Stimpson JP (2020) Association of maternal characteristics with latino youth health insurance disparities in the United States: a generalized structural equation modeling approach. Abstract BMC Public Health 20(1). https://doi.org/10.1186/s12889-020-09188-1
5. Anselin L (1995) Local indicators of spatial association—LISA. Geogr Anal 27(2):93–115. https://doi.org/10.1111/j.1538-4632.1995.tb00338.x
6. Arashi M, Rounaghi M (2022) Analysis of market efficiency and fractal feature of NASDAQ stock exchange: Time series modeling and forecasting of stock index using ARMA-GARCH model. Futur Bus J, 8(1). https://doi.org/10.1186/s43093-022-00125-9
7. Barber C, Lamontagne JR, Vogel RM (2020) Improved estimators of correlation and R2 for skewed hydrologic data. Hydrol Sci J 65(1):87–101. https://doi.org/10.1080/02626667.2019.1686639
8. Breslow NE, Clayton DG (1993) Approximate Inference in Generalized Linear Mixed Models. J Am Stat Assoc 88(421):9. https://doi.org/10.2307/2290687
9. Callahan ST, Cooper WO (2004) Gender and uninsurance among young adults in the United States. Pediatrics 113(2):291–297. https://doi.org/10.1542/peds.113.2.291
10. Cameron AC, Trivedi PK (2013) Regression analysis of count data. In AC. and Cameron TPK (Ed), Regres Anal Count Data. Cambridge University Press
11. Casella George, Berger RW (2024) Statistical inference. Chapman & Hall
12. Centers for Disease Control and Prevention (CDC). (2024) About PLACES: local data for better health. PLACES. CDC. https://www.cdc.gov/places/about/index.html
13. Cohen J, Cohen P, West SG, Aiken LS (2013) Applied multiple regression/correlation analysis for the behavioral sciences, third edition. Taylor and Francis. https://doi.org/10.4324/9780203774441
14. DeGroot M (1986) Probability and statistics (DeGroot M, Schervish M, Eds, 2nd ed). Addison-Wesley. http://thuvienso.thanglong.edu.vn/handle/TLU/11620
15. Draper NR, Smith H (1998) Applied regression analysis. In John Wiley & Sons, Inc. (3rd ed, 326)
16. Efeizomor RO (2023) A comparative study of methods of remedying multicolinearity. Am J Theor Appl Stat, 12(4), 87–91. https://doi.org/10.11648/J.AJTAS.20231204.14
17. Engle RF (1982) Autoregressive conditional heteroscedasticity with estimates of the variance of United Kingdom inflation. Econ 50(4):987. https://doi.org/10.2307/1912773
18. Fatima S, Uddin M (2022) On the forecasting of multivariate financial time series using hybridization of DCC-GARCH model and multivariate ANNs. Neural Comput Appl 34(24):21911–21925. https://doi.org/10.1007/S00521-022-07631-5

19. Finegold K, Conmy A, Chu R, Bosworth A, Sommers B (2010) Trends in the U.S. uninsured population, 2010–2020 (Issue Brief). U.S. Department of Health and Human Services, Office of the Assistant Secretary for Planning and Evaluation. https://aspe.hhs.gov/reports/trends-us-uninsured-population-2010-2020
20. Flom P, Cassell D (2007) Stopping stepwise: Why stepwise and similar selection methods are bad, and what you should use. NorthEast SAS users group Inc 20th annual conference, 11–14
21. Gareth J, Daniela W, Trevor H, Robert T (2013) An introduction to statistical learning: with applications in R
22. Griffith DA (2003) Spatial Filtering. In Spat Autocorrelation Spat Filtering Adv Spat Sci (pp 91–130). Springer, Berlin, Heidelberg. https://doi.org/10.1007/978-3-540-24806-4_4
23. Grubesic TH, Matisziw TC (2006) On the use of ZIP codes and ZIP code tabulation areas (ZCTAs) for the spatial analysis of epidemiological data. Int J Health Geogr 5(1):1–15. https://doi.org/10.1186/1476-072X-5-58/FIGURES/7
24. Haight FA (1967) Handbook of the poisson distribution. John Wiley & Sons
25. Hanks EM, Schliep EM, Hooten MB, Hoeting JA (2015) Restricted spatial regression in practice: geostatistical models, confounding, and robustness under model misspecification. Wiley Online Library 26(4):243–254. https://doi.org/10.1002/env.2331
26. Hosmer DW, Lemeshow S (1980) Goodness of fit tests for the multiple logistic regression model. Commun Stat-Theory Methods 9(10):1043–1069. https://doi.org/10.1080/03610928008827941
27. Jacob BG, Izureta R, Bell J, Parikh J, Loum D, Casonova J, Gates T, Murray K, White L, Aceng JR (2023) Approximating non-asymptoticalness, skew heteroscedascity and geo-spatiotemporal multicollinearity in posterior probabilities in bayesian eigenvector eigen-geospace for optimizing hierarchical diffusion-oriented COVID-19 random effect specifications geosampled in uganda. Am J Math Stat 13(1):1–43
28. James G, Witten D, Hastie T, Tibshirani R (2023) Linear regression. In An Introd Stat Learn: Appl Python (pp. 69–134). Springer International Publishing. https://doi.org/10.1007/978-3-031-38747-0_3
29. James G, Witten D, Hastie T, Tibshirani R, Taylor J (2023) An introduction to statistical learning. Springer International Publishing. https://doi.org/10.1007/978-3-031-38747-0
30. Kennedy J, Wood E, Frieden L (2017) Disparities in insurance coverage health services use and access following implementation of the affordable care act: a comparison of disabled and nondisabled working-age adults INQUIRY. J Health Care Finance 54. https://doi.org/10.1177/0046958017734031
31. Kind A, Buckinghamd W (2018) Making neighborhood-disadvantage metrics accessible—the neighborhood atlas. N Engl J Med, 278(26), 2456. https://www.ncbi.nlm.nih.gov/pmc/articles/PMC6051533/
32. King G (2016) Unifying Political Methodology. University of Michigan Press, In Unifying Political Methodology. https://doi.org/10.3998/MPUB.23784
33. Konishi S, Kitagawa G (2008) Information Criteria and Statistical Modeling. Springer, New York. https://doi.org/10.1007/978-0-387-71887-3
34. Krieger N, Waterman P, Chen JT, Soobader MJ, Subramanian SV, Carson R (2002) Zip code caveat: Bias due to spatiotemporal mismatches between zip codes and US census-defined geographic areas - The public health disparities geocoding project. Am J Public Health 92(7):1100–1102. https://doi.org/10.2105/AJPH.92.7.1100
35. Lee DC, Liang H, Shi L (2021) The convergence of racial and income disparities in health insurance coverage in the United States Abstract. Int J Equity Health 20(1). https://doi.org/10.1186/s12939-021-01436-z
36. Lepot M, Aubin JB, Clemens FHLR (2017) Interpolation in time series: an introductive overview of existing methods, their performance criteria and uncertainty assessment. Water 9, 796, 9(10), 796. https://doi.org/10.3390/W9100796
37. Marks R, Rios-Vargas M (2021) Improvements to the 2020 census race and hispanic origin question designs, data processing, and coding procedures
38. MedPAC (2018) Report to the congress: medicare and the health care delivery system

39. Montez JK, Cheng KJ (2022) Educational disparities in adult health across U.S. states: Larger disparities reflect economic factors. Frontiers in Public Health, 10, 966434. https://doi.org/10.3389/FPUBH.2022.966434/BIBTEX
40. Moran PAP (1950) Notes on continuous stochastic phenomena. Biometrika 37(1/2):17. https://doi.org/10.2307/2332142
41. Payne EH, Hardin JW, Egede LE, Ramakrishnan V, Selassie A, Gebregziabher M (2015) Approaches for dealing with various sources of overdispersion in modeling count data: Scale adjustment versus modeling. Stat Methods Med Res 26(4):1802–1823. https://doi.org/10.1177/0962280215588569
42. Rodriguez-Aguilar M, Marmolejo-Saucedo L, Rodriguez-Aguilar R, Marmolejo-Saucedo JA (2024) Volatility in private health services and supplies in Mexico, 2002–2023. EAI/Springer Innov Commun Comput, 157–168. https://doi.org/10.1007/978-3-031-67440-2_13
43. Ross SM (2014) Introduction to Probability Models, Eleventh. Eleventh Edition. Elsevier Inc., In Introduction to Probability Models. https://doi.org/10.1016/C2012-0-03564-8
44. Rowland M, Peterson-Besse J, Dobbertin K, Walsh ES, Horner-Johnson W (2014) Health outcome disparities among subgroups of people with disabilities: a scoping review. Disabil Health J 7(2):136–150. https://doi.org/10.1016/j.dhjo.2013.09.003
45. Sheather S (2009) A modern approach to regression with R. Springer Sci & Bus Media
46. Shi L, Stevens GD (2005) Disparities in access to care and satisfaction among U.S. children: The roles of race/ethnicity and poverty status. Public Health Reports, 120(4), 431–441. https://doi.org/10.1177/003335490512000410
47. Shrestha N (2020) Detecting multicollinearity in regression analysis. Am J Appl Math Stat 8(2):39–42
48. Spielman S, Folch D, Nagle N (2014) Patterns and causes of uncertainty in the American Community Survey. Appl Geogrpahy, 46, 147–157. https://doi.org/10.1016/j.apgeog.2013.11.002
49. Stanley NB, Burus T, Oh DL, Tribby CP, Mcnair C, Delnero P, Tossas KY (2024) Identifying and addressing the needs of catchment area data experts: Insights from the inaugural catchment area data conference (CADC23). Taylor & Francis 2025(1):2416192. https://doi.org/10.1080/28322134.2024.2416192
50. Sørensen J, Sorenson O, Winship C, Western B (2016) Multicollinearity and model misspecification. Sociol Sci, 3, 627–649. https://doi.org/10.15195/V3.A27
51. US Census Bureau (2023) ZIP code tabulation areas (ZCTAs). U.S. census surveys and progams. https://www.census.gov/programs-surveys/geography/guidance/geo-areas/zctas.html
52. Yabroff K, Zhao J, Halpern M, Fedewa S, Han X, Nogueira L, Zheng Z, Jemal A (2021) Health insurance disruptions and care access and affordability in the U.S. Am J Prev Med 61(1):3–12. https://doi.org/10.1016/j.amepre.2021.02.014
53. Zuur AF, Ieno EN, Walker N, Saveliev AA, Smith GM (2009) Mixed effects models and extensions in ecology with R. Springer, New York. https://doi.org/10.1007/978-0-387-87458-6

Anticipatory Innovation for Strengthening Pandemic Preparedness and Response: Tech Enabled Predictive Pandemic Intelligence for Capability Planning

Edwin Michael and Anthony J. Masys

Abstract COVID-19 was not a black swan event. The emergence and spread of COVID-19 have dramatically highlighted the vital interest for the US and the global community to counter and manage the persistent risk of the biosecurity threats posed by infectious diseases, whether arising naturally, through accidents with high-consequence pathogens, or via deliberate release of harmful bioagents (Miley in Global health security: recognizing vulnerabilities, creating opportunities. Springer Publishing, 2020). For years stemming from our experience with SARS, H1N1 and Ebola, Public Health and foresight experts have been calling attention to the global and national security impacts of global outbreaks. COVID-19 has in fact stress tested our national and global security revealing inherent vulnerabilities within our societal systems and infrastructure in addition to vulnerabilities and shortcomings associated with our mitigation, preparedness and response capabilities. Given the impacts of COVID-19 on national and global security and the risks pertaining to the increasing frequency and severity of upcoming pathogens, there is an urgent demand for novel approaches and capabilities to detecting and responding to pandemic threats. As described in Tõnurist and Hanson (Anticipatory innovation governance: Shaping the future through proactive policy making. OECD working papers on public governance, no 44. OECD Publishing, Paris, 2020. https://doi.org/10.1787/cce14d80-en.) '…even though foresight tools are increasingly integrated into policymaking, governments often lack a practical understanding of how to anticipate uncertain futures but also how to act on them today to achieve desired outcomes'. Our premise is that assessing and tackling the future risks stemming from pandemic threats requires innovative, collaborative approaches that draw on diverse capacities in the areas of AI-driven disease sensing, large-scale data-driven epidemic modelling, multi-scaled projections of socio-ecological risks, big data, systems and futures thinking, policy analysis, human resource development, and knowledge exchange.

E. Michael · A. J. Masys (✉)
College of Public Health, University of South Florida, Tampa, USA
e-mail: tmasys@usf.edu

A. J. Masys
Global and National Security Institute (GNSI), University of South Florida, Tampa, USA

B. Jacob et al. (eds.), *Tech Enabled Global Health Security*, Advanced Sciences and Technologies for Security Applications, https://doi.org/10.1007/978-3-031-86997-6_9

This chapter presents a **complexity-informed inquiry** challenging traditional mindsets and practices to support anticipatory innovation that enables next generation pandemic preparedness and response capability-based planning.

Keywords Pandemic preparedness · Complexity · Digital Twin · Agent based models · Anticipatory innovation

1 Introduction

COVID-19 was not a black swan event. For years stemming from our experience with SARS, H1N1 and Ebola, Public Health and foresight experts have been calling attention to the global and national security impacts of global outbreaks. Apart from causing over 14.8 million deaths globally and continuing long-term morbidity [32], the pandemic also underscored the vast societal damage that such threats can generate—with studies indicating that the '…COVID-19 pandemic triggered the sharpest downturn in the world economy since the Great Depression, with global GDP declining 3.0% in 2020 [9]'.

The risk of another large-scale pandemic occurring in the next few decades is ever-present and is increasing. COVID-19 revealed the existing vulnerabilities of the global and the US's pandemic surveillance, mitigation, preparedness, and response capabilities. These challenges mean that strengthening our capabilities to assess and address the security and health threats posed by these diseases also requires recognition that these risks constitute complex collective intelligence (CI) challenges surrounding surveillance and predictive pandemic intelligence (rooted in a capability framework).

As described in Tõnurist and Hanson [31] '…even though foresight tools are increasingly integrated into policymaking, governments often lack a practical understanding of how to anticipate uncertain futures but also how to act on them today to achieve desired outcomes', hence the requirement for anticipatory innovation.

Anticipatory innovation [24] is the act of creating and implementing new, potentially value-shifting innovations in environments of deep uncertainty, particularly for the purpose of exploration and with emergent issues that might shape future priorities and future commitments. Our premise is that assessing and tackling the future risks stemming from pandemic threats requires innovative, collaborative approaches that draw on diverse capacities in the areas of AI-driven disease sensing, large-scale data-driven epidemic modelling, multi-scaled projections of socio-ecological risks, big data, systems and futures thinking, policy analysis, human resource development, and knowledge exchange.

This chapter presents a **complexity-informed inquiry** challenging traditional mindsets and practices to support anticipatory innovation that enables next generation pandemic preparedness and response capability. This is realized through the development of an Anticipatory Innovation Framework, harnessing of new developments in Tech-enabled Intelligence, and a Next Generation Capability-based Planning Framework.

2 Pandemic Threats: An Issue of National Security and Biodefence Capabilities

The COVID-19 pandemic has highlighted the inherent fragility in our health security systems [22] pointing to the necessity for redesigned capabilities to be better prepared to cope with multiple and diverse health threats. A Chatham House report, 'Preparing for High Impact, Low Probability Events', found that governments and businesses remain unprepared for such events [15]. Helbing [11: 51] argues that:

> Many disasters in anthropogenic systems should not be seen as 'bad luck', but as the results of inappropriate interactions and institutional settings. Even worse, they are often the consequences of a wrong understanding due to the counter-intuitive nature of the underlying system behaviour.

As described in Houser [13] 'The terror attacks of September 11, 2001 uncovered various weaknesses in the national public health infrastructure that have persisted into the current pandemic'. With this in mind, there is an urgent need to re-examine our exposure to security risks and vulnerabilities, in order to transform our understanding and perception of security. The ability to improve preparedness of health security systems under VUCA (Volatility, Uncertainty, Complexity, Ambiguity) conditions relies on the development of capabilities to 'future proof' health security systems against biothreat scenarios. Achieving such strategic objectives associated with biosecurity call upon the application of Capability-Based Planning (CBP). Emerging from the defence community CBP '…aims to ensure that military forces are equipped with the right mix of capabilities to effectively respond to a wide range of potential threats and challenges. By adopting a CBP approach, military organizations aim to be more flexible, adaptable, and responsive to changing circumstances' [23]. A capability is defined as:

> The ability to achieve a desired effect under specified standards and conditions through a combination of means and ways across the Doctrine, Organization, Training, Materiel, Leadership and Education, Personnel, and Facilities (DOTMLPF) to perform a set of tasks to execute a specified course of action. Capabilities result from a combination of tasks. (cited in Samaras and Willis [28])

From a capability mindset perspective, key COVID-19 lessons learned described in Williams et al. [32] include:

- Pandemic preparedness requires comprehensive thinking, from the earliest stages of research to enable effective development of countermeasures, to monitoring pandemic threats, and to coordination of pandemic response efforts in real-time. The time for action is during non-pandemic times, enabling a more effective response in an emergency.

- Bolster pandemic pathogen intelligence: Detecting novel pathogens as they arise allows for the earliest possible response, so surveillance systems should be expanded and more extensively leveraged to better detect and respond to infectious disease outbreaks in real time.
- Optimize and de-risk earlier pandemic interventions: Non-pharmaceutical interventions (NPIs) saved lives but are not without limitations and consequences. Persistent long-term reliance on NPIs can be challenging because people grow tired and apathetic toward them.

Specifically, the pandemic has highlighted that health security systems must be redesigned to 'future-proof' society from multiple and diverse health threats (both well-known and unexpected). Such capability can be expressed through a complexity-informed tech enabled approach that draws on diverse capacities in the areas of AI-driven disease sensing, large-scale data-driven epidemic modelling, multi-scaled projections of socio-ecological risks, big data, systems and futures thinking, policy analysis, human resource development, and knowledge exchange.

3 Futures Thinking: Anticipatory Posturing

As described in Tonurist et al. [31: 31] 'Anticipation does not mean predicting the future, but rather it is about asking questions about plausible futures so that we may act in the present to help bring about the kind of futures we decide we want…It is a capacity connected to engaging with alternative futures, based on sensitivity to weak signals, and an ability to visualize their consequences, in the form of multiple possible outcomes …'. As presented in Masys [16], '…anticipatory innovation emerges from a complexity-informed inquiry at the intersection of futures thinking (exploring the possibility space), systems thinking (understanding the complex global health security landscape) and design thinking (a problem-solving approach that is human centered, possibility driven, option focused)'. Within the context of pandemics and emerging diseases in shaping national security capabilities, anticipation supports an opportunity to assess mental models, practice, rehearse, develop and validate capabilities.

The next section highlights the leveraging of tech enabled scenario analysis as a backdrop to the development of a CBP framework.

4 The Digital Twin (DT) as an Intelligence Tool for Pandemic Planning and Response

Over the past few decades, the world has become increasingly interconnected, creating all kinds of complex adaptive systems [3, 14]. Such global coupling has led to many advantages but has also resulted in the formation of distributed risks that can emerge and propagate rapidly through interacting international and regional social and physical networks, nodes, feedback loops and synchronizing factors [14]. Pandemics, such as the ongoing COVID-19, highlight the risk that is possible with the increased coupling and complexity of these multi-layer interacting networks, with increasing work showing how risk flows can interact with node or state features to either amplify or diffuse the emergent shocks both globally and locally [10].

The need to address the effects of the above features that drive global risks, such as that embodied by a pandemic, means that simulation tools are required to capture the combined effects of the many intertwined human as well as the physical networks and their interactions that govern the patterns and intensity of pathogen spread through space and over time in a population. The complexity and emergent properties of the local interactions that occur among multiple elements of such a system, including its complex behaviour, adaptivity, creativity and response to infectious risk, implies that computational systems that can provide insights into the driving mechanisms and the collective outcome of these interactions are needed if better intelligence for pandemic planning and response is to be generated [3]. Purely top-down equilibrium-based models do not suffice, as they cannot easily capture the myriad mechanisms by which this type of collective system behaviour appears and manifests itself over space and time [19]. They also are not able to characterize fully how top-down control can interact with bottom-up evolution to provide better solutions to the problems produced by pathogen emergence or importation and propagation among the multi-faceted components that underlie pandemic transmission dynamics in different communities.

4.1 The Digital Twin as an Enabling Technology for Simulating Complex Phenomena

The Digital Twin (DT) is a relatively new technology that has gained growing interest for the study of infectious diseases since the start of the COVID-19 pandemic [4, 25]. In simple terms, it can be defined as a virtual replica of a physical object of any scale and all its relevant interacting components and characteristics that also allow the transmission of real-time data between the two (ie. physical and virtual) entities [6, 8, 26, 30]. The definition of the DT has evolved since the concept was first proposed in 2003 in the manufacturing industry, with NASA extending it to model complex systems in 2011. Thus, it was first proposed as a simple way to replicate product life cycle management, then it was implemented to simulate aerodynamic vehicles as complex systems by NASA, and currently it is a technology used to model and predict the performance of any physical system of high complexity in up to five dimensions [21, 26]. It is now becoming the main technology used to analyse cities as complex systems with various interacting sub-systems in order to foster understanding of real-time urban dynamics and the fate of smart cities [2].

4.2 CitySEIRCast

The above features of place digital twins (pDTs), particularly their ability to capture in real-time the collective impacts of the myriad physical and human networks that govern the operation and dynamics of complex phenomena in a populated setting, has indicated that coupling the appropriate network-informed disease transmission models with such DTs can provide a new environment for improving the real-time simulation of pathogen propagation in real-world settings [1, 4, 25]. Towards this end, we have developed CitySEIRCast, a place DT coupled with an agile agent-based model (ABM) of epidemic transmission, to serve as a virtual environment for modeling the emergence/importation, spread and management of a pandemic among the diverse subpopulations and locations of an urban locality, focussing on COVID-19 in the first instance as an example [5]. Figure 1 illustrates the building blocks of CitySEIRCast. First, it comprises the relevant models for the physical elements of a place and the disease under consideration. This includes models for creating the virtual environment of the study setting based on the concepts of city information modelling (CIM) [29], and algorithms for generating synthetic populations (see details given in Bilal et al. [5]). The third model is the ABM-based simulation framework for COVID-19 which describes the states, processes, and parameters governing infection progression through individuals [5]. The second block is a focus on capturing and managing evolving data on locations, routes and movement, human behaviour and epidemiological data pertinent to tracking the epidemic and calibration of the ABM model. This focus is critical because these conditions are likely to change over time and thus will not only be key for supporting real-time disease spread assessment and policy decisions, but also for improving understanding of the epidemic impact of systemic changes in disease drivers and management responses over longer time scales. A key need here is the need for agile data technologies, data management, and modeling protocols for facilitating dynamic updates to the simulation models so that these can be adapted and changed as physical and disease conditions change [1, 5]. Such dynamic correspondence linking the DT architecture with models and data makes it possible not only to monitor and synchronise the state of the DT with the physical environment and human population being mirrored but also to assess how these changes can in turn affect pandemic transmission. Third, CitySEIRCast is founded on a central entity-data broker to integrate the various components together and execute tasks such as data acquisition, mangling, exchange, model calibration, validation and simulation. This includes a data registry for administering different sources of information (Fig. 1), a data catalogue for storing descriptions of data sources in the form of schemas, knowledge graphs, and metadata, an event module for handing movement and behavioural data and to execute calibrations and simulations, and a security module for user authentication and access management. Finally, execution of the ABM and physical entity co-simulations are handled using high performance computing based on MPI and Open MPI parallel computing standards and libraries [5]. The fourth element of CitySEIRCast is a scalable web interface

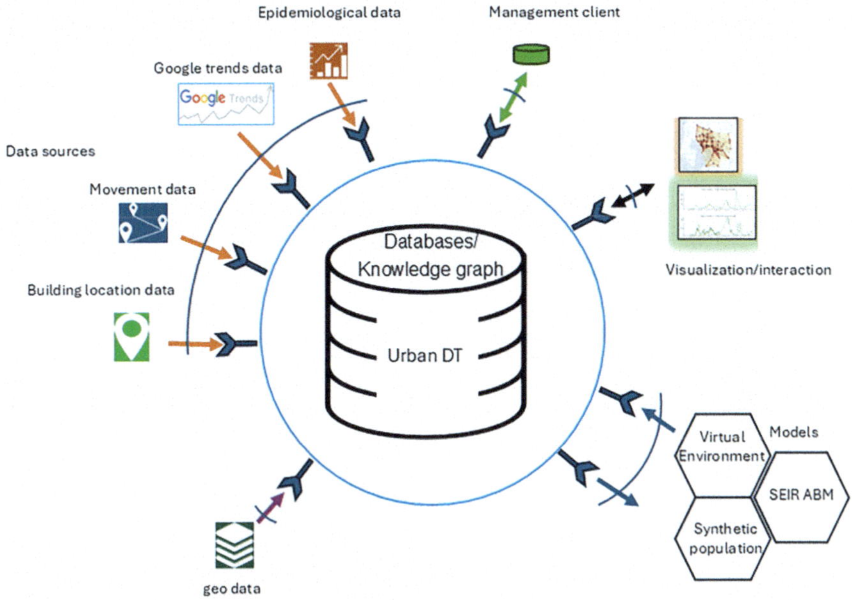

Fig. 1 Architecture of CitySEIRCast onto which different data sources, models, visualization clients, and other components connect

for analysts and users to set up, run, visualize, and analyse computer experiments (Fig. 1).

4.3 Performing Complexity-Informed Pandemic Simulations, Forecasts, and Management.

Figure 2 shows how the City Information Modeling Module of CitySEIRCast is able to faithfully create a 3-D model of the urban environment representative of Hillsborough County, while Fig. 3 indicates how the synthetic population generator is able similarly to closely mimic the County's real population. CitySEIRCast further uses an origin–destination (OD) matrix and distance kernels derived from both Tom Tom's real time Floating Car Data and public transportation usage data to inform the mobility of agents in the DT [5]. The OD information is used to assign workplaces and schools to the fractions of employed adults and school-going children within and beyond their original resident zipcodes, whereas the estimated distance kernel measures are used to model movement to community spaces. Finally, as these movements are used by the ABM to simulate the motion of agents in the virtual environment through a simulation time step of 6 h, we further divide each day into four time segments (morning, afternoon, evening and night) and synthesize agent mobility patterns for each time period. This allows us to simulate the daily activity of agents according to the time of the day as well as when different categories of buildings are open at different times during the day.

Figure 4 shows a sample of movements made by agents to different locations as they carry out their normal activities in a section of Hillsborough County, highlighting the differential agent movement patterns that can be expected in that particular location during the course of a day.

The evolution of daily COVID-19 cases and hospitalizations predicted (filled curves) for Hillsborough County by the DT-ABM taking full account of the social protective measures, vaccinations, and changes in behavioural and mobility responses that occurred during the course of the pandemic in the county is shown in Fig. 5 compared to observed data (black curves). The results indicate that CitySEIRCast

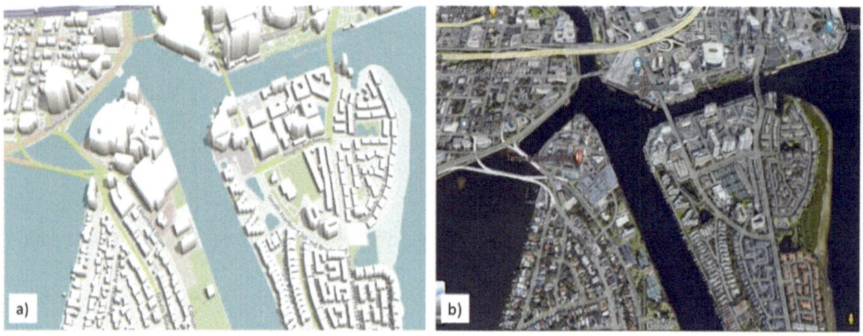

Fig. 2 **a** A section of Hillsborough County showing 3D building locations, shapes and sizes. **b** The satellite image of the same area

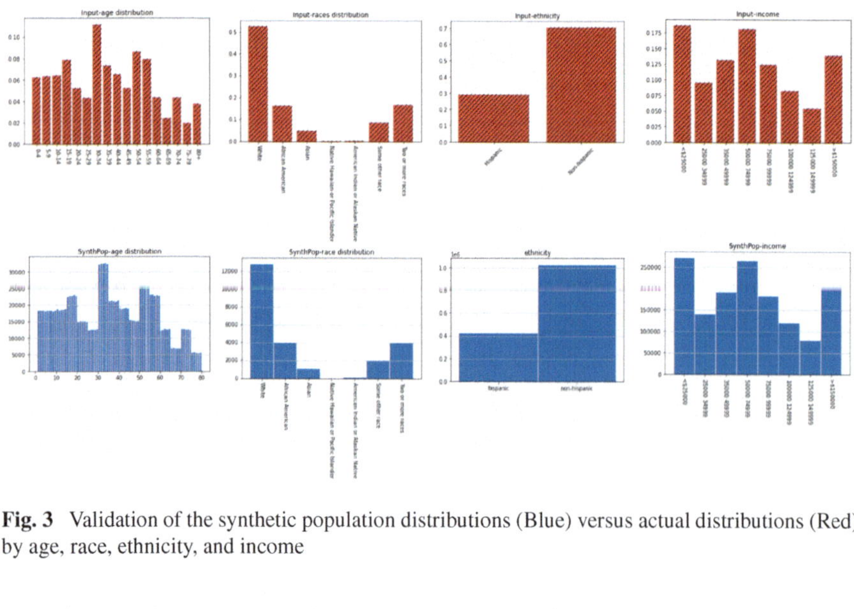

Fig. 3 Validation of the synthetic population distributions (Blue) versus actual distributions (Red) by age, race, ethnicity, and income

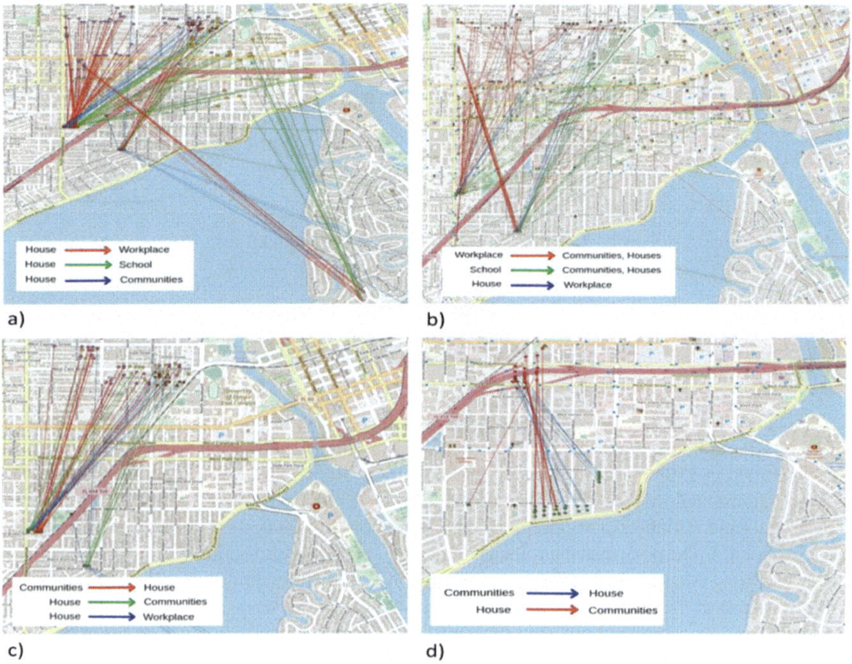

Fig. 4 Agent movement from and to different places during 6-h time blocks. **a** 7 a.m.–12 p.m., **b** 1 p.m.–6 p.m., **c** 7 p.m.–12 a.m., **d** 12 a.m.–6 a.m. 12 p.m., **b** 1 p.m.–6 p.m., **c** 7 p.m.–12 a.m., **d** 12 a.m.–6 a.m.

was able to simulate the average as well as the variance in the course of cases and hospitalizations caused by SARS-CoV-2 across the zipcodes of the County reasonably well. Furthermore, given that it is also able to capture and track the geographic spread of the outbreak in the County, it provides the facility to address the spatial patterns of evolution of the pandemic. Figure 6 portrays the spatial distribution of the cases predicted at the zipcode level and indicate that significant heterogeneity or spatial asynchrony between zip codes emerged early during the pandemic, which persisted over time (high infection zipcodes remaining highly infected) throughout the duration of the pandemic. Figure 7 demonstrates another important utility of employing a place DT to run a disease ABM, viz. that it also allows investigation of the population attributes that may underlie the spatial distribution of disease transmission. The figure plots the fractions of agents exhibiting different attributes that were found to be associated with COVID-19 transmission in a sample of high incidence versus low incidence zipcodes. The results show that among the agent attributes, the higher rate of transmission observed in the high incidence zip codes is associated with individuals belonging to >18 year groups, the Hispanic ethnic group, poorer income brackets and greater travel outside the resident zip code.

A key feature afforded by the DT-ABM is the ability to support the performance of scenario analyses helpful for guiding policy choices [7, 8, 27]. An example of this

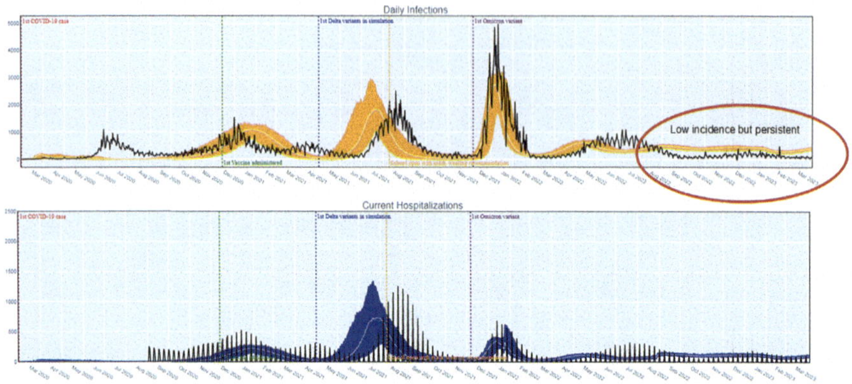

Fig. 5 Daily infected cases and hospitalizations across Hillsborough County. An ensemble of 10,000 sets of parameters from prior distributions were simulated and a selection criterion based on RMSE resulted in 159 best fitting (shaded region) curves. Reported daily infection/hospitalization cases are represented by black lines. Vertical dashed lines reflect the dates of arrival of the major virus variants and initiation of vaccinations (shaded region) curves against reported daily infection cases (black line). The prior distribution were simulated and a selection criterion by RMSE resulted in 159 best fitting (shaded region) curves against reported daily infection cases (black line). Vertical dashed lines reflect the dates of arrival of the major virus variants. Shaded blue region indicates the 99 percent confidence interval obtained from ensemble of simulations using parameter priors as in Table 1. An ensemble of 10,000 sets of parameters from the prior distribution were simulated and a selection criterion by RMSE resulted in 159 best fitting (shaded region) curves against reported daily infection cases (black line). Vertical dashed lines reflect the dates of arrival of the major virus variants

functionality related to vaccination strategy is depicted in Fig. 8, in which outcomes of two hypothetical preferential vaccination scenarios, viz. either vaccinating only low-income groups (those below the US Federal Poverty Level (FPL)) or vaccinating workers alone, are compared. In both scenarios, the preferential vaccination causes an initial drop in the daily cases due to the protective effect of vaccination on those contributing most to the transmission dynamics due to their movement and mixing. However, due to the faster waning of vaccine induced immunity relative to the waning rate of natural immunity, both groups will lose their immunity status, and upon the release of lockdown measures, will facilitate the earlier rise of the subsequent epidemic wave. The wave created by the workers, however, will be larger than the FPL groups since they represent a larger fraction of the population. This result, and the fact that the differential impact on cases in the two groups will dissipate as natural immunity in the entire population builds up (see predictions beyond Jan 2022), indicates that random vaccination may have a better outcome than preferential vaccination.

These findings, overall, highlight how by acting as a digital mirror of the local environment and populations through continuous monitoring, DT-ABMs may provide a powerful technology for both tracking the complex course and distributional effects

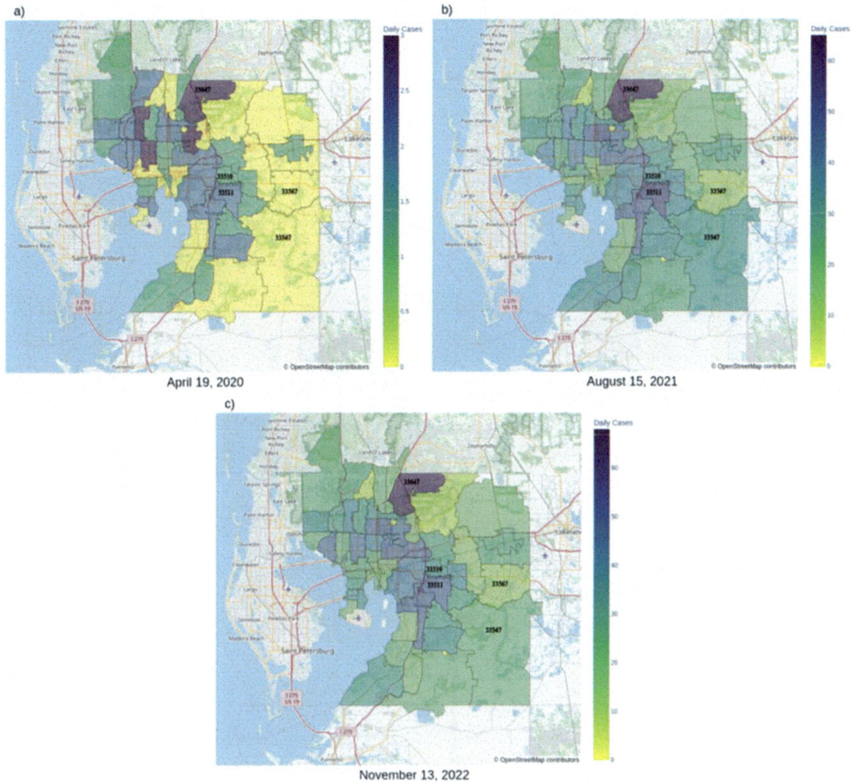

Fig. 6 Maps of infection counts across zip codes on three dates (19th April 2020, 15th August 2021, and 13th November 2022), showing spatial asynchrony between zip codes

of pandemics and for testing of strategies in conditions closely reassembling the real-world pandemic environment. Additionally, it is also clear that DT-based epidemic modelling systems can enable the development and operation of more effective multi-actor collaborations. As pandemic and local setting dynamics are visualized, these co-simulation systems have the potential to coalesce stakeholders, including ordinary citizens, around the challenges in planning and responding to epidemics, and define and cocreate solutions before testing them for impact in a transparent manner. DT-ABMS can thus act as a civil tech for promoting participation, understanding roles of complexity and uncertainty, improving response and planning capacities, and guiding collective assessment and exploration of emergent biothreat risks and their management. Such fostering of collective intelligence could be key to supporting the development of more systemic preparedness approaches and strategies for lowering the vulnerability and increasing resilience of communities to ongoing and future biothreat risks.

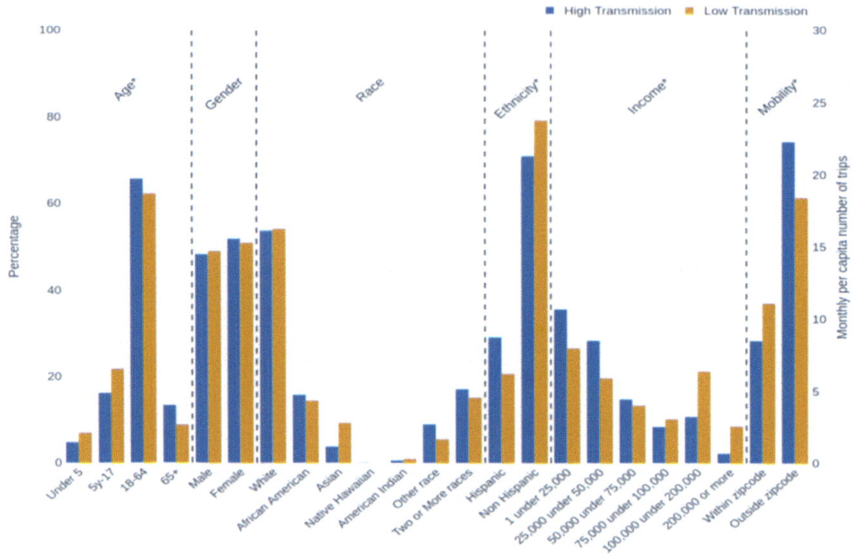

Fig. 7 Population characteristics of low versus high transmission zip codes

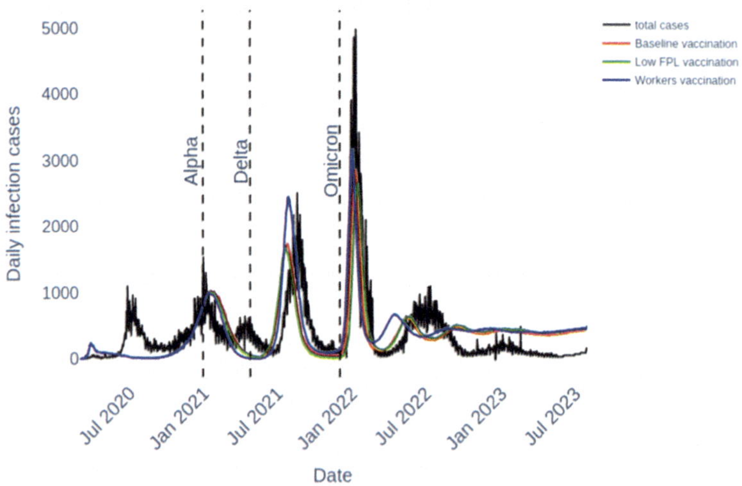

Fig. 8 Daily infected cases across the Hillsborough County for vaccination scenarios. The red curve indicates the simulation curves for baseline vaccination while green and blue indicated scenarios where preferential low FPL and workers are preferentially vaccinated before the rest of the population. The reported data is given by the black curve

5 Anticipatory Innovation Framework

The example of the *CitySEIRCast* case study illustrates the predictive, exploratory and 'intervention flight simulator' capability of such a tech-enabled scenario approach. In planning, it is essential to consider a broad range of future challenges within a 'scenario space' (a cone of plausibility) as illustrated in Fig. 9.

The scenario space is examined through a complexity lens that leverages systems thinking, futures thinking and design thinking [16]. This approach supports a red-teaming mindset [17] that can be used to understand a situation from an alternative perspective and explore the range of possible reactions to a policy implementation and the introduction of a capability. In so doing, this approach can:

- uncover hidden biases;
- challenge assumptions and beliefs;
- identify flaws in logic;
- widen scope of information searches;
- identify different options and alternatives; and
- stress-test a plan [20].

This 'scenario space' is explored via the ABM/DT applications described earlier. In this way, the ABM/DT becomes an 'implementation strategy flight simulator' to explore options in the biodefence capabilities. As described in Hodicky et al. [12] 'Capability-based planning as an approach to defense planning is an almost infinitely complex engineered system with countless nodes and layers of interdependency, influenced by state and non-state diplomatic activities, information, military and economic actions creating secondary and third order effects. The main output of capability-based planning is the set of capability requirements needed to achieve the expected end-state. One revitalized qualitative technique that allows us to gain

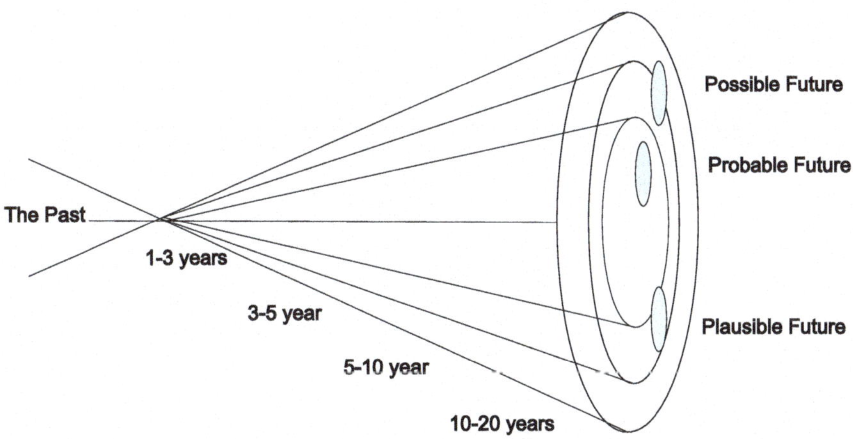

Fig. 9 Cone of plausibility [16]

insights into unstructured'. The ADM/DT is a unique digitization of the scenario space to facilitate an analysis of fuzzy problems in the complex landscape associated with a biothreat. The scenario space lends itself to policy development and 'flight-testing' of policy deployments thereby supporting anticipatory innovation (Fig. 10).

Strengthening the capabilities to assess and address the security and health threats posed by pandemics requires recognition that these risks constitute complex collective intelligence (CI) challenges surrounding surveillance and predictive pandemic intelligence. Figure 11, illustrates the key features of what constitutes a Capability Framework for Pandemic Preparedness. The capability framework is operationalized through an Anticipatory Innovation Framework (Fig. 10) that enables policy makers and decision makers to plan and execute under conditions of Volatility, Uncertainty, Complexity and Ambiguity (VUCA).

As illustrated by COVID-19 experience, a newly emerged pathogen can grow from a disease outbreak to pandemic proportions. The capability framework above highlights the complexity-informed approach to biothreat preparedness to include: AI-driven disease sensing, large-scale data-driven epidemic modelling, multi-scaled projections of socio-ecological risks, big data, systems and futures thinking, policy

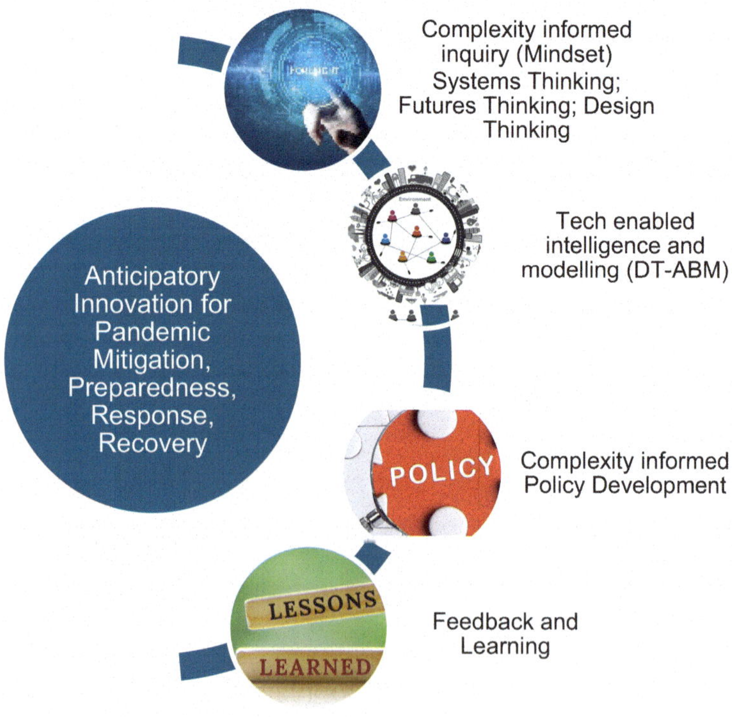

Fig. 10 Anticipatory innovation framework for pandemic mitigation, preparedness, response and recovery

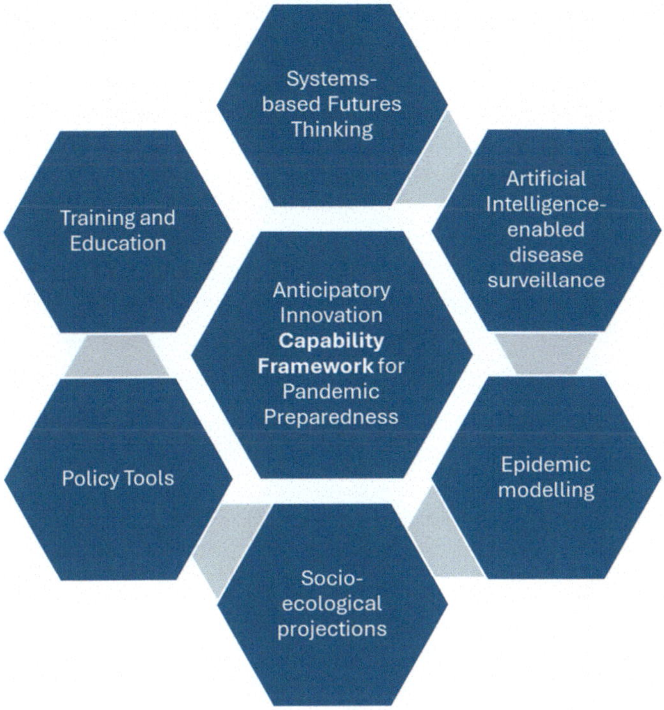

Fig. 11 Capability framework for pandemic preparedness

analysis, human resource development, and knowledge exchange. A capability analysis is conducted across Doctrine, Organization, Training, Materiel, Leadership and Education, Personnel, and Facilities (DOTMLPF) to perform a set of tasks to execute a specified course of action. Capabilities result from a combination of tasks (cited in Samaras and Willis [28]).

6 Conclusion

As described in Zawadzki et al. [33: 78] '... the pandemic has highlighted that health security systems must be redesigned to cope with multiple and diverse health threats, not only well-known diseases, but also unexpected and challenging emerging health threats'. To address this national and global security threat, our premise is that assessing and tackling the future risks stemming from pandemic threats requires innovative, collaborative approaches that draw on diverse capacities in the areas of AI-driven disease sensing, large-scale data-driven epidemic modelling, multi-scaled projections of socio-ecological risks, big data, systems and futures thinking, policy analysis, human resource development, and knowledge exchange. We highlight here the role that place DTs coupled with appropriate disease ABMs can play in performing complexity-driven scenario analyses of outbreak and spread potential as well as impacts of management options to guide policy making. The ability to improve preparedness of health security systems under VUCA (Volatility, Uncertainty, Complexity, Ambiguity) conditions relies on the development of capabilities to 'future proof' health security systems against biothreat scenarios. Achieving such strategic objectives associated with biosecurity call upon the application of next generation tech-enabled Capability-Based Planning (CBP). CBP thereby enables contemplating extreme uncertainties and emerging threats, rather than being prepared for traditional threats.

References

1. Agrawal S, Bhandari S, Bhattacharjee A, Deo A, Dixit NM, Harsha P, Juneja S, Kesarwani P, Swamy AK, Patil P, Rathod N, Saptharishi R, Shriram S, Srivastava P, Sundaresan R, Vaidhiyan NK, Yasodharan S (2020) City-scale agent-based simulators for the study of non-pharmaceutical interventions in the context of the covid-19 epidemic. J Ind Inst Sci 100(4):809–847. https://doi.org/10.1007/s41745-020-00211-3
2. Ammara U, Rasheed K, Mansoor A, Al-Fuqaha A, Qadir J (2022) Smart cities from the perspective of systems. Systems 10(3):77. https://doi.org/10.3390/systems10030077
3. Arcaute E, Barthelemy M, Batty M, Caldarelli G, Gershenson C, Helbing D, Moreno Y, Ramasco J, Rozenblat C, Sánchez A (2021) Future cities: why digital twins need to take complexity science on board. https://www.researchgate.net/publication/354446988_Future_Cities_Why_Digital_Twins_Need_to_Take_Complexity_Science_on_Board
4. Barat S et al (2021) An agent-based digital twin for exploring localized non-pharmaceutical interventions to control covid-19 pandemic. Trans Ind Nat Acad Eng 6(2):323–353. https://doi.org/10.1007/S41403-020-00197-5
5. Bilal S, Zaatour W, Otano YA, Saha A, Newcomb K, Soo K, Kim J, Groen D, Michael E (2023) CitySEIRCast: an agent-based city digital twin for pandemic analysis and simulation. https://doi.org/10.1101/2023.12.22.23300481.
6. Deng T, Zhang K, Shen ZJM (2021) A systematic review of a digital twin city: a new pattern of urban governance toward smart cities. J Manage Sci Eng 6(2):125–134 (2021). https://doi.org/10.1016/J.JMSE.2021.03.003
7. Deren L, Wenbo Y, Zhenfeng S (2021) Smart city based on digital twins. Comput. Urban Sci. 1:4. https://doi.org/10.1007/s43762-021-00005-y
8. Fuller A, Fan Z, Day C, Barlow C (2020) Digital twin: enabling technologies, challenges and open research. IEEE Access 8:108952–108971. https://doi.org/10.1109/ACCESS.2020.2998358
9. Gagnon JE, Kamin SB, Kearns J (2023) The impact of the COVID-19 pandemic on global GDP growth. J Jpn Int Econ 68:101258. https://doi.org/10.1016/j.jjie.2023.101258. Epub 2023 Mar 22. PMID: 37012983; PMCID: PMC10030258
10. Guilien M, Ontiveros E (2012) Global turning points: the challenges for business and society in the, 21st Century. Cambridge University Press, Cambridge
11. Helbing D (2013) Globally networked risks and how to respond. Nature 497:51–59
12. Hodicky J, Prochazka D, Baxa F, Melichar J, Krejčík M, Křížek P, Stodola P, Drozd J (2020) Computer assisted wargame for military capability-based planning. Entropy 22:861. https://doi.org/10.3390/e22080861

13. Houser RS (2022) The role of public health emergency management in biodefense: a COVID-19 case study. Disaster Med Public Health Prep. https://doi.org/10.1017/dmp.2022.113
14. Le Coze JC (2023) Coupling and complexity at the global scale: flows, networks, interconnectedness and synchronicity (e.g. Covid-19). Saf Sci 165:106193
15. Lee B, Preston F, Green G (2012) Preparing for high-impact, low-probability events lessons from Eyjafjallajökull: A Chatham House Report. January 2012. https://www.chathamhouse.org/sites/default/files/public/Research/Energy,%20Environment%20and%20Development/r0112_highimpact.pdf
16. Masys AJ (2022) Transforming future defence capabilities through anticipatory innovation. NATO Symposium STO-MP-MSG-197
17. Masys AJ (2012) Black swans to grey swans—revealing the uncertainty. Int J Disas Prev Manage 21(3):320–335
18. Miley K (2020) Global health biosecurity in a vulnerable world—an evaluation of emerging threats and current disaster preparedness strategies for the future. In: Masys AJ, Izurieta R, Reina Ortiz M (eds) Global health security: recognizing vulnerabilities, creating opportunities. Springer Publishing
19. Miller JH, Page SE (2007) Complex adaptive systems: an introduction to computational models of social life. Princeton University Press
20. MoD (2021) Red teaming handbook. UK Development Concepts and Doctrine Centre: London, UK. https://assets.publishing.service.gov.uk/media/61702155e90e07197867eb93/20210625-Red_Teaming_Handbook.pdf
21. Negri E, Fumagalli L, Macchi M (2017) A review of the roles of digital twin in CPS-based production systems. Procedia Manuf 11:939–948. https://doi.org/10.1016/j.promfg.2017.07.198
22. OECD (2020) A systemic resilience approach to dealing with Covid-19 and future shocks. In: New approaches to economic challenges (NAEC) 28 April 2020
23. Okromtchedlishvili I (2023) Capability-based planning and program budgeting approaches: enhancing sustainability and utility of defense forces. J Def Resour Manage Buchar 14(2):21–32
24. OPSI (2021) Public sector innovation facets ANTICIPATORY INNOVATION October 2021. https://oecd-opsi.org/wp-content/uploads/2021/10/OECD-Innovation-Facets-Brief-Anticipatory-Innovation-2021.pdf
25. Phung S (2021) Digital twin in epidemic: how does a smart city cope with novel challenges through digital simulation, automation and visualisation. The University of Auckland. Conference contribution. https://doi.org/10.17608/k6.auckland.13578224.v2
26. Qi Q et al (2021) Enabling technologies and tools for digital twin. J Manuf Syst 58:3–21. https://doi.org/10.1016/J.JMSY.2019.10.001
27. Raes et al (2022) DUET: a framework for building interoperable and trusted digital twins of smart cities. IEEE Internet Comput 26(3):43–50. https://doi.org/10.1109/MIC.2021.3060962
28. Samaras C, Willis HH (2013) Chapter 2: using joint capability areas to inform installation energy security decisions. In: Capabilities-based planning for energy security at department of defense installations. RAND Corporation. https://doi.org/10.7249/j.ctt2jc9g1
29 Souza L, Bueno C (2022) City information modelling as a support decision tool for planning and management of cities: a systematic literature review and bibliometric analysis. Build Environ 207:108403. https://doi.org/10.1016/j.buildenv.2021.108403
30. Tao F, Zhang H, Liu A, Nee AYC (2019) Digital Twin in industry: state-of-the-art. IEEE Trans Ind Inf 15(4):2405–2415. https://doi.org/10.1109/TII.2018.2873186
31. Tõnurist P, Hanson A (2020) Anticipatory innovation governance: shaping the future through proactive policy making. In: OECD working papers on public governance, no 44. OECD Publishing, Paris. https://doi.org/10.1787/cce14d80-en
32. Williams BAB, Jones CH, Welch V, True JM (2023) Outlook of pandemic preparedness in a post-COVID-19 world. NPJ Vaccines 8:178
33. Zawadzki M, Montibeller G (2023) A framework for supporting health capability-based planning: identifying and structuring health capabilities. Risk Anal 43:78–96. https://doi.org/10.1111/risa.14014

Tech Enabled Approaches to Support Access and Equity in Public Health Education

Ann Joyce, Danielle Secker, Mary Keyhani, Silvia Moreno, Neil Bleiweiss, Alyscia Quichocho, and Anthony J. Masys

Abstract The effect of the COVID-19 pandemic on the education system and learners was palpable. It was considered to be '…the most significant disruption of education systems in human **history**, affecting nearly 1.6 billion learners in more than 200 countries' (Pokhrel and Chhetri in High Educ Future 8:133–141, [12]) cited in (Gopika and Rekha in Int J Educ Reform, 105678792311733, [5]):2). In response to the pandemic, the closure of schools, institutions, and other learning spaces brought about significant changes in educational delivery methods and student learning impacting more than 94% of the world's student population (Pokhrel and Chhetri in High Educ Future 8:133–141, [12]). By the same token it also ushered in innovative strategies to support remote learning thereby transforming education and creating new opportunities for knowledge access and equity across the learning community. With the advent of Artificial Intelligence (AI) becoming mainstream and accessible, it is now part of the e-learning pedagogy creating opportunities for enhanced student engagement and AI literacy. This chapter focuses on the development of e-learning strategies to support public health education in post-secondary institutions as well as support adult learning, workforce development and lifelong learning.

A. Joyce (✉) · D. Secker · M. Keyhani · S. Moreno · N. Bleiweiss · A. Quichocho · A. J. Masys
College of Public Health, University of South Florida, Tampa, FL, USA
e-mail: ajoyce@usf.edu

D. Secker
e-mail: secker@usf.edu

M. Keyhani
e-mail: marytoth@usf.edu

S. Moreno
e-mail: smoreta@usf.edu

N. Bleiweiss
e-mail: nbleiweiss@usf.edu

A. Quichocho
e-mail: alyscia@usf.edu

B. Jacob et al. (eds.), *Tech Enabled Global Health Security*, Advanced Sciences and Technologies for Security Applications, https://doi.org/10.1007/978-3-031-86997-6_10

Keywords E-learning · Pedagogy · COVID-19 · Education · Artificial Intelligence (AI) · Workforce development · Public health

1 Introduction

The COVID-19 pandemic profoundly impacted the education system and learners worldwide. It has been described as "the most significant disruption of education systems in human history, affecting nearly 1.6 billion learners in more than 200 countries" (Pokhrel and Chhetri [12], as cited in Gopika and Rekha [5]:2). The widespread closure of schools, institutions, and other learning spaces led to significant changes in educational delivery methods, affecting over 94% of the global student population [12]. At the same time, the pandemic spurred innovative approaches to remote learning, transforming education and creating new opportunities for knowledge access and equity across learning communities.

With the mainstream adoption of Artificial Intelligence (AI), e-learning pedagogy has evolved to enhance student engagement and promote AI literacy. Recent technological advancements have enriched e-learning content, tools, and methods, making them more tailored to students' needs (Ivanović et al. [6], as cited in Wang et al. [15]:5088). E-learning democratizes access to critical knowledge and skills development, addressing the growing demand for expertise in a world characterized by VUCA conditions—volatility, uncertainty, complexity, and ambiguity—particularly in the rapidly evolving public health landscape.

The integration of digital pedagogy and e-learning is transforming the workforce by emphasizing digital fluency, adaptability, and collaboration in virtual environments [2]. As automation and digital tools become pervasive, the workforce is increasingly oriented toward a high-skills knowledge economy, prioritizing specialized expertise in areas like data science, machine learning, and AI-driven technologies [1]. This shift is reshaping traditional educational pathways, directing attention toward equipping students with relevant technical skills, career competencies, and specialized knowledge. Emerging trends such as micro-certificates and digital badges are providing innovative strategies to prepare public health students for an increasingly dynamic and complex job market [4].

Over the past decade, the public health workforce development infrastructure has faced significant challenges, with the pandemic highlighting critical gaps, particularly in continuing education and professional training [3]. The rising adoption of alternative credentials, the expansion of e-learning, and the growing demand for a skilled workforce underscore the need for a multi-faceted approach to career and professional development. Establishing robust online platforms for public health professionals ensures that educational programs remain responsive to the sector's evolving needs, supporting the reconstruction of public health infrastructure and preparing the workforce for future challenges [13].

2 E-Learning

As part of the distance educational landscape of delivery modes, the popularity of e learning has increased significantly. For example as reported by the [16]:

> In 2016, 21 million students registered for Coursera's online courses, a number that increased annually by around 7 million over the next two years. But the switch to remote working as the pandemic hit triggered a three-fold increase in new registrations, bringing the figure to 71 million in 2020, and 92 million in 2021... These increases reflect growing global acceptance of online teaching, including increases in remote learners taking higher education courses and those from vulnerable or remote communities.

The e-learning environment provides learners with a variety of learning resources such as images, audio, video and AI plug in to enhance the learning experience. As cited in Nazim and Mukherjee ([11]:136–137), 'E-learning is defined as "an instructional process that gives online learners access to a wide range of resources—teachers, other learners, and content such as readings and exercises—independently of place and time" (Xu and Wang [17]:442). A game changer in education, an e-learning system facilitates knowledge transfer, sharing and creation allowing the user to learn self-paced and receive feedback.

Facilitating on-demand self-paced learning, e-learning eliminated many time and space restrictions that characterize traditional learning modalities. The flexibility and adaptability of the e-learning content delivery platform helps to reduce '...the effort and travel expenses and other expenses that accompany traditional learning. E-learning reduced significantly the administrative effort, preparation and lectures recording, attendance, and leaving classes' [9]. In support of adult learners, workforce development and lifelong learner, the online environment with its synchronous and asynchronous delivery methods, encourages learning from anywhere which is especially relevant during times of poor weather and in difficult circumstances (post hurricane) that prevent them from reaching universities and schools.

3 Global Health/Public Health Context

As [10] observed, COVID-19 stress-tested "our national and global societal systems" and exposed critical vulnerabilities within these frameworks. Taleb [14] aptly noted, "not seeing a tsunami or an economic event coming is excusable, building something fragile to them is not." The pandemic, along with recent public health crises, highlighted substantial deficiencies in preparedness, response, and recovery efforts at local, regional, and global levels, underscoring the urgency of advancing public and global health education and skill development.

COVID-19 also revealed gaps and opportunities in teaching competencies that ensure public health students and professionals are workforce-ready. When travel is restricted—whether due to pandemic lockdowns, financial limitations, logistical challenges, or visa restrictions—the shift to e-learning has proven transformative.

Luacus et al. [8] emphasize that "virtual approaches to global health electives are impactful for learner development and foster greater diversity of participants. … Virtual environments can potentially change the face of global health education, reaching those previously excluded from in-person experiences and mitigating the power asymmetries that have historically plagued the field."

The rise of digital health technologies, including telemedicine, wearable devices, and AI-driven diagnostics, has fueled demand for innovative, accessible learning opportunities. Unlike traditional degree programs or conferences, micro-certificates offer a flexible, affordable pathway to credentialing. These programs cater to busy professionals, requiring no prerequisites or applications, allowing participants to immediately enroll and build targeted skills. Each micro-certificate consists of short, competency-based courses completed at an individualized pace and delivered entirely online. Upon completion, participants earn digital badges to enhance resumes, LinkedIn profiles, or portfolios.

Micro-certificates not only offer immediate value but also act as steppingstones for further education and career growth [18]. Credits earned through these programs contribute to CPH license maintenance and recertification requirements. As demand for alternative credentials grows, the Learning and Leadership Academy (LLA) remains at the forefront, providing career-focused learning solutions that bridge education and workforce needs, ensuring public health professionals are equipped to meet evolving challenges [19].

4 Discussion

The COVID-19 pandemic brought about significant changes in educational delivery methods and student learning. E-learning has transformed the healthcare education system by providing healthcare professionals with training and development opportunities, regardless of their location. Incorporating AI as part of the experiential learning strategies to support Public Health education as well as develop AI literacy.

The University of South Florida's College of Public Health's Lifelong Learning Academy (LLA) is part of the College's service mission to provide educational opportunities to people around the world and create innovative approaches to teaching and learning beyond the degree program. The LLA provides an array of lifelong learning and educational opportunities designed to augment the public health practitioner's knowledge, skills, competence, performance, and attitude to improve public health practice.

The LLA consists of an interdisciplinary team of learning designers, a project manager, director, and academic dean. Their mission and vision embody four core values: inquiry, innovation, integrity, and inclusion. With a growing need for innovative approaches to public health workforce development, the COPH Lifelong Learning Academy has two main goals: workforce development and interdisciplinary knowledge promotion.

With a focus on alternative credentials, the LLA created a web portal designed to massify educational content and resources tailored to public health professionals, including lectures, resources, and networking opportunities. The LLA equips public health professionals with essential education and skills for a competitive global job market, maintaining their licenses, and developing new public health research.

The LLA promotes new knowledge across interdisciplinary programs, creating a virtual learning environment to deliver training courses, administer and monitor them, and award digital badges. This platform enables researchers from all USF colleges and departments to integrate public health into collaborative research projects, fostering innovative practices and innovation in specific areas of interest.

5 Hallmark Program: Disease Intervention Specialist Training Program

The Lifelong Learning Academy (LLA) has played a pivotal role in the development of the Disease Intervention Specialist Training Program (DISTA), a groundbreaking alternative credential program designed to enhance the skills and retention of Disease Intervention Specialists (DISs). DISTA represents a multidisciplinary approach to public health workforce training, addressing the need for comprehensive and collaborative preparation for DISs, who play a critical role in managing and responding to disease outbreaks.

DISTA is designed with a tiered structure that mirrors a career ladder, offering specialized training that supports DIS professionals in advancing their careers. The program aims to address the challenges of DIS retention by providing a clear, structured pathway that not only equips new DISs with the skills needed for their complex roles but also offers continuous professional development opportunities for those already in the field. This approach ensures that DISs are well-prepared to manage the demanding responsibilities of their positions while also increasing the overall capacity of the public health system to respond effectively to disease outbreaks.

The DISTA program is the result of a strategic collaboration between the Department of Health and four colleges at the university, bringing together a diverse range of expertise to create a training program that is both comprehensive and relevant. This multidisciplinary approach ensures that the training provided is not only grounded in the latest public health practices but also incorporates perspectives from various fields, enhancing the overall learning experience for participants.

One of the most significant advantages of the DISTA program is its accessibility. By launching the program on the LLA platform, it has become available to DISs across the nation, breaking down geographic and financial barriers that might otherwise prevent public health professionals from accessing such valuable training. The LLA open payment system further eliminates barriers by providing a flexible and accessible way for participants to enroll in the program.

As the first program of its kind, DISTA is tailored specifically for public health professionals outside of traditional degree programs and represents a significant innovation in the field of alternative credentials. The program offers a micro-certificate that not only serves as a career advancement tool but also provides DISs with a digital badge that they can publicly display to showcase their new skills. This digital badge is a modern, recognized credential that adds value to the participant's professional profile, making it easier for them to demonstrate their competencies to employers and colleagues.

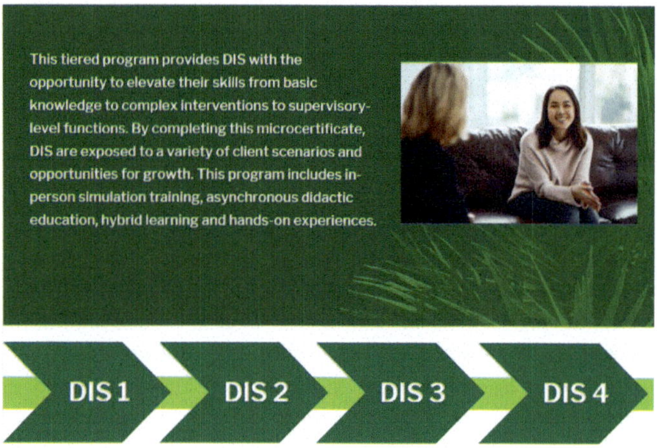

Through the unveiling of the DISTA Program, the LLA demonstrates the eight Hallmarks of Excellence: Advocacy in Leadership, Entrepreneurial Initiative, University-to-Business Stakeholder Engagement, The Faculty Experience, The Learner Experience, Digital Technology, External Advocacy and Leadership beyond the university, and Professionalism.

The Lifelong Learning Academy (LLA) demonstrates Advocacy in Leadership by incorporating a dynamic and forward-thinking approach that involves championing new possibilities for their institution. The LLA advocates for innovative changes that propel USF into the future. If the LLA did not advocate for innovative technology, new business models, and community partnerships, the DISTA program would not have manifested.

The LLA demonstrates University-to-Business Stakeholder engagement particularly through initiatives like applying for grants to support workforce development, bridging the gap between academic institutions and industry needs, and addressing continuing education through alternative credentials. By focusing on continuing and professional education, universities can enhance their impact on the workforce and the broader economy, ensuring that professionals remain competitive and up to date with industry trends. By applying for grants focused on workforce development, these institutions can fund programs that directly address these gaps, providing tailored training that equips workers with the necessary skills to thrive in their respective industries. This alignment of educational offerings with industry needs not only benefits individual learners but also strengthens the local and national economy by creating a more skilled and adaptable workforce. The use of University-to-Business Stakeholder engagement is the foundation of the DISTA program.

The LLA exemplifies a strong commitment to enhancing the Faculty and Learner Experience by creating an environment that fosters multidisciplinary inquiry, promotes excellence in teaching and learning, and nurtures collegiality across the university and beyond. Through innovative approaches such as alternative credentials and a culture of open dialogue, the LLA plays a crucial role in advancing both the academic and professional growth of faculty and learners alike. By promoting alternative credentials, the LLA not only enhances professional development opportunities but also strengthens connections within the academic community and with external partners. The LLA worked closely with the faculty to ensure the learners journey was seamless and enhanced learner engagement.

In the rapidly evolving educational landscape, technology plays a pivotal role in transforming how institutions operate, teach, and engage with students and the broader community. The LLA's commitment to incorporating innovative digital technology is evident when it pioneered the adoption of cutting-edge technologies to enhance learning outcomes, streamline operations, and create more immersive and interactive educational experiences. By advocating for the integration of advanced digital platforms, the LLA facilitated more personalized and accessible learning experiences. These platforms can include AI-driven learning tools, virtual reality simulations, and online collaboration spaces that cater to diverse learning needs and styles. Through revenue generated by the DISTA program, the LLA was able to purchase Vyond, Camtasia, Canva, and the external version of Canvas, which houses a secure payment system so external learners can attend courses at the University.

Strong community partnerships are a hallmark of effective advocacy in leadership beyond the university. The LLA positioned itself as a hub for collaboration, bringing together various stakeholders—including local businesses, nonprofit organizations, government entities, and community members—to work towards shared goals. By advocating for and developing programs that directly benefit the community, such as the DISTA program, the LLA strengthened its ties with the local population and created a positive social impact. Additionally, through the LLA, the DISTA program shows advocacy for adult education, workforce development, or health and wellness initiatives.

Last, the LLA exemplifies a deep commitment to professionalism through all its endeavors, setting a standard of excellence that permeates every aspect of its operations. This commitment is reflected in the way the LLA approaches communication, marketing, project management, course development, and team building, ensuring that every interaction and initiative is managed with the utmost care, integrity, and expertise.

6 Conclusion

The implications of innovative e-learning opportunities for public health are profound and far-reaching. E-learning is emerging as a transformative tool for addressing disparities in the dissemination of public and global health knowledge, tools, and methodologies. Career readiness, lifelong learning, and skill development are becoming cornerstones of e-learning as it evolves. Flexible, short-term educational models—such as bootcamps, online courses, and certifications—are increasingly complementing or supplementing traditional degree programs. By incorporating modern teaching methodologies like micro-certifications and expanding e-learning offerings, academic institutions can better align workforce development with industry requirements.

As the demand for public health professionals continues to grow, it is essential for educational institutions to design programs that target workforce competencies. Policymakers should prioritize funding and support for alternative credentialing initiatives that bridge gaps between academic outcomes and workforce demands. Programs like the Learning and Leadership Academy (LLA) exemplify how curricula and training can be tailored to meet the evolving needs of public health. These approaches provide students with focused, career-oriented education, preparing them to thrive in a digital and data-driven world [7].

The COVID-19 pandemic underscored the need for a well-trained, adaptable, and ready public health workforce. Traditional degree programs often fall short of addressing the immediate demands of the field, while micro-certificates offer a flexible, cost-effective alternative for professional development. Policies such as reimbursement incentives, stipends, and the recognition of micro-credentials for promotions and job qualifications can encourage professionals to engage in lifelong learning and skill advancement.

Tech-enabled global and public health education is an evolving domain that requires innovative methods of knowledge transfer to overcome challenges and disparities in access. Moving beyond theoretical knowledge to practical implementation demands scalable solutions. Initiatives like the LLA are pioneering such efforts through advanced software tools, micro-certifications, and digital badges. These forward-thinking strategies reshape public health education, enhance workforce readiness, and empower professionals to address the complex challenges of community health and safety in a rapidly changing landscape.

Implications for E-Learning in Public Health:

1. **Enhanced Workforce Readiness and Competency Alignment**
   E-learning enables the rapid development and deployment of training programs that align with emerging public health challenges and workforce demands. By offering flexible, targeted, and competency-based learning pathways—such as micro-certifications and digital badges—public health professionals can quickly acquire relevant skills to address immediate needs in areas like epidemiology, data analysis, and health equity.
2. **Increased Accessibility and Equity in Education**

E-learning reduces barriers to education by making public health training accessible to a broader audience, including individuals in underserved or remote areas. Virtual platforms facilitate global participation, diversify the learner base, and mitigate logistical challenges such as travel restrictions, financial constraints, and visa requirements, promoting equitable access to knowledge and skills development.

3. **Integration of Digital Health Technologies**
   The rise of telemedicine, wearable health devices, and AI-driven diagnostics demands a digitally literate workforce. E-learning platforms can integrate simulations, interactive modules, and case studies that incorporate these technologies, preparing public health professionals to leverage digital tools effectively in practice and contributing to modernized, technology-driven public health systems.

By embracing these trends, public health education and workforce development can better prepare professionals to navigate the dynamic demands of the field while addressing critical health disparities.

References

1. Anderson V (2020) A digital pedagogy pivot: re-thinking higher education practice from an HRD perspective. Hum Resour Dev Int 23(4):452–467. https://doi.org/10.1080/13678868.2020.1778999
2. Bećirović S (2023) Digital pedagogy : the use of digital technologies in contemporary education (1st edn.). Springer Nature Singapore Pte Ltd. https://doi.org/10.1007/978-981-99-0444-0
3. Dean HD, Myles RL, Spears-Jones C, Bishop-Cline A, Fenton KA (2014) A strategic approach to public health workforce development and capacity building. Am J Prev Med 47(5):S288–S296. https://doi.org/10.1016/j.amepre.2014.07.016
4. Gish-Lieberman JJ, Tawfik A, Gatewood J (2021) Micro-credentials and badges in education: a historical overview. TechTrends 65(1):5–7. https://doi.org/10.1007/s11528-020-00567-4
5. Gopika J, Rekha R (2023) Awareness and use of digital learning before and during covid-19. Int J Educ Reform:105678792311733. https://doi.org/10.1177/10567879231173389
6. Ivanovic M, Milicevic AK, Aleksic V, Bratic B, Mandic M (2018) Experiences and perspectives of technology-enhanced learning and teaching in higher education–Serbian case. Procedia Comput Sci 126:1351–1359. https://doi.org/10.1016/j.procs.2018.08.086
7. Kulik PKG, Alperin M, Todd Barrett KS, Bekemeier B, Documet PI, Francis KA, Gloria CT, Healy E, Hileman R, Kenefick HW, Lederer AM, Leider JP, McCormick LC, Prechter L, Reynolds KA, Rogers MH, Rose B, Scallan Walter EJ, Walkner LM, Power LE et al (2024) The need for responsive workforce development during the pandemic and beyond: a case study of the regional public health training centers. J Public Health Manag Pract 30(1):46–55. https://doi.org/10.1097/PHH.0000000000001835
8. Luaces MA, Cochran MS, Finocchario-Kessler S, Connelly K, Polivka B, Young R, Anguyo G, Nwobu C, Evert J (2023) Impacts, learner diversity, and curricular framework of a virtual global health elective catalyzed by the covid-19 pandemic. Ann Global Health 89(1):32, 1–10. https://doi.org/10.5334/aogh.4060
9. Maatuk M, Elberkawi EK, Aljawarneh S, Rashaideh H, Alharbi H (2022) The COVID-19 pandemic and E-learning: challenges and opportunities from the perspective of students and instructors. J Comput High Educ 34:21–38. https://doi.org/10.1007/s12528-021-09274-2

10. Masys AJ (2021) Crisis leadership and sensemaking. In: Masys AJ (ed) Sensemaking in security. Springer Publishing
11. Nazim M, Mukherjee B (2016) Chapter 6 knowledge management tools for libraries. In Knowledge management in libraries. Chandos Publishing, pp 115–147
12. Pokhrel S, Chhetri R (2021) A literature review on impact of COVID-19 pandemic on teaching and learning. High Educ Future 8(1):133–141. https://doi.org/10.1177/2347631120983481[DOI]
13. Sherrer KJ, Prelip ML (2019) A multifaceted approach to public health career and professional development training. Health Promot Pract 20(6):932–940. https://doi.org/10.1177/1524839918783744
14. Taleb NN (2014) Antifragile: things that gain from disorder. Random House, New York
15. Wang Y-M, Wei C-L, Chen W-J, Wang Y-S (2024) Revisiting the E-learning systems success model in the post-covid-19 age: the role of monitoring quality. Int J Human Comput Interact 40(18):5087–5102. https://doi.org/10.1080/10447318.2023.2231278
16. WEF (2022) These 3 charts show the global growth in online learning, https://www.weforum.org/stories/2022/01/online-learning-courses-reskill-skills-gap/
17. Xu D, Wang H (2011) Integration of knowledge management and e-learning. In: Schwartz DD, Te'eni D (eds) Encyclopedia of knowledge management (2nd edn), pp 442–451. Information Science Reference, Hershey, NY
18. Young D, West RE, Nylin TA (2019) Value of open microcredentials to earners and issuers: a case study of national instruments open badges. Int Rev Res Open Distrib Learn 20(5):104–121. <Go to ISI>://WOS:000502555700007
19. Zemmel DJ, Kulik PKG, Leider JP, Power LE (2022) Public health workforce development during and beyond the covid-19 pandemic: findings from a qualitative training needs assessment. J Public Health Manag Pract 28(5):S263–S270. https://doi.org/10.1097/phh.0000000000001524

Leveraging AI and Genealogy for Family Reunification During Humanitarian Crises

Alyssa Durfey

Abstract Millions of people are affected by humanitarian crises each year. The process of aiding these individuals is complex, but has lasting physical, emotional, and social effects. These effects are further exacerbated by familial separation in displacement. Understanding the importance of genealogy across the globe, a better strategy for tracking records and reunification is critical to expediting healing from trauma. By leveraging existing genealogical platforms presently available in many countries worldwide, the reunification process could be significantly simplified and expedited. This application holds the potential to save lives and years of unnecessary trauma with only minor additions to the existing framework. Barriers to implementation include stringent monitoring of data use, storage, and security in hopes of addressing institutional mistrust and censorship. The sustainable life cycle of this intervention depends on the support of current technology platforms, but growth is nearly unlimited with the right global acceptance and application.

Keywords Refugees · Artificial Intelligence · Mental health · Genealogy

1 Background

More than 110 million people world-wide have been forced to flee their homes due to violations of human rights, conflict, and violence (USA for UNHCR 2024). These individuals are separated from loved ones and all familiarity, left to simply hope for reunification in the distant future, often unsure how or where their loved ones are coping physically, mentally, and geographically. In an age of unbelievable technological advancement and rapidly available cloud data, reuniting displaced individuals with their loved ones should not take years, or even months. It is imperative for the health and wellbeing of refugees that the trauma of displacement be eased by the

A. Durfey (✉)
College of Public Health, University of South Florida, Tampa, FL, USA
e-mail: alyssadurfey@gmail.com

B. Jacob et al. (eds.), *Tech Enabled Global Health Security*, Advanced Sciences and Technologies for Security Applications, https://doi.org/10.1007/978-3-031-86997-6_11

reconnection and reunification of individuals, regardless of their geographic location. To understand the future of reunification, one must first understand the driving forces of displacement and genealogy.

1.1 Displacement

Individuals seeking asylum may flee their home, or are forced to leave, due to conflict, human rights atrocities, persecution, or natural disaster, all of which expose them to significant trauma as they navigate or avoid violence, abuse, brushes with law enforcement, or exploitation [5]. Forced displacement is a humanitarian cause supported by billions of dollars in donations annually, much of which addresses urgent needs and helps restore dignity, community, and independence while establishing and maintaining support for initiatives involving shelter, health care, nutrition, sanitation, and potable water. These programs funded by the European Commission help refugees rebuild their lives by providing access to education, livelihood support, or legal protection to asylum seekers (2024). The International Refugee Assistance Project (IRAP) aims to assist in the family reunification process, making that their primary route of resettling refugees (n.d.). These legal professionals help families navigate the intricacies of refugee and reunification applications when seeking asylum across the United States and several European countries. The importance of such work is evident in global policies. For example, the United Nations High Commissioner for Refugees (UNHCR) establishes that family unity is a universal right as a fundamental and natural societal unit [18].

1.2 Genealogy

Genealogy is the scientific study of an individual's direct relatives. It is arguably similar to the study of Family History, which is a more broad study of accessory non-direct family, as well as the cultural, traditional, and contextual life of ancestors, but genealogy typically focuses specifically on major events like birth, death, and marriage [12]. The Board for Certification of Genealogists blurs the line between these two fields of study by encouraging the recording of "aspects of identity, relationship, event, or situations…artifacts, authored narratives, oral histories…and other sources concerning agriculture, demographics, DNA, [and] economies" by genealogic professionals [12]. The fields have separate applications, utility, and propriety, but the two will be encompassed under "genealogy" for the purposes of this recommendation.

Genealogy is a centuries-old tradition, evidenced by class and caste systems, pedigree tracing, and family history societies [16]. The Bible, the Book of Mormon, the Quran, the Upanishads, and countless other sacred texts of religions worldwide describe historically relevant genealogy through records of succession, legitimacy, and lineage continuity in many instances. The concept of genealogy is similarly relevant in the historical tracing of religious movements, human migration, and social constructs, scientific applications of natural selection or inheritance in natural history as well as modern laboratory medicine and multi-omics, or even through the registration of a family pet to the American Kennel Club and tracing their pedigree. This incredibly broad nature of this concept indicates that every person is likely familiar with genealogy or some version of its application.

A humanitarian crisis is a major disruptor in the life of impacted individuals and communities as a whole. Oftentimes, the trauma of displacement is exacerbated by the fear, unknowns, and grief associated with family separation. The importance of recording family structure is enmeshed in human evolution, and has been a major theme of historic and cultural literature for hundreds of years. The purpose of this chapter is to propose the application of Artificial Intelligence (AI) and genealogical work to expedite reconnection and reunification of families during and following a humanitarian crisis.

2 Humanitarian Concerns

Every year, millions of people are forced to abandon safety and familiarity to seek asylum. The estimated number of individuals seeking asylum is ever-growing. With an anticipated surge of 20 million people between 2023 and 2024, the total estimated forced displacement worldwide will be 130 million by the end of this year [5]. Further, while refugees come from several different locations, 75% of those forced from their homes will go to low-income or middle-income countries where they hope to find solace. In reality, however, the European Commission further explains that in their search for survival essentials like food, shelter, and basic services, these individuals are wholly reliant on their host country which puts a significant burden on local communities and infrastructure, landing 2 out of 3 refugees in poverty (2024).

Refugee status is a complex legal and ethical dilemma heavily influenced by the home country, host country, an individual's legal status, and the involvement of humanitarian organizations [8]. When considering refugees, it is easy, perhaps due to Western media coverage or technology, to imagine migrant families traveling as a unit. However, often, for one reason or another, a single family member is often sent to seek refugee status, and if it is granted, to then apply for family reunification [10]. The timeframe and approval status of asylum or family reunification applications are fraught with unknowns and ambiguity, leaving many families without thorough guidance or expectations. Families approach the ambiguity with pessimism, often sending one member of the family ahead to prepare the way for the rest of the family.

In the cases where families are separated, there remains a great deal of uncertainty. Family reunification is not a straightforward or easy process as it presents practical, legal, and financial obstacles [18]. The International Refugee Assistance Project (IRAP) shared the story of a young boy, left with his aunt when his parents fled retaliatory action and conscription. The aunt traveled to a neighboring African country with the boy, following his parents' guidance. She was intending to reunite the boy with his mother, only to discover that his mother was already transferred to Europe. After years of legal proceedings, the boy was reunited with his mother and siblings in France (n.d.). Other stories from IRAP detail years of legal struggles and duress for sibships separated between their home and asylum. An unfortunate reality, however, is that many families will never reunite [18]. Hvidtfeldt, Petersen, & Efstathiou found this to typically be fathers in Europe [10], but Giannopoulou et al. recounted cases where a child was sent first and a humanitarian disaster ensued for years to come [9].

In 2016, the Grecian government faced major scrutiny when a migration route was closed due to inhospitable climate, leaving many children stranded in Greece, though they had no desire or intention to stay [9]. This study, involving 90 of the affected children, found that most had been attempting reunification and had applied for asylum in a central or northern European country [9]. The majority of these children exhibited significant trauma responses, especially related to war and violence, death, kidnapping, and theft during migration. There were major concerns raised about the mental health comorbidities found in this study, especially in regards to the psychosocial burden of comorbid Post-Traumatic Stress Disorder (PTSD) and depression. Giannopoulou et al. concluded that it is the responsibility of clinicians to recognize and empathetically treat the immense psychological toll displacement can have on children; but it is up to politicians and governments to work to address the root cause of this trauma, noting that severe damage can be done mentally when children are left in uncertain upheaval, and this continued sense of displacement is an infringement on the foundational welfare concepts of a child's best interest and family unity [9]. Experiencing domestic or gang violence adds to risk factors for anxiety, depression, and PTSD, and such experiences early in development can affect a child negatively for the rest of their lives (NCTSN, n.d.).

Feelings of powerlessness, guilt, and distress are discussed extensively in literature, most often reported by the family members that sought asylum and left their family behind, presumably in danger [10]. It is evident that reunification is a journey of complex procedures with short actionable periods followed by long wait times, financial responsibilities in some countries, different definitions of "family", protection and documentation obstacles, and embassy visits that can be costly and dangerous [18]. These obstacles cannot overshadow the reality of the impacts of separation. It is incredibly detrimental to the health and wellbeing of young children to experience separation from their families. The National Child Traumatic Stress Network (NCTSN) cites intrusive dreams or thoughts, avoidant or antisocial behavior, negative beliefs and emotions, hyperarousal and sensitivity, and inappropriate behavior as potential symptoms of such trauma (n.d.). They elaborate that prolonged separation from primary attachments, such as parents, especially following chaotic, scary, or sudden changes in a child's environment affect emotional and physical development and the importance and impressionability of this population means that separation can incapacitate a child forever by hindering their ability to self-soothe, trust, and develop relationships well into adulthood. This exposes the children to cognitive, emotional, and physiologically impactful trauma symptoms that may persist even after reunification.

Children are not the sole people affected, though. Lange, et al. concluded that women were affected more than men in regards to trauma, when compared to their family members following reunification (2023). The same study concluded that though refugees and their families exhibited similar baseline psychopathology, the sociodemographic impact of displacement affected refugees, and effects of the trauma experienced impacted individuals even postmigration. These results demonstrate the need for comprehensive mental health support during and after displacement and relocation. Further supporting this need, poor mental health has been associated with poor work performance, which could perpetuate poverty, poor physical health, and the detriment of refugee children educationally and emotionally [10]. A 24-year longitudinal study in Denmark found significant PTSD in fathers separated from their families during displacement, with a positive relationship between mental health hazard risk and the waiting period between arrival in host country and family reunification, as well as increased hazard risk in those waiting for reunification versus those reunified [10]. Policies for reunification and documentation vary by location and region, which presents a significant barrier in family reunification, especially in cases of death [8]. The psychological burden of unknowns has been discussed previously, but there are additional legal implications to displacement in these cases. A verified death certificate can award family members guardianship of minors, status as a widow(er), or establish orphanhood, any of which could provide domestic and international protections, rights, and benefits [8]. Therefore, a displaced family could be separated, and if one of the parents dies, the other parent has few options for confirming death versus loss of communication means. Once death is confirmed, though, the process to receive official documentation is tenuous, at best. Add to that the concerns for child welfare if the deceased parent has one of the family's children with them, and the rest are with the other parent. That now unaccompanied minor is likely fleeing without documentation as an easy target for abusers or traffickers, and has no ability to locate or make contact with their living relative, all while coping with the psychological distress of displacement and the grief of losing a parent. The surviving parent must navigate a foreign legal system for their minor while simultaneously doing the same for themselves or their other children in the same or another country. An alternative scenario is when one parent goes to seek asylum, intending to request family reunification, but simultaneously, violence makes the home country too dangerous, and the family they are seeking asylum for are forced to seek asylum elsewhere. Clearly, countless variations in the map of displacement are possible and a single answer will not resolve all obstacles, but improvement of the process of reunification is a critical step in human welfare.

Immigration policy changes have extended the reunification process significantly, with some countries temporarily suspending reunification, and others requiring DNA testing to allow for reunification, all of which is counter-intuitive to the risks of depression, anxiety, suicide, and PTSD associated with displacement [10]. These concerns outline the desperate need for a better way to manage and understand family relationships, extended kinship, legal documentation and vital records, and keep families connected throughout displacement.

3 Current Landscape

3.1 Reunification

Reunification policies vary by country, even in the same geographic regions. For example, Canada's immigration laws for family reunification includes dependents and family members, defined as legal or common-law spouses, and biological or adopted, unmarried children younger than 22 years [3]. Further, they clarify that applicants must be either settled in Canada via a sponsor or by applying for refugee status, and they must apply for family reunification within one year of Canadian arrival. In contrast, refugees to the United States have a two year window to apply for family reunification, and children must be unmarried and under 21 years old, but with several timeline and age waivers available [17]. These differences demonstrate the nuances between countries where families may seek reunification. However, this is merely one obstacle in the current process.

Identifying kinship is another significant obstacle to reunification, though its documentation remains a keystone of human rights, dignity, and protection, according to Franceschetti et al. [8]. European law states that with complete parental consent, a minor can exercise their right to family reunification [8]. This further supports the need for adequate and expeditious record keeping in a humanitarian crisis. The right to family life is supported by European law, allowing children to be granted asylum in the country their parents find refuge in, and in cases where the parents' legal status is in question, the welfare of the child is taken into the highest account, and the premise of encouraging an ongoing relationship with both parents is ideal [8]. These legal stipulations provide great flexibility, but also great ambiguity for those caught in a crisis. The balance of family rights and immigration enforcement is a seemingly endless battle to be faced.

3.2 Genealogy and Artificial Intelligence

There is an emerging trend of DNA collaboration with genealogical study, contributing to the success of commercial services like 23andMe, AncestryDNA, FamilySearch, and MyHeritage. These companies have leveraged AI to aid in their processing of historical documents through new handwriting deciphering technology, extensive cataloging, relationship prediction, and identification of missing data by automatically rifling through thousands, if not millions, of records in no time [1, 15]. Artificial Intelligence is a computer science application where a machine is used to identify information, interpret it, and generate easily accessible summaries and statistics with little or no human intervention [15]. Various algorithms are used to extract data points like dates and locations, scan records to identify missing information about a relative of the person of interest, scan photos to identify people, and is now commonly utilized to compare genetic markers and suggest relationships between people that would likely have never known they were related [15].

4 Recommendation

Countless genealogy applications are available online to track family history records, genealogy, and even DNA connections. The largest publicly available shared family tree in the world is FamilySearch, which is conveniently partnered with AncestryDNA, the world's largest consumer genetic testing database [5–11, 14]. These services are available in 128 countries, and with the platform already having completed more than 25 million genetic tests around the globe, it is widely accepted by users [2]. This well-supported partnership has the infrastructure to support a global initiative, or is at least the platform most likely to succeed in a global health endeavor. With minor adjustments to the current user portal, FamilySearch could add an optional Refugee badge to an individual's profile. This badge could enable searchability across the platforms' worldwide database. By entering life events to the dropdown menus, such as "Migration from...", and entering the year, as completed with currently available life events in the system, users would be able to search fellow refugees that migrated from the same place, to the same place, and countless other search parameters. Current user profiles are limited to ages 13 and up, or with parental consent from 8 to 12 years old [7]. A system bypass could be created for the profiles of younger children with a refugee badge if they are unaccompanied minors, or be created by parents for their younger children. This unified searchable portal already utilizes AI to predict family relations, so if a displaced mother is separated from her husband, and they both created free profiles in the system, AI would predict a match based on refugee badge, similar "Migrated from..." choices, and potentially with photographs of refugees. AI in the Ancestry platform can already map dates and locations (MacEntee, n.d.), so additional measures to accommodate refugee status should be limited [13].

The use of AI on these platforms can address record loss in the refugee population since vital records are scanned and suggested as matches for different individuals. Uploading documents like birth or death certificates strengthens the data in the system, but also could inform family members of someone's death, allowing them to address the legal implications. The platforms allow users to connect with relatives, so in the cases described by IRAP referenced previously, the siblings could connect sooner rather than later, and the aunt could have gone directly to France instead of taking a dangerous journey to a different country, one that her sister was not even in.

Enrollment in a genealogy platform that allows for searching for relatives that have been displaced would likely be easiest to implement as people enter a refugee camp, but could be advertised more broadly because the platform already has mobile device compatibility. The FamilySearch and Ancestry platforms are user-friendly and available in several languages, making them widely accessible for many refugees. If personal devices are not available, registration upon camp entry would likely be the most effective way to register as many participants as possible. As this approach would limit access, assuming few or no personal devices are available, it would be beneficial to create workstations in a central location. Another idea is to use this reunification platform within a mental health resource center, bringing more residents in, destigmatizing the center, and ideally, exposing them to the mental health resources available at the camp.

A program like this addresses several major concepts in public health. It supports policy development by tracing migration patterns and trends, encourages legal and regulatory action by guiding individuals to the appropriate channels for where they want to settle, and it can be used as a basis for community information and education. It could aid in assessment practices by better tracing deaths, transmission paths, and root causes for a genetic abnormality being introduced to a new population. It addresses assurance by fostering collaboration and community building, provides work opportunities for genealogy experts, and it could improve processes and research by closing the loop of documentation practices and record keeping.

4.1 *Potential Barriers to Implementation*

4.1.1 Data Use

A 2018 article released by Canada's CBC News quoted a Canada Border Patrol Services Agency representative explaining that they use DNA testing and genealogy websites to investigate migrant nationality in efforts to reduce long-term incarceration of non-citizens. The statement raised concerns over immigration policies and highlighted the sudden influx of refugees under certain government figures. This statement was released following interviews with an immigration attorney defending a client being investigated for lying about his nationality because his application claimed Liberia, but his DNA indicated Nigerian descent (Reuters, 2018). This example in mind, an open-access portal to track refugees welcomes the opportunity for the targeting of refugees. Whether for marketing, scams, or discrimination, having an online roster of refugees would require significant data use, storage, and security measures to ensure safety of all people involved. Those signing up would have to be well-informed on how their host country can access or use that data, they must understand that their contributions will become publicly available information, and there must be guidelines on the types of communication shared across the platform. The use of that information for criminal investigation, though creative, poses a threat to the efficacy of such a project as this because it will inevitably allow or encourage people to falsify information. That understood, data verification is a necessary improvement to the system that should be investigated further as soon as it is feasible.

4.1.2 Distrust

Literature extensively demonstrates the effects of trauma. An aspect not reviewed so much herein is the erosion of trust that comes from experiencing a humanitarian crisis. Many individuals will have trauma that is unimaginably difficult to process, and if that is stemming from governmental unrest or gang activity, the reliance on a rostered database may cause various levels of distrust in how any shared information will be utilized.

4.1.3 Data Storage

Cloud storage of the magnitude necessary to accommodate such a huge refugee population is, admittedly, a major obstacle in this program's implementation. Such data storage and the appropriate security measures is a burden that should be shared internationally. Perhaps grants, partnerships, and collaborations would make such a project possible, as long as all funders are acknowledging that the social responsibility of caring for the marginalized should be shared for the benefit of mankind.

4.1.4 Censorship

In 2021, President Museveni outlawed certain social media platforms in Uganda, allegedly due to censorship concerns [4]. A representative from a humanitarian organization in Uganda's capital, who requested to remain unnamed, stated the publicized context of this ban was incomplete (2024). According to the unnamed source at a nonprofit in Uganda, news articles outside of Uganda fail to mention that the sudden eradication of the platform was actually in response to the platform banning the President's own personal account in light of his continued threats of violence and inciting civil unrest through the platform (R.G., personal correspondence, 15 March 2024). This is one of many reports of censorship globally, identifying the disconnect between life in a developed country versus a developing country, highlighting the importance of culturally appropriate, accessible, and unobtrusive intervention methods. To establish an effective intervention in developed and developing countries across the globe, the platform must remain relatively neutral politically, economically, and religiously to establish the largest footprint. This challenge is a precarious one and must be addressed carefully, but support may rely on inherently biased sources, so checks and balances should be implemented with this in mind. To that end, the joint platforms discussed herein, Ancestry and FamilySearch, are heavily funded by The Church of Jesus Christ of Latter-Day Saints (LDS), and while the LDS church values diversity, inclusion, and humanitarian relief, a significant religious affiliation could affect receptivity to the platform in some regions.

4.2 Program Evaluation, Development, and Upkeep

Successful implementation of genealogy in reunification will have several measurable outcomes. First, the speed at which reunification occurs is a quantitative measure that can be assessed by time, but could also be compared between countries. The number of connections correctly made can be tracked, and by adjusting accordingly, those trends being identified will aid the machine learning algorithm of AI to better predict future connections, similar to the learning models discussed by Pillai in 2023. Lastly, extensive user influx will aid current users of genealogic platforms by establishing a more broad database of historical documents and records with more human connections, effectively tracking human movement and activities more accurately for generations to come. Data mining such extensive collections of information is an important piece of AI that improves AI in itself, but this information can also be used extensively in research, planning, and humanitarian endeavors.

This technology could additionally be used to reunite displaced communities and rebuild social support. It is not uncommon to see stories in mass media about cherished items or pets left behind in a humanitarian crisis or natural disaster. By uniting displaced individuals with their previous neighbors or community members, communities could be restored, items once believed to be lost forever could be returned to their rightful owners, and displaced individuals can begin the process of reflecting and healing from their trauma.

Big data is an actively evolving field. The continual changes in data security laws, data retention practices, and availability of information is shifting at a near constant rate. By employing this approach, there will surely be obstacles to address in coming years, but the benefits of reunification surely outweigh the cost of flexibility and development to compensate for necessary updates and changes. In addition to the ambiguous future of big data, AI faces many challenges as its utility and applications evolve. MacEntee discusses copyright infringement on sources and AI-generated content, citation ambiguity, ethical concerns, high cost of deployment, bias, false information, violations of privacy, and informed consent or lack thereof as major concerns for the emergence of AI (n.d.) [13].

5 Conclusion

Displacement in a humanitarian crisis is an incredibly traumatic event. The impact of displacement is significantly worsened by the asylum seeking and family reunification process. The current landscape for reunification is a drawn out legal process that differs from place to place on different timeframes. Stories from refugees indicate a loss of connectivity, lack of records, and legal procedures that severely complicate reunification. By employing AI and publicly available genealogical software around the globe, refugees have the possibility of retaining communication with family members, tracking vital records, and identifying where loved ones have settled. These improvements to the current state of affairs have potentially life-altering impacts for all individuals involved emotionally, physiologically, and mentally. As the world faces the largest refugee movement in history, integrating technological advances has the potential to restore communities and families in ways previously unimaginable.

References

1. Ancestry (2023) Ancestry® to apply AI-driven handwriting recognition technology to create a searchable index of the 1931 census of Canada. Ancestry®. https://www.ancestry.com/corporate/blog/ancestry_-to-apply-ai-drive-handwriting-recognition-technology-t
2. AncestryDNA (2024) Countries where AncestryDNA® is available. Ancestry®. https://support.ancestry.com/s/article/Countries-Where-AncestryDNA-is-Available?language=en_US
3. Canadian Council for Refugees (n.d.) Family reunification—options for refugees. https://ccrweb.ca/en/family-reunification-options-refugees#:~:text=Canada%27s%20options%20for%20family%20reunification,beyond%20the%20capacity%20of%20many
4. Dahir AL (2021) Uganda blocks Facebook ahead of contentious election. The New York Times. https://www.nytimes.com/2021/01/13/world/africa/uganda-facebook-ban-elections.html
5. European Commission (2024) Forced displacement. European Civil Protection and Humanitarian Aid Operations. https://civil-protection-humanitarian-aid.ec.europa.eu/what/humanitarian-aid/forced-displacement_en
6. FamilySearch (2024a) How do I create a free FamilySearch account? https://www.familysearch.org/en/help/helpcenter/article/how-do-i-create-a-free-familysearch-account#:~:text=Anyone%20over%2013%20years%20of,username%20or%20reset%20a%20password
7. FamilySearch (2024b) How do I set up my ancestry account? https://www.familysearch.org/en/help/helpcenter/article/how-do-i-set-up-my-ancestry-account#:~:text=FamilySearch%20collaborates%20with%20Ancestry%20on,your%20own%20account%2C%20visit%20Ancestry
8. Franceschetti L, Mazzarelli D, Ragni C, Paltenghi F, Pecoraro A, Albani D, Giuffrida R, Siccardi C, Polizzi N, Di Simone P, D'Apuzzo A, Mandrioli D, Buscemi M, D'Amico M, Viarengo I, Cattaneo C (2024) Why identification matters: an explorative study on six cases of family reunification. Int J Leg Med 138(3):1187–1192. https://doi.org/10.1007/s00414-024-03163-w
9. Giannopoulou I, Mourloukou L, Efstathiou V, Douzenis A, Ferentinos P (2022) Mental health of unaccompanied refugee minors in Greece living "in limbo." https://www.psychiatrikijournal.gr/documents/psychiatry/33.3-EN-2022.pdf
10. Hvidtfeldt C, Petersen JH, Norredam M (2022) Waiting for family reunification and the risk of mental disorders among refugee fathers: a 24-year longitudinal cohort study from Denmark. Soc Psychiatry Psychiatr Epidemiol 57(5):1061–1072. https://doi.org/10.1007/s00127-021-02170-1
11. Jaffe J, Leamey T (n.d.) Best DNA test for 2024. CNET. https://www.cnet.com/health/medical/best-dna-test/#:~:text=AncestryDNA%20says%20its%20database%20contains,the%20DNA%20test%20kit%20services
12. Koch A (2023) Genealogy versus family history: definitions and examples of each. Family Tree Mag. https://familytreemagazine.com/general-genealogy/what-is-genealogy-family-history/
13. MacEntee T (2023) A genealogist's guide to artificial intelligence. Family Tree Mag. https://familytreemagazine.com/resources/software/ai-and genealogy/
14. Morton SJ (2019) The world's largest shared family tree. FamilySearch. https://www.familysearch.org/en/blog/online-family-tree
15. Pillai A (2023) AI revolutionizes genealogy: discovering family history and relationships with data-driven insights. LinkedIn. https://www.linkedin.com/pulse/ai-revolutionizes-genealogy-discovering-family-history-pillai/

16. Sharpe M (2015) The roots of genealogy. The History Press. https://thehistorypress.co.uk/article/the-roots-of-genealogy/
17. U.S. Citizenship and Immigration Services (USCIS) (2024) I-730, refugee/Asylee relative petition. USCIS Website. https://www.uscis.gov/i-730
18. USA for UNHCR (n.d.) Refugee statistics: USA for UNHCR. The Un Refugee Agency. https://www.unrefugees.org/refugee-facts/statistics/

The manufacturer's authorised representative in the EU is Springer Nature Customer Service Centre GmbH, Europaplatz 3, 69115 Heidelberg, Germany. If you have any concerns regarding our products, please contact ProductSafety@springernature.com

Printed and bound by CPI Group (UK) Ltd, Croydon, CR0 4YY
05/10/2026
02228127-0010